Explainable Artificial Intelligence for Intelligent Transportation Systems

Artificial Intelligence (AI) and Machine Learning (ML) are set to revolutionize all industries, and the Intelligent Transportation Systems (ITS) field is no exception. While ML, especially deep learning models, achieve great performance in terms of accuracy, the outcomes provided are not amenable to human scrutiny and can hardly be explained. This can be very problematic, especially for systems of a safety-critical nature such as transportation systems. Explainable AI (XAI) methods have been proposed to tackle this issue by producing human interpretable representations of machine learning models while maintaining performance. These methods hold the potential to increase public acceptance and trust in AI-based ITS.

FEATURES:

- Provides the necessary background for newcomers to the field (both academics and interested practitioners)
- Presents a timely snapshot of explainable and interpretable models in ITS applications
- Discusses ethical, societal, and legal implications of adopting XAI in the context of ITS
- Identifies future research directions and open problems

Amina Adadi is an assistant professor of Computer Science at Moulay Ismail University, Morocco. She has published several papers including refereed IEEE/Springer/Elsevier journal articles, conference papers, and book chapters. She has served and continues to serve on executive and technical program committees of numerous international conferences such as IEEE IRASET, ESETI, and WITS. Her research interests include Explainable AI, Data Efficient Models: Data Augmentation, Few-shot learning, Self-supervised learning, Transfer Learning, Blockchain, and Smart Contracts.

Afaf Bouhoute holds a Ph.D., a Master's degree in information systems, networking, and multimedia, and a bachelor's degree in computer science, all from the faculty of science, Sidi Mohamed Ben Abdellah University, Fez, Morocco. She regularly serves in the technical and program committees of numerous international conferences such as ISCV, WINCOM, ICECOCS, and ICDS. She also served as a co-chair of the First International Workshop on Cooperative Vehicle Networking (CVNET 2020), which was organized in conjunction with EAI ADHOCNETS 2020. Her research interests span different techniques and algorithms for modeling and analysis of driving behavior, with a focus on their application in cooperative intelligent transportation systems.

Explainable Artificial Intelligence for Intelligent Transportation Systems

Edited by
Amina Adadi
Moulay Ismail University, Morocco
Afaf Bouhoute
Sidi Mohamed Ben Abdellah University, Morocco

CRC Press is an imprint of the
Taylor & Francis Group, an **informa** business

First edition published 2024
by CRC Press
6000 Broken Sound Parkway NW, Suite 300, Boca Raton, FL 33487-2742

and by CRC Press
4 Park Square, Milton Park, Abingdon, Oxon, OX14 4RN

CRC Press is an imprint of Taylor & Francis Group, LLC

ISBN: 978-1-032-34457-7 (hbk)
ISBN: 978-1-032-34853-7 (pbk)
ISBN: 978-1-003-32414-0 (ebk)

DOI: 10.1201/9781003324140

Typeset in LM Roman font
by KnowledgeWorks Global Ltd.

Publisher's note: This book has been prepared from camera-ready copy provided by the authors.

Contents

Preface

Artificial Intelligence (AI), particularly Machine and Deep Learning, has been significantly advancing Intelligent Transportation Systems (ITS) research and industry. Due to their ability to recognize and to classify patterns in large datasets, AI algorithms have been successfully applied to address the major problems and challenges associated with traffic management and autonomous driving, e.g., sensing, perception, prediction, detection, and decision-making. However, in their current incarnation, AI models, especially Deep Neural Networks (DNN), suffer from the lack of interpretability. Indeed, the inherent structure of the DNN is not intrinsically set up for providing insights into their internal mechanism of work. This hinders the use and acceptance of these "black-box" models in systems of a safety-critical nature like ITS. Transportation usually involves life-death decisions; entrusting such important decisions to a system that cannot explain or justify itself presents obvious dangers. Hence, explainability and ethical AI are becoming subjects to scrutiny in the context of intelligent transportation.

Explainable Artificial Intelligence (XAI) is an emergent research field that aims to make AI models' results more human-interpretable without sacrificing performance. XAI is regarded as a key enabler of ethical and sustainable AI adoption in transportation. In contrast with "black-box" systems, explainable and trustworthy intelligent transport systems will lend themselves to easy assessments and control by system designers and regulators. This would pave the way for easy and continual improvements leading to enhanced performance and security, as well as increased public trust.

Given its societal and technical implications, we believe that the field of XAI needs an in-depth investigation in the realm of ITS, especially in a post-pandemic era. This book aims at compiling into a coherent structure the state-of-the-art research and development of explainable models for ITS applications. The contributions are organized in three topical parts:

1. Toward explainable ITS: This part aims to round up reference material for newcomers and researchers willing to join the dynamic area of explainable intelligent transportation systems by introducing them to fundamental concepts and topics.

2. Interpretable methods for ITS applications: This part provides interested and more experienced researchers with a timely holistic view of existing and potential methods and techniques for enabling and improving the interpretability of transportation systems.

3. Ethical, social, and legal implications of XAI in ITS: This part discusses ethical and social implications of XAI in the ITS context.

In Chapter 1, the editors provide a comprehensive survey on the topic of XAI and its application to ITS. This survey aims to provide the fundamental knowledge required of researchers who are interested in explainability in ITS. The second part begins by introducing a theoretical design of an XAI-based autonomous driving system in Chapter 2; the theoretical framework is projected to a practical case study. Afterward, some inherent interpretable methods are explored for different ITS application, namely Chapter 3 proposes a graph feature selection-based interpretability for predicting accident injury severity. In Chapter 4, a decision tree-based model is used to classify crash effects in an attempt to evaluate the impact of the COVID-19 pandemic on traffic crash injuries in the city of Barcelona. A survey on explainable reinforcement learning applied to ITS is proposed in Chapter 5. Chapter 6 deals with missing data estimation for traffic flow in Morocco using linear regression. The rest of the chapters focus on ad-hoc explainable techniques. Chapter 7 studies the explainability of the GNN model using two methods: SHAP and Zorro method. The aim of the study is to estimate the importance of intersections and their impact on the times of waiting on red signals. The third survey proposed by the book investigates machine learning and XAI techniques for vessel traffic service. In Chapter 9, an explainable model for the detection and recognition of traffic road signs is proposed using heatmapping techniques (Saliency Map and Grad-cam). An ensemble machine learning-based model is proposed in Chapter 10 to classify transportation modes; SHAP is used to enhance the interpretability of the model. Finally, Chapter 11 explores the potential of combining XAI and Blockchain for better trust in autonomous vehicles. Explainability is not a purely technological issue; it entails trust, fairness, and ethical considerations. In the last part of this book, a study of the public preference for autonomous vehicles in moral dilemma situations is presented. The chapter presents a questionnaire survey in order to study and analyze the moral choices of the public under different perspectives.

To the best of our knowledge, this is the first dedicated edited book that focuses on XAI in the context of ITS. We believe that the rich portfolio of the XAI techniques and the diverse ITS applications discussed in this book will help to bridge AI and transportation communities, and will contribute to the progress toward explainable, more trusted, and ethical intelligent transportation.

We would like to avail this opportunity to express our thanks to the contributing authors for their precious collaboration. But for their contributions, this initiative could not have become a reality. We would also like to thank Taylor and Francis / CRC Press as publication partner for their guidance and assistance; a special thanks to Elliott Morsia and Simran Kaur for their kind help throughout the book-editing process.

Dr. Amina Adadi, Moulay Ismail University, Morocco
Dr. Afaf Bouhoute, Sidi Mohamed Ben Abdellah University, Morocco

Contributors

Rahim Ali Abbaspour
School of Surveying and Geospatial Eng
University of Tehran, Iran

Amina Adadi
SIC Research Team, L2ISEI Laboratory
Moulay Ismail University, Meknes, Morocco

Ahmad Aiash
Civil Engineering School
UPC– BarcelonaTech, Barcelona, Spain

Yassine Akhiat
Department of Informatics
LPAIS Laboratory, Fez, Morocco

Shahin Atakishiyev
Department of Computing Science
University of Alberta, Edmonton, Canada

Abderrahim Bajit
ISA Laboratory, ENSA
Ibn Tofail University, Kenitra, Morocco

Anass Barodi
ISA Laboratory, ENSA
Ibn Tofail University, Kenitra, Morocco

Mohammed Benbrahim
ISA Laboratory, ENSA
Ibn Tofail University, Kenitra, Morocco

Mohammed Berrada
LIASSE Laboratory
ENSA Fez, USMBA, Morocco

Dominik Bogucki
TensorCell, Poland
Poland

M. Latif Bolum
TensorCell, Poland
Poland

Younes Bouchlaghem
Department of Informatics, UAE
Tetouan Morocco

Afaf Bouhoute
Sidi Mohamed Ben Abdellah University
Faculty of Science, Fez, Morocco

Mohamed Chahhou
Department of Informatics, UAE
Tetouan Morocco

Alireza Chehreghan
Faculty of Mining Eng
Sahand University of Technology, Iran

Meng Joo Er
Dalian Maritime University
Dalian, Liaoning, China

Wenxiao Gao
Dalian Maritime University
Dalian, Liaoning, China

Randy Goebel
Department of Computing Science
University of Alberta, Edmonton, Canada

Huibin Gong
Dalian Maritime University
Dalian, Liaoning, China

Mouna Jiber
Ministry of Equipment and Water
Fez, Morocco

Chuang Ma
Dalian Maritime University
Dalian, Liaoning, China

Ouassima Markouh
LIASSE Laboratory
ENSA Fez, USMBA, Morocco

Abdelilah Mbarek
LISAC Laboratory, FSDM
USMBA, Fez, Morocco

Rudy Milani
Universitaet der Bundeswehr Muenchen
Neubiberg, Germany

Maximilian Moll
Universitaet der Bundeswehr Muenchen
Neubiberg, Germany

Kareem Othman
Civil Engineering Department
University of Toronto, Canada

Stefan Pickl
Universitaet der Bundeswehr Muenchen
Neubiberg, Germany

Francesc Robusté
Civil Engineering School
UPC– BarcelonaTech, Barcelona, Spain

Abdelouahed Sabri
LISAC Laboratory, FSDM
USMBA, Fez, Morocco

Sajjad Sowlati
School of Surveying and Geospatial Eng
University of Tehran, Iran

Mohammad Salameh
Huawei Technologies Canada Co., Ltd
Edmonton, Canada

Ahmed Tamtaoui
SC Department, INPT-Rabat
Mohammed V University, Rabat, Morocco

Ali Yahyaouy
LISAC Laboratory, FSDM
USMBA, Fez, Morocco

Hengshuai Yao
Department of Computing Science
University of Alberta, Edmonton, Canada

Ahmed Zinedine
Department of Informatics
LPAIS Laboratory, Fez Morocco

Abdelkarim Zemmouri
ISA Laboratory, ENSA
Ibn Tofail University, Kenitra, Morocco

I

Toward Explainable ITS

CHAPTER 1

Explainable Artificial Intelligence for Intelligent Transportation Systems: Are We There Yet?

Amina Adadi
ISIC Research Team, L2ISEI Laboratory, Moulay Ismail University, Meknes, Morocco

Afaf Bouhoute
Sidi Mohamed Ben Abdellah University, Faculty of Science, Fez, Morocco

CONTENTS

Artificial Intelligence (AI) and Machine Learning (ML) are set to revolutionize all industries. Intelligent Transportation Systems (ITS) field is no exception. However, being a safety-critical system, transportation needs explanation and justification in order to ensure that AI-based decisions and results are made fairly and without errors. Nevertheless, due to the block-box nature of ML especially deep

DOI: 10.1201/9781003324140-1

learning models, the outcomes provided by these models are not amenable to human scrutiny. Explainable AI (XAI) methods have been proposed to tackle this issue by producing human interpretable representations of ML models while maintaining performance. These methods hold the potential to increase public acceptance and trust in AI-based ITS.

This chapter investigates the use of XAI in smart transportation, it aims to (i) provide the necessary background regarding XAI, its main concepts, and algorithms; (ii) identify the need of explanations for intelligent systems in transportation; (iii) examine efforts deployed to enable and/or improve ITS interpretability, by reviewing the literature related to explanations for intelligent transportation problems; and finally (iv) explore potential challenges and possible research avenues in the emergent field of Explainable ITS.

1.1 INTRODUCTION

AI is everywhere, in everything from the mobile voice chatbot to self-driving cars. Indeed, aided by advances in AI and automation, autonomous driving and intelligent transportation, in general, have seen rapid progress and maturation in recent years. However, from a public perspective, autonomous vehicles and intelligent traffic systems are still perceived as futuristic and even dangerous technology. Much of the discussion and critics of self-driving vehicles have been concerned with issues of safety and trust. Transportation usually involves life-death decisions, and entrusting such critical decisions to an algorithm may cause some frustration. A transparent algorithm with an "if-then" structure is relatively accepted since its behavior is known and predictable ahead. However, in the case of a black-box Deep Neural Networks (DNN) model, even its creator cannot predict its outcomes. Indeed, the inherent structure of the DNN is not intrinsically set up for providing insights into their internal mechanism of work. This hinders the use and acceptance of these "black-box" models in systems of safety-critical nature like transportation. Interestingly though, DNNs are proven to be the best models in terms of performance for tasks like perception, prediction, and object recognition, which are considered the core tasks in the intelligent transportation industry. Making DNN models' results more human-interpretable without scarifying performance has become then an intriguing research frontier. As a response to this challenge, explainable AI field emerged. XAI studies and designs methods producing human-understandable explanations of black box models. This technology is regarded as a key enabler of ethical and responsible AI adoption in transportation.

The purpose of this work is to highlight the potential of using XAI in the context of intelligent transportation. Despite the fact that the extent of research on interpretable and explainable models is quickly expanding, a holistic review and a systematic classification of these research works within the transportation field are missing from the literature. Few works have attempted to review the issue [4,54], but the scope was limited only to autonomous vehicles. To the best of our knowledge, this is the first study that focuses on XAI in the context of transportation broadly including sea, road, air, and rail transportation. This work may offer a helpful starting point for researchers and practitioners who are interested in implementing interpretability in transport intelligent systems.

The rest of the chapter is organized as follows. Section 1.2 provides background on XAI and projects the concept into the transportation field. Section 1.3 presents a literature review of XAI for smart transport systems based on recent works. Section 1.4 discusses current applications and future research directions, and finally, Section 1.5 concludes the paper.

1.2 BACKGROUND

1.2.1 Interpretability: Concepts and Terminology

Interpretable AI or Explainable AI refers to a suite of techniques that produce more human interpretable models while maintaining high prediction accuracy [2]. Interpretability in AI is not a new issue. In the mid-1970s researchers studied explainability for expert systems. Because the systems were based on symbolic knowledge representations, it was relatively straightforward to generate symbolic traces of their execution [58]. However, the pace of progress toward resolving interpretability has slowed down once ML and DNN, with their spectacular advances, have entered the game. Recently, there has been a resurgence of interest in XAI as researchers and practitioners seek to make their algorithms more understandable in order to comply with social, ethical, and legal obligations.

While there is general acknowledgment that it is important for intelligent systems to be explainable/interpretable, there is no general consensus over what is meant by "explainable" and "interpretable." Two perspectives should be taken into account while approaching the notion of explanation, namely (i) the social science perspective and (ii) the etymology/terminology perspective.

In his work, Miller [48] explored bodies of research in philosophy, psychology, and cognitive science to investigate how people define, generate, and present explanations. He concluded that:

1. Explanations are contrastive: people do not ask why event P happened, but rather why event P happened instead of some event Q.

2. Explanations are selected: humans are adept at selecting one or two causes from a sometimes-infinite number of causes to be the explanation.

3. Explanations are social: explanations depend on the social and cultural background of both the explainee and the explainer.

4. Explanations are post hoc: generally, the explanation process is triggered after the reasoning process produces a decision. That means that the brain generates first a decision and then looks for a contextual and convincing explanation, if one is required.

All these four characteristics should be considered in developing XAI techniques in order to generate a human-like explanation.

From a terminology perspective, in the interpretable ML community, many terms are used interchangeably. However, they refer to different meanings, which led to terminology confusion. Murdoch et al. [51] define Interpretable AI as the extraction of relevant knowledge from a machine-learning model concerning relationships either contained in data or learned by the model. AI interpretable systems become explainable if their operations can be understood by humans. Transparent AI refers to white-box models that are inherently interpretable without alteration. On the other hand, Trustworthy AI is when an intelligent system succussed to generate the trust of its users. The guidelines for trustworthy AI published by the EU High-Level Expert Group on AI [16] list seven key requirements that AI systems should meet in order to be trustworthy (1) human agency and oversight, (2) technical robustness and safety, (3) privacy and data governance, (4) transparency, (5) diversity, non-discrimination and fairness, (6) environmental and societal well-being, and (7) accountability. It should be noted that while trustworthiness should be a property of every explainable system, this does not mean that every trustworthy system is also explainable. Finally, XAI implies Responsible AI, a methodology for the large-scale implementation of AI methods in real organizations with fairness, ethics, and accountability at its core [8].

1.2.2 Need for Interpretability

Broadly speaking, there are three key reasons why interpretability needs to be integrated into the design of AI-based systems [2], namely (i) to justify, (ii) to debug, or (iii) to discover.

- Justification: Intelligent models' outcomes need to be justified, particularly when unexpected decisions are made. This ensures an auditable way to prevent unfair and/or unethical algorithmic decisions. From a legal point of view, models need to be explainable to meet adherence to regulatory requirements. Indeed, many international regulations require now "the right of explanation." For instance, in the European General Data Protection Regulation (GDPR)[1], it is stated that in case an automated decision-making system is used, "data subjects" have the right to obtain explanations to ensure fair and transparent processing. Algorithmic Accountability Act 2019[2] requires companies to provide an assessment of the risks posed by the automated decision system to the privacy or security and the risks that contribute to inaccurate, unfair, biased, or discriminatory decisions impacting consumers. Washington Bill 1655[3] established a guideline for the use of automated decision systems to protect consumers, improve transparency, and create more market predictability. Massachusetts Bill H.2701[4] established a commission on automated decision-making, transparency, fairness, and individual rights. Finally, Illinois House Bill 3415[5] stated predictive data analytics determining credit worthiness or hiring

[1] https://gdpr.eu/what-is-gdpr/
[2] https://www.congress.gov/bill/116th-congress/house-bill/2231/all-info
[3] https://legiscan.com/WA/bill/HB1655/2023
[4] https://malegislature.gov/Bills/192/H2701
[5] https://www.ilga.gov/legislation/101/HB/PDF/10100HB3415lv.pdf

decisions may not include information that correlates with the applicant race or zip code. Specifically in transportation, the reason to justify is very critical. The lack of reasons by AI systems leads to muted trust and limited mainstream adoption. Furthermore, legal institutions are interested in justifications for liability and accountability purposes, especially when a self-driving system is involved in a car accident. Recent events from autonomous cars illustrate well this fact. In 2018, a self-driving Uber killed a woman in Arizona. It was the first known fatality involving a fully autonomous vehicle. The information reported by anonymous sources who claimed the car's intelligent driving system registered an object in front of the vehicle, but treated it in the same way it would a plastic bag or tumbleweed carried on the wind [9]. Many crashes of the type were reported afterward. Only an explainable system can help investigate and clarify the circumstances of such accidents and eventually prevent them from occurring.

- Debuggability: A model that is not scrutable can lead to incorrect conclusions, even with correct outcomes. In this case, explanation methods could not only help in understanding but also debugging and validating the behavior of a learned model. Lapuschkin et al. [40] demonstrated in their study how recent techniques for explaining ML-based decisions can help evaluate whether the learned strategy is valid and generalizable or whether the model has based its decision on a spurious correlation in the training data. A pretrained DNN fine-tuned on PASCAL VOC image dataset, shows an excellent accuracy on categories such as "ship" and "train." However, inspecting the basis of the decisions with explainable methods reveals substantial divergence for certain images. For example, in the ship and train images, the heatmap pointed respectively at the sea and the rail as the most relevant features (see Figure 1.1). These features are not characteristic of these two objects. For instance, a ship or a boat image could possibly not include the sea, in this case, the algorithm will fail to recognize the object. Therefore, the fact that an algorithm produces a good result does not mean that it is valid and generalizable. XAI can contribute to control and adjust and eventually improve intelligent models. In the ITS realm, XAI can play a major in testing and validation operations, particularly for autonomous vehicles.

- Knowledge Discovery: One of the most spectacular accomplishments of AI is for the intelligent machine AlphaGo to win against the world champion of Go game. At the time, Fan Hui,[6] a renowned go expert stated that "It's not a human move. I've never seen a human play this move" refereeing to the last move, which allowed AlphaGo to win the match. It is incredible for a machine to win against a human player. However, it would be more useful if the machine could explain its game tactics for knowledge sharing. It is believed that systemizing expandability in the intelligent technology pipeline would help discover new

[6]https://www.wired.com/2016/03/googles-ai-viewed-move-no-human-understand/

Figure 1.1 Images of the classes "boat" and "train," processed by DNN models and heatmapped using LRP [40].

facts and realities provided by machines. XAI models can teach us about new and hidden facts in driver behavior and machinery dynamics.

1.2.3 Interpretability Stakeholders

Explainability could mean different things depending on the recipient. The stakeholders involved in the interpretability process can be divided into three broad categories [54]:

- Developers/data scientists: industrials and/or researcher actors who develop intelligent models for ITS. The primary motive for seeking explainability/interpretability for this community is quality assurance, meaning system testing, debugging, and evaluation, and to improve the performance of their models [58].
- Regulators and policy-makers: this category includes auditors, insurance agents, and managers who inspect and control ITS design processes and operations in order to ensure fairness, unbiased behavior, and compliance with regulations.
- End-Users: people who interact with ITS; this could include drivers, passengers, pedestrians, and other road agents. Members of the user community need explanations to help them decide whether to accept, trust, and thus act according to the machine outputs, or not.

Accordingly, adaptation and personalization are seen to be essential for the generation of meaningful and useful explanations. A creator, a controller, a user, or simply an observer, naturally need different representation of explanation. While a simple user without a technical background may be satisfied with a user-friendly explanation,

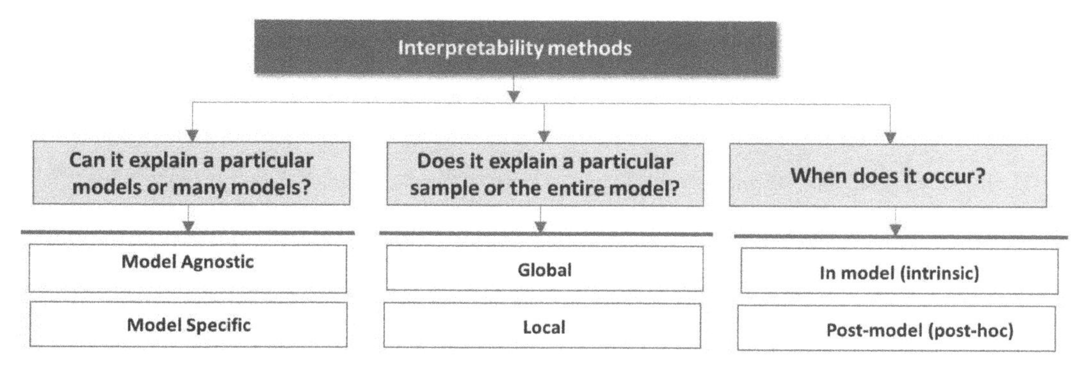

Figure 1.2 Interpretable methods categories.

developers would require a technical explanation that will support the design and validation of the intelligent system.

1.2.4 Interpretability Techniques

At a high level, interpretable methods could be categorized depending on three aspects: the scope, the stage, and the applicability [2, 31] (Figure 1.2).

1. ***Scope of interpretability:***
   Global: the methods with a global scope explain the functionality of the whole model.
   Local: the methods with local scope are used to explicitly explain a single instance (outcome or prediction). Local methods are the most used in the real-world scenarios as they could provide the instance level justification required by regulations.

2. ***Stage of interpretability:***
   Intrinsic: methods which have interpretability built-in, meaning that from the beginning the model is designed to generate both the perdition and its related explanation. Generally, this class includes transparent models that are inherently interpretable, such as fuzzy models and tree-based models.
   Post hoc: in this case, the system includes two components; the predictive model (typically a DNN) and an external or surrogate model. The predictive model remains intact, while the external model mimics the predictive model's behavior to generate an explanation. Most recent works done in XAI field belong to post hoc class.

3. ***Applicability of interpretability:***
   Model specific: interpretable methods that work only for a specific model, generally this class refers to inherently interpretable models (which are specific to themselves).
   Model agnostic: post hoc methods that provide explanations undependably to the type of the predictive model. Most of the state-of-art XAI methods belong to this class.

Table 1.1 describes the most popular interpretability techniques in the literature. The input data of an interpretable model vary following the problem to be solved and the predictive model in use. In transportation, as in other domains, inputs of interpretable methods can be in form of images, times series, text, or graphs. As for XAI method outputs, as aforementioned, their format depends on the expertise of the end-users. The most common forms of explanations are numeric, rules, textual, visual, and mixed [31]. Furthermore, it should be noted that implementing interpretability comes at a cost. Generally, a tradeoff is observed between accuracy and interpretability. The more the model is interpretable, the less it is accurate and vice versa [2]. The tradeoff accuracy/explanation is an important factor to take in account in transportation context to make choice of appropriate interpretable techniques. Though, ITS is a safety-critical field, so trust and security come naturally first, however, in some tasks that do not involve safety, accuracy could be favored.

1.3 REVIEW

We conducted an extensive literature review by examining papers from two major academic databases, namely SCOPUS and Web of Science, in addition to Google Scholar and preprints posted on arXiv. Valuable literature of 154 articles was retrieved and scanned. Next, we will present a detailed review of the relevant works by approaching them from a use case perspective. Indeed, additionally to investigating XAI methods that have been used within the field, we would like also to identify and categorize the niche applications of XAI in smart transportation. Hence, we suggest describing the XAI efforts along the four main Transportation Modes: (i) Road Transport (including road traffic and autonomous vehicles), (ii) Air Transport, (iii) Maritime Transport, and (iv) Railway Transport.

1.3.1 XAI in Autonomous Vehicles

Autonomous vehicles (AVs) are systems capable of driving themselves with little to no human intervention. Specifically, self-driving vehicles sense their environment and map such sensing data to real-time driving decisions using intelligent technologies. The Society of Automobile Engineers (SAE) defines six levels of automation from zero to five [14]. Level zero automation indicates no automation or fully manual, meaning the driver takes over all driving tasks himself. While level five describes full and complete automation. This means that the vehicle is capable of performing all driving maneuvers and tasks in a completely autonomous way, the only time a driver takes control of the car is when they request it. While AV is arguably the next significant disruptive innovation in transportation, widespread production and adoption are still some time away from anything higher than level two. In order for autonomous cars to enter the mainstream, they must fulfill several criteria, one of which is trust and safety. Multidisciplinary research is very active to address trust and safety roadblocks. As a trust-enabling technology, interpretable AI has stimulated a vibrant research activity in this field resulting in valuable literature about when and how interpretability should be incorporated in AV design and development.

TABLE 1.1 XAI Techniques

Class of technique	Description	Techniques	Scope		Applicability		Stage	
			Local	Global	Model Specific	Model Agnostic	Intrinsic	Post-hoc
Surrogate models	Methods that mimics (approximates) predictive model to generate explanations	Model Distillation		X		X	X	
		LIME	X			X		X
Feature relevance explanation	Methods that evaluate feature importance that influenced the prediction	SHAP	X	X		X		X
		LRP	X	X		X		X
Visualization Explanation	Curves and plots that give insights of the predictive models by internal pattern	PDP		X		X		X
		ICE	X			X		X
Pixel attribution	A special case of feature attribution, but for images. It also provides Visual Explanations	Grad-CAM	X					
		DeepLIFT	X	X		X		X
Self-Explanatory Neural Networks	Methods that generate in-model interpretability by making DNN more transparent	Capsule Network PINN KPRN Attention Mechanism	X	X	X		X	
Example based explanation	Select particular instances of the dataset to explain the behavior of the predictive mode	Contrefactuels	X			X		X

In this subsection, we propose to explore this literature by reviewing, organizing, and discussing the recent work on explainability for AV. Only road AVs are considered in this section, other self-driving vehicles from other modes of transport such as aerial, maritime, and railway will be discussed in the next sections.
Reviewed AV studies fall under two main application categories, (i) Environment Perception and (ii) Action Planning.

(a) Environment perception: AV needs to perceive its environment to have the necessary data based on which it makes autonomous decisions in real-world driving scenarios. If the perception is wrong, the consequences could be harmful. Hence, a significant line of work discussed interpretability in this context.

Environment perception rely heavily upon reliable object detection techniques. Many models have been described in the literature for the detection of various objects relevant to the tasks of self-driving cars or driver assistance systems. Most of them used variations of CNN architectures (e.g., YOLO and Faster R-CNN), this is the model of choice for detection and classification of objects from images with excellent performance. Accordingly, many methods explaining CNN opaque outcomes have been discussed in the literature. Nowak et al. [53] detected charging stations from a moving electric bus under varying lighting and weather conditions. The interpretation was achieved by visualization of attention heat maps. Mankodiya et al. [45] proposed a segmentation model for road detection, Grad-CAM, and saliency map were used to explain the segmentation maps generated from DL models. In Marathe et al. [47] study, the problem of image restoration was addressed. Weather corruptions can hinder the object detectability. Thus, there is a need for efficient denoising, deraining, and restoration techniques. The authors used Generative Adversarial Networks (GAN) model to this aim. Once more, Grad-CAM attention maps were used to visualize the results.

One of the earliest applications where explainability was discussed in transportation was traffic sign detection. Back in 2016, Lu et al. [43] proposed a multi-modal tree-structure embedded multi-task learning for traffic sign recognition. The tree structure of the model made the proposed approach inherently explainable. In the same vein, deep reinforcement learning was used by Rizzo et al. [59] on the traffic light control optimization problem, neural network agents were trained to select the traffic light phase. SHAP was used to estimate the effect of the road detector state on the agent selected phases.

Another class of object detection applications is intention recognition, meaning detecting non-static, uncontrollable moving objects (pedestrians and vehicles). Li et al. [42] developed a vision-based system to detect and identify various objects in order to recognize the intention of pedestrians in the traffic scene. The system is based on YOLO v4 model. In this study, Randomized Input Sampling for Explanation (RISE) algorithm is applied to make the final classification explainable by generating the saliency map. The intention of other vehicles was recognized in [18] to perform a lane change maneuver. Layerwise Relevance Propagation (LRP) was used as heatmapping technique to emphasize the salient areas of the image.

Although, most of Deep Neural Perception works rely on Visual Explanations. Few works have attempted to generate Textual Explanations. Kim et al. [37]

introduced an attention-based video-to-text model to produce Textual Explanations and descriptions of model actions. Chen et al. [12] proposed a framework imitating the learning process of human drivers by jointly modeling the visual input images and natural language, while using the language to induce the visual attention in the image.

(b) Action Planning: Decision or action planning refers to the process of making purposeful decisions based on collected data in order to achieve the AV's order goals. In this category, we find mainly studies related to planning (i) paths/trajectories, (ii) behaviors, and (iii) other motions. The goal of these works is to justify the selected class of actions by providing necessary explanations.

AVs are expected to navigate complex urban environments safely. Therefore, interpretable trajectory planning and prediction is crucial for safe navigation and has therefore received much interest. Sun et al. [63] proposed human-like trajectory planning at dynamic highway scenes. The model is based on inverse reinforcement learning which started to be used recently to improve interpretability of deep driving models by learning the cost maps which are visually understandable. Neumeier et al. [52] introduced the Descriptive Variational Autoencoder (DVAE) for predicting vehicle trajectories that provide partial interpretability. By introducing expert knowledge within the decoder part of the autoencoder, the encoder learns to extract latent parameters that provide a graspable meaning in human terms. Such an interpretable latent space enables explanations and thus validation by expert defined rule sets. Rjoub et al. [60] designed an explainable federated deep reinforcement learning model to improve the trustworthiness of the trajectory decisions for newcomer AVs. The basic idea behind this approach is that when newcomer AV seeks help for trajectory planning, the edge server launches a federated learning process to train the trajectory and velocity prediction model in a distributed collaborative fashion among participating AVs. To resolve the problem of selecting the appropriate AVs that should participate in the federated learning process, the authors capitalized on SHAP method to assess the importance of the features contributed by each vehicle to the overall solution in order to derive trust scores for them. Then, a trust-based deep reinforcement learning model is put forward to make the selection decisions.

On the other hand, numerous studies have been conducted in the area of behavior planning. AV operates in an open environment alongside other road users with whom it will interact regularly. When humans drive, they continually predict the behaviors of other vehicles based on their prior knowledge. In order for a self-driving car to be driven safely, it needs to rely on interpretable data-driven algorithms to be able to plan its actions according to the predicted behavior of other traffic agents. Zhou et al. [81] predicted situation awareness during the takeover transition period in conditionally automated driving using eye-tracking and self-reported data. First, an ensemble ML model called light gradient boosting machine tree (LightGBM) was used to predict situation awareness. Then, in order to understand what factors influenced situation awareness and how, SHAP values of individual predictor variables in the LightGBM model were calculated. Hossain et al. [29] explored pedestrian behavior prediction model as a prerequisite for AV path planning. The authors proposed SFMGNet, a hybrid approach combining social force model, a classical simulation model for the dynamic behavior of pedestrians extended by pedestrian group behavior modeling with a neural network for predicting pedestrian motion. The model is proved to be partially interactable. Kalatian et al. [34] predicted pedestrian wait time behavior using immersive virtual reality and data-driven Cox Proportional Hazards (CPH) model, in which the linear combination of the covariates is replaced by a flexible non-linear DNN. A SHAP-based interpretability is provided to understand the contribution of different covariates to the time pedestrians wait before crossing.

Even though trajectory and behavior planners are the most reported applications in the literature, growing research interest in explaining other actions/motions has been noted. Generally, deep reinforcement learning agents are used to learn complex driving tasks and motions. AVs operate in a continuous action space in which these types of models are specialized. Studies proved that these agents lead to a smooth human-like driving experience, but they are limited by the lack of interpretability. In this sense, Gangopadhyay et al. [19] proposed a hierarchical program-triggered reinforcement learning, which uses a hierarchy consisting of a structured program along with multiple agents, each trained to perform a relatively simple task. The proposed architecture is proved to leading to a significantly more interpretable implementation as compared to a complex reinforcement learning agent. Furthermore, Gyevnar et al. [24] proposed a transparent, human-centric explanation generation method for autonomous vehicle motion planning and prediction based on an existing white-box system called IGP2. The method integrates bayesian networks with context-free generative rules and can give causal natural language explanations for the high-level driving actions of AV. Mankodiya et al. [46] used decision tree-based random forest to plan cybersecurity related actions. The model is set to detect malicious AV in vehicular adhoc networks. The authors used a particular model interface of the evaluation metrics to explain and measure the model's performance.

While the separation between perception and action planning is clear in modular driving systems, recent end-to-end neural networks blur the lines and perform both operations simultaneously and in an integrated way. Indeed, some recent works have been found to provide end-to-end autonomous driving systems while ensuring the interpretability of the overall solution. For instance, Chen et al. [11] proposed an interpretable deep reinforcement learning method for end-to-end autonomous driving, which is able to handle complex urban scenarios. The model introduced a latent space that enables an interpretable explanation of how the policy reasons about the environment by decoding the latent state to a semantic bird eye mask. Sadat et al. [61] proposed a novel end-to-end learnable network that performs joint perception, prediction, and motion planning for self-driving vehicles and produces interpretable intermediate representations. Similarly, Zeng et al. [77] also introduced a neural motion planner for learning to drive autonomously in complex urban scenarios. The authors designed a holistic model that takes as input raw LIDAR data and a HD map and produces interpretable intermediate representations in the form of 3D detections and their future trajectories, as well as a cost volume defining the goodness of each position that the self-driving car can take within the planning horizon.

1.3.2 XAI in Road-Traffic Management

Road traffic management and optimization are widely regarded as essential component of the modern transportation ecosystem. Complementary to the avant-garde AV technology, road traffic management seeks to resolve problems such as flow forecast ing, road safety, parking optimization, and managing traffic congestion. This allows traffic managers to anticipate and take the proper decisions in order to improve the security, safety, and reliability of the overall road transport system.

Road traffic management appears to be a particularly promising application area for explainable models. Efforts in this domain are mainly oriented to ward the development of prediction models for short-term traffic forecast ing. The earliest work on interpretable smart transportation found in the literature studied precisely traffic flow prediction. In the 2014 study by Xu et al. [73], the authors introduced interpretable spatiotemporal Bayesian multivariate adaptive-regression splines to predict short-term freeway traffic flow. The model assists traffic managers by extracting the most related road segments that have the greatest contribution to the future traffic state of the target road. Later on, Kruber et al. [39] developed a modified version of the random forest algorithm for the categorization of traffic situations. The algorithm generates a proximity matrix that contains a similarity measure. This matrix is then reordered with hierarchical clustering to achieve a graphically interpretable representation. In the studied literature, DNNs were introduced for traffic forecast ing in Barredo-Arrieta et al. [5] work, where SHAP was applied to extract additional knowledge from two black-box models widely used in traffic forecast ing namely the classic model random forest and the emergent model Recurrent Neural Network (RNN) falling within the DNN family. Traffic data involve spatial–temporal features, which made RNN and, particularly, LSTM the best prediction models. In the recent traffic prediction literature, LSTM was widely used with an attention mechanism to weigh the contribution of each traffic sequence toward the output predictions. When visualized, the weight values provide insights into the decision-making process of the LSTM and consequently produce explainable outputs [1, 41]. Always in traffic flow prediction, Wang et al. [70] proposed a deep polynomial neural network combined with a seasonal autoregressive integrated moving average model for predicting short-term traffic flow. The model has a good predicting accuracy as well as enhanced clarity on the spatiotemporal relationship in its deep architecture. Gu et al. [22] exploited interpretable patterns for flow prediction in dockless bike sharing systems. Specifically, by dividing the urban area into regions according to flow density, spatiotemporal bike flows are modeled between regions with graph regularized sparse representation, where graph Laplacian is used as a smooth operator to preserve the commonalities of the periodic data structure. Then, traffic patterns from bike flows use subspace clustering with sparse representation to construct interpretable base matrices. Recently, Zhai et al. [78] proposed a two-step interpretability evaluation framework with land-use probing tasks to facilitate understanding the internal functions of a given DL-based traffic prediction model. the interpretability of three models namely Auto-encoders, Variational Auto-encoders and Graph Convolutional Auto-encoders is evaluated using Partial Dependence Plot (PDP).

In public transportation applications, the interpretability of passenger flow prediction models was often studied alongside the model accuracy. For instance, Monje et al. [49] designed a model for bus passenger forecast ing in Spain, the model is based on LSTM networks. The authors obtained an interpretable model from the LSTM neural network using a global surrogate model based on regression tree and the two-tuple fuzzy linguistic model, which improves the linguistic interpretability. To discover the underlying passenger flow patterns over all stations in the Hong Kong metro system, Wang et al. [68] proposed a multivariate functional principal component analysis (MFPCA) method. To improve interpretability, the MFPCA is formulated as a minimization problem with both sparsity and robustness penalty term. A computationally efficient algorithm is developed accordingly to obtain the eigenvectors. The derived sparse and smooth eigenvectors can be well interpreted as empirically meaningful passenger flow patterns.

In the same vein, methods were proposed to improve the interpretability of travel time prediction. Fiosina et al. [17] introduced an explainable federated learning based system for taxi travel time prediction in Germany. Since the explainability of geographically distributed federated deep learning is not typically studied in the literature. This work proposed to investigate how exiting XAI methods could explain a federated models. Earlier, Zhang et al. [80] employed a gradient boosting regression tree method (GBM) to analyze and model freeway travel time, the model improves inherently the interpretability.

In incident management, in order to develop appropriate safety countermeasures, it is useful to identify and quantify the influential factors in road crashes. Many works shed light on this type of application. Ma et al. [44] studied how to prioritize influential factors for freeway incident clearance time prediction using the gradient boosting decision trees method that provides an inherent explainability. Arteaga et al. [3] proposed a text mining approach to analyze crash narratives in order to identify contributing factors to injury severity. The interpretability of the results is enabled by LIME method. On the hand, Movsessian et al. [50] used SHAP to explain the results of their damage detection model.

Other traffic road management applications were less studied from an interpretability point of view. For example, parking optimization and energy efficiency. The study of Parmar et al. [57] integrated LIME with LSTM for modeling parking duration in urban land-use. While both Jiang et al. [33] and Wang et al. [69] addressed the energy-management issue using an explainable reinforcement learning model.

1.3.3 XAI in Air-Traffic Management

Air Traffic has increased dramatically in the past few years. The International Civil Aviation Organization (ICAO) estimates that the number of domestic and international travelers is expected to reach six billion by 2030 [56]. Sustained growth of the aviation industry is putting enormous pressure on Modern Air Traffic Management (ATM) required to operate quickly but safely and efficiently. These constraints prompt the need for increased and advanced automation in order to increase system performance and decrease workload. AI and ML technologies provide opportunities for a fundamental change in the ATM automation landscape. Currently, highly automated ATM systems rely on AI algorithms for tasks related to airspace management (e.g., managing resources and optimizing the structure of the sectors), Air Traffic flow and capacity management (e.g., predicting the network delay), and Air Traffic controller (e.g., conflict detection and avoidance) [15]. AI is also actively guiding Unmanned Aerial Vehicles (UAVs) in making decisions without human intervention [6]. However, despite the promising potential of AI-driven automation, ATM industry remains inherently conservative. High-level of automation may compromise safety. Thus, adoption of AI within the sector will likely be limited to specific use cases involving relatively little uncertainty and risks [38]. The tradeoff between automation and safety in ATM field is expected, in return, to contribute to developments in explainable AI which may untimely change the foreseen premise and enable mainstream adoption of AI in ATM.

By scanning the literature on explainable models for ATM applications, it is clear that XAI is still an as-yet untapped area of research in the sector. The few works that address the interpretability issue, deal with it as a secondary goal of the work or as a perspective to undertake in future works.

Xie et al. [71] used XGBoost model to predict the risk of flight incidents and accidents based on meteorological data, which allows the implementation of a real-time situation assessment tool for air-traffic advisory services. The prediction results were explained through SHAP and LIME models. In [35], Adaptive Neuro-Fuzzy Inference System (ANFIS) models were used to give explainable features of UAV decisions. The system predicts the outputs regarding the weather conditions and enemy positions whenever UAV deviates from the predefined path. In this same vein, He et al. [27] developed an explainable deep reinforcement learning -based path planner for UAV. To get a better understanding of the model outcomes, both visual and Textual Explanations were provided as local explanations for non-expert users. SHAP was used to measure the feature attribution. Activation map was drawn to show the detailed visual feature extracted by the CNN part. Hossain [28] employed many ML models including Random Forest, Gradient Boost, and XGBoost to predict flight delay, the interpretability of the model was assured by using LIME explanations. Goh et al. [21] proposed a system for approach and landing accidents prediction using tunnel Gaussian process (TGP) model. The authors claimed that TGP can provide both the generative probabilistic description of landing dynamics and interpretable tunnel views of approach and landing parameters.

For the aerospace field, Cao et al. [10] considered both the model interpretability requirements and the low data constraint, and proposed an enhanced version of SYNDY method to ensure the achievement of interpretable and accurate predictions under limited data. In the same vein, Yadam et al. [74] investigated different visualization techniques for airfoil self-noise prediction. Based on NASA's open-source dataset, the authors inspected the significance of each airfoil parameter leveraged in predicting the sound pressure level by using different feature importance plots including Individual Conditional Expectation (ICE) plot, SHAP, swarm plots, and tree interpreter.

As shown by the literature analysis, XAI has been modestly studied in the ATM field. The few spotted works cover a diverse application related essentially to Air Traffic controller tasks. However, most of them are still in their early stage of development and simulation, their results are not yet tangible on the ground. Fortunately, a growing literature is urgently calling for more research and industrial projects on explainable ATM systems [15,38,62]. Such works recognize XAI as a necessity to ride the strong tailwind of growth and pave the way forward for a smart, sustainable, and fully automated aviation industry.

1.3.4 XAI in Maritime Transport

Marine regions cover about two-thirds of the earth. Consequently, Maritime Transport (MT) is considered one of the major links in the global transport chain. More than 80% of global trade is handled by the shipping industry [64]. Having vessels as means of transport, MT research is conventionally divided into two categories, namely the shipping side and the port side. And like in other modes of transport safety is crucial. In recent years, emerging technologies, notably ML and deep learning models have been receiving wider attention to solve practical problems on both sides [76]. On the shipping side, AI is used to address problems related to ship trajectory prediction, ship risk prediction and safety management, and ship energy efficiency prediction. On the port side, examples of AI applications include ship destination and arrival time prediction, and port condition prediction. Naturally, works on interpretability within the field would follow the same research trajectory. While reviewing the related literature, we noted that most works fall under the maritime surveillance issue, specifically (i) vessel types classification and (ii) ship detention identification, two vital operations for civil and military applications.

Vessel types classification was studied by Burkat et al. [9], the authors introduced a model that classifies ship vessel types based on their trajectory. They illustrated the explainability approaches by means of an explanatory case study, evaluation was based on human expert opinion. An explainable attention network for fine-grained ship image classification was proposed by Xiong et al. [72] by using remote-sensing images obtained by satellite sensors. In an interesting study, Veerappa et al. [66] explored LIME, SHAP, and path integrated gradient (PIG) for explaining the results of a ResNet based classifier for multivariate time series data representing ship trajectories, ship's positional information together with speed and other features. In a previous work [65], the authors used association rule mining methods for the same purpose which are inherently explainable. The interesting thing about these two studies is that the evaluation aspect was adequately discussed and addressed. To assess the quality of the XAI method's explanations, the authors used perturbation and sequence analysis techniques.

He et al. [26] addressed ship detention identification. They proposed an interpretable ship detention decision-making model based on XGBoost algorithm and synthetic minority oversampling technique (SMOTE) algorithm. The interpretability of the model is assured by using SHAP explainer. Yan et al. [75] viewed ship detention as a type of anomaly and developed an isolation forest model for detention prediction. The prediction results were also explained based on SHAP in this case.

Another line of work studied water traffic accidents and the related risk assessment. Even though the rate of maritime accidents is relatively low compared to other kinds of accidents, a shipping accident can cause a huge loss of property, loss of life, and environmental pollution. Transparent intelligent models can help in preventing such fatalities. Kim et al. [36] developed an interpretable maritime accident prediction system using XGBoost. PDP was used as an interpretation visualization tool, and both LIME and SHAP were used as explainer techniques. In their study, Huang et al. [30] proposed an automatic collision avoidance system called HMI-CAS whose decision-making process is interpretable and interactive for human operators. The focus of this study is much on hybrid intelligence where machine and human share their intelligence in solving collision avoidance problem. Generalized velocity obstacle (GVO) algorithm has been applied which allows visualizing the solution space, so human operators can easily detect rule-violation behavior and intervene.

Interpretability is also studied in the Autonomous Marine Vehicles (AMV) context. The ultimate aim of AMV is to remove human operators in the control loop, which is still unrealistic. However, XAI is a prerequisite in order to approach this goal. In a theoretical study, Veitch et al. [67] investigated the following research question: can a human-centered design approach to XAI contribute to align Autonomous Surface Vehicles (ASV) technology toward real-world stakeholders? The work contributes to construct and promote the concept of "human-centered XAI" to address the distinct needs of end user interactions for ASVs. It underlies that as ASVs scale for widespread deployment, design practices are orienting towards end user interaction design. However, this is occurring without the guidance of a fully formulated conceptualization of what this design practice entails. In Gjaerum et al. [20] work, linear model trees are used to approximate the deep reinforcement learning model controlling an ASV in a simulated environment and then run in parallel with the deep model to give explanations. Visualization of the docking agent, the environment, and the feature attributions are proposed in an adaptative way. Indeed, two different visualizations were suggested for two different users—the developer and the seafarer/operator. Chowdhury et al. [13] developed a multi-objective planner for optimal path planning of autonomous agents in stochastic dynamic flows. The planner finds optimal routes in stochastic dynamic ocean environments with dynamic obstacles. The authors suggested that their solution can serve as the benchmark for other approximate AI algorithms such as reinforcement learning and help improve the explainability of those models.

Like in ATM literature, the study of interpretability is notably limited in waterborne transport. However, unlike ATM applications, research clusters can be identified in the maritime context. Driven by the maritime safety goal, most research addressed applications related to maritime surveillance, collision avoidance, and autonomous vessel related operations.

1.3.5 XAI in Railway Transport Systems

AI is influencing almost every bit of transportation. Railway is no exception. AI techniques would be most appropriate to tackle the challenges associated with modern smart railways. However, the literature clearly shows that AI is still in its very infancy stage for the railway sector compared to the other transport modes. As such, it comes as no surprise that interpretability has been much less investigated in railway transportation. In a recent survey where Besinovic et al. [7] introduced the basic concepts and possible applications of AI to railway academics and practitioners. The authors strongly stressed the particular importance of explainability for railways, especially in safety and security.

It is true that railways are typically among the safest form of transport. Thus, accident prediction and avoidance may not be the first AI application within the field. However, railway operations are so vulnerable to delays. AI, and thus interpretability, could be a game changer in rail traffic planning and management, railway maintenance, passenger mobility and untimely in autonomous train driving. Recently, some of these issues are starting to attract some research attention.

Zhang et al. [79] proposed a station delay prediction model based on graph community neural networks and time-series fuzzy decision tree. In the interpretability modeling process, this work used a multi-objective optimization of time series fuzzy decision tree based on error rate pruning algorithm to improve the model's interpretability and transparency. Jiang et al. [32] combined the interpretability of logistic regression models with the accuracy of random forest models to create a hybrid model predicting punctuality for the Sweden railway. The work of Hak et al. [25] aims at estimating passengers' preference between local and express trains based on XGBoost model. SHAP is used to interpret the importance of individual features. Earlier, Oneto et al. [55] investigated the problem of analyzing train movements in large-scale railway networks for the purpose of understanding and predicting their behavior. They focused on predicting the running time of a train between two stations, the dwell time of a train in a station, the train delay, the penalty costs associated with a delay, and the train overtaking between two trains which are in the wrong relative position on the railway network. For this, they used an interpretable decision-tree based model relying on the knowledge of the network and the experience of the operators as well as on historical data about the network. To support smart maintenance decisions, Bukhsh et al. [8] developed a model to predict maintenance need, activity type and trigger's status of railway switches. The predictive model is based on the decision tree, random forest, and gradient boosted trees. To facilitate the model's interpretability, LIME was used to provide a detailed explanation of models' predictions by features importance analysis and instance level details. Zhu et al. [82] proposed hybrid PVAR-NN prediction methods to predict passenger flow in the railway system. The model combined a hybrid linear and nonlinear time series analysis model, which uses the panel vector autoregression (PVAR) and neural networks. The advantage of the PVAR method is providing an accurate analysis of the marginal effects of influencing factors, which enhances interpretability.

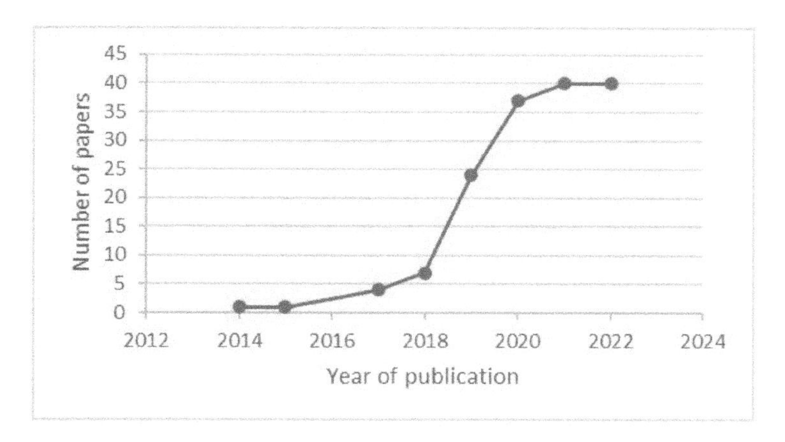

Figure 1.3 Temporal distribution of the reviewed papers.

1.4 DISCUSSION

A bird's eye view of the above literature exposes the clear trend of the transportation community to enter the race towards designing intelligent systems incorporating both accurate predictive and useful explainable models. Figure 1.3 shows the distribution of the included studies by year of publication. An upward trend in publications is noted in the last four years. 91% of the reviewed studies have been published since 2019. However, as a relatively conservative industry, we noted that transportation is slower than other fields in accepting advanced AI innovations, consequently, the pace of research on interpretability is a bit slow compared to other safety-critical field such as healthcare. A quick query on Scopus returns more than twenty review works on XAI in the healthcare field. Each analyzed hundreds of papers on the topic. The increasing number of recent survey papers in the area evidences the increasing emphasis on XAI in this field. While in transportation, reviews on the matter are absent from the literature. This alone is an indicator that although is growing, the body of research on interpretability still has not had the expected impact on the transportation community.

Down into the ITS categories, a wide variation in the response to the interpretability requirement is observed among transportation subfields. As depicted in Figure 1.4, developments in road transport are the most advanced. Specifically, most researches are conducted in the road autonomous vehicles field. The situation is different in air, maritime and rail transportation. only 17% of the studies belong to these subfields, which is marginal compared to road transport. In fact, the extent of attention given to interpretability is directly correlated with the AI implication in a given subfield. AI is an integral part of the self-driven car industry. Naturally, model interpretability should be addressed as part of the system validation. However, for railway transport systems interpretability is barely addressed because the overall industry does not rely on AI.

Another notable fact is that clearly, XAI is not the main target of the general transportation community. In the studied literature, explainability is mostly approached as a secondary or complementary to the work's main goals. That means

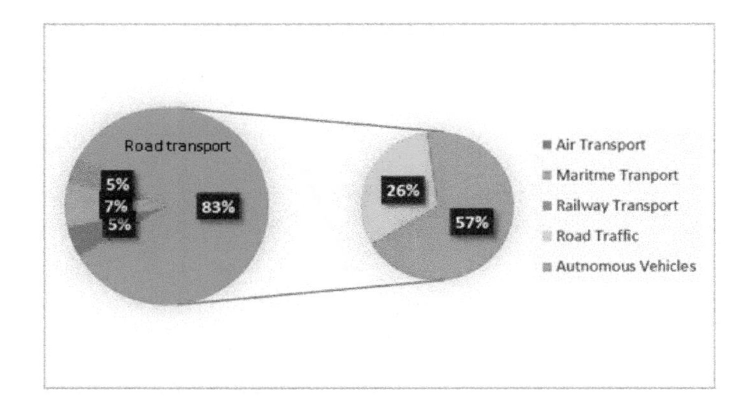

Figure 1.4 Classification of papers based on the transportation sectorss.

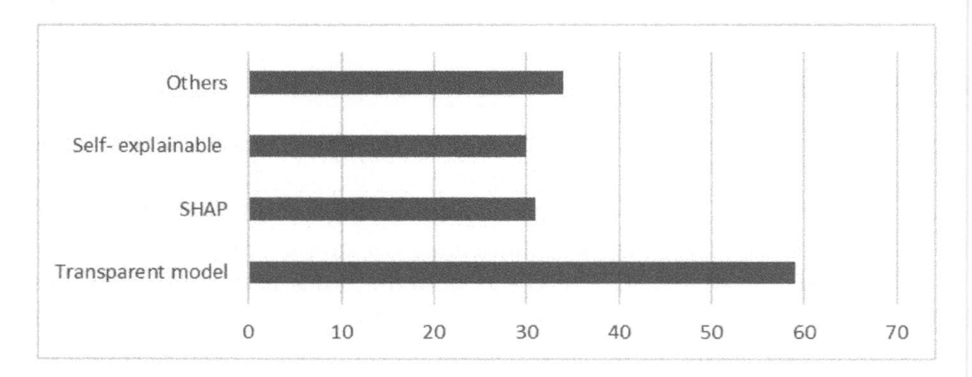

Figure 1.5 Classification of papers based on the XAI model used.

that contributions essentially aim at proposing predictive models, and as a part of performance evaluation, researchers highlight the capability of their models to interpret as extra propriety. Very recently, some works began to address interpretability as the main purpose. For instance, without studying the predictive models and by focusing exclusively on explainability, Gyevnar [23] proposed "Explainable Autonomous Vehicle Intelligence (XAVI)," a typical architecture of a system that delivers relevant and accurate explanations about its behavior in a natural, conversational manner.

Furthermore, analyzing the interpretable strategies adopted by the reviewed works reveals that in the current transportation literature, we can distinguish between four types of XAI techniques, (i) transparent models, (ii) self-explainable models, (iii) SHAP techniques, and (iv) others techniques. Figure 1.5 describes the distribution of the reviewed paper per the XAI technique used.

1.5 CONCLUSION

Inscrutable AI systems are difficult to trust, especially if they are used in high-stakes scenarios like transportation. Interpretable AI holds the potential to engender the sought trust and potentially mitigate various kinds of bias. While the potential benefits are immense, as matter of fact, interpretability is not yet attracting so much attention in intelligent transportation. In this chapter, we have provided an extensive review of the state-of-art of interpretable AI in smart transportation applications. Results reveal, that, overall, interpretability research is still in its early stage compared to other domains such as healthcare. Specifically, the extent of research varies according to the transport modes, most research has been made in road transport, while fewer explainable models have been proposed in railway, marine, and air transport. The data also shows that transparent models and SHAP are the most widely used XAI techniques in transportation applications.

Given its societal and technical implications, interpretability needs to be more often addressed in transportation literature. Explaining intelligent models needs to be set as the main target and not just as complementary propriety. In this sense, more efforts should be deployed toward (i) developing self-explainable DNN models, (ii) integrating human stakeholders in the design and validation of ITS solutions, and (iii) systemizing the quality assessment and evaluation of the produced explanations.

Bibliography

[1] Amr Abdelraouf, Mohamed Abdel-Aty, and Jinghui Yuan. Utilizing attention-based multi-encoder-decoder neural networks for freeway traffic speed prediction. *IEEE Transactions on Intelligent Transportation Systems*, 23(8):11960–11969, 2021.

[2] Amina Adadi and Mohammed Berrada. Peeking inside the black-box: a survey on explainable artificial intelligence (xai). *IEEE Access*, 6:52138–52160, 2018.

[3] Cristian Arteaga, Alexander Paz, and JeeWoong Park. Injury severity on traffic crashes: A text mining with an interpretable machine-learning approach. *Safety Science*, 132:104988, 2020.

[4] Shahin Atakishiyev, Mohammad Salameh, Hengshuai Yao, and Randy Goebel. Explainable artificial intelligence for autonomous driving: An overview and guide for future research directions. *arXiv: 2112.11561*, 2021.

[5] Alejandro Barredo-Arrieta, Ibai Lana, and Javier Del Ser. What lies beneath: A note on the explainability of black-box machine learning models for road traffic forecasting. In *2019 IEEE Intelligent Transportation Systems Conference (ITSC)*, pages 2232–2237. IEEE, 2019.

[6] Lidia Maria Belmonte, Rafael Morales, and Antonio Fernandez-Caballero. Computer vision in autonomous unmanned aerial vehicles—a systematic mapping study. *Applied Sciences*, 9(15):3196, 2019.

[7] Nikola Besinovic, Lorenzo De Donato, Francesco Flammini, Rob M. P. Goverde, Zhiyuan Lin, Ronghui Liu, Stefano Marrone, Roberto Nardone, Tianli Tang, and Valeria Vittorini. Artificial intelligence in railway transport: Taxonomy,

regulations, and applications. *IEEE Transactions on Intelligent Transportation Systems*, 23(9):14011–14024, 2021.

[8] Zaharah Allah Bukhsh, Aaqib Saeed, Irina Stipanovic, and Andre G Doree. Predictive maintenance using tree-based classification techniques: A case of railway switches. *Transportation Research Part C: Emerging Technologies*, 101:35–54, 2019.

[9] Nadia Burkart, Marco F Huber, and Mathias Anneken. Supported decision-making by explainable predictions of ship trajectories. In *15th International Conference on Soft Computing Models in Industrial and Environmental Applications (SOCO 2020) 15*, pages 44–54. Springer, 2021.

[10] Rui Cao, YuPing Lu, and Zhen He. System identification method based on interpretable machine learning for unknown aircraft dynamics. *Aerospace Science and Technology*, 126:107593, 2022.

[11] Jianyu Chen, Shengbo Eben Li, and Masayoshi Tomizuka. Interpretable end-to-end urban autonomous driving with latent deep reinforcement learning. *IEEE Transactions on Intelligent Transportation Systems*, 23(6):5068–5078, 2021.

[12] Sikai Chen, Jiqian Dong, Runjia Du, Yujie Li, and Samuel Labi. Reason induced visual attention for explainable autonomous driving. *arXiv preprint arXiv:2110.07380*, 2021.

[13] Rohit Chowdhury, Atharva Navsalkar, and Deepak Subramani. Gpu-accelerated multi-objective optimal planning in stochastic dynamic environments. *Journal of Marine Science and Engineering*, 10(4):533, 2022.

[14] SAE On-Road Automated Vehicle Standards Committee et al. Taxonomy and definitions for terms related to driving automation systems for on-road motor vehicles. *SAE International: Warrendale, PA, USA*, 2018.

[15] Augustin Degas, Mir Riyanul Islam, Christophe Hurter, Shaibal Barua, Hamidur Rahman, Minesh Poudel, Daniele Ruscio, Mobyen Uddin Ahmed, Shahina Begum, Md Aquif Rahman, et al. A survey on artificial intelligence (ai) and explainable ai in air traffic management: Current trends and development with future research trajectory. *Applied Sciences*, 12(3):1295, 2022.

[16] digital strategy.ec.europa.eu. Ethics guidelines for trustworthy ai. `https://digital-strategy.ec.europa.eu/en/library/ethics-guidelines-trustworthy-ai`.

[17] Jelena Fiosina. Explainable federated learning for taxi travel time prediction. In *VEHITS*, pages 670–677, 2021, Scitepress.

[18] Oliver Gallitz, Oliver De Candido, Michael Botsch, and Wolfgang Utschick. Interpretable feature generation using deep neural networks and its application to lane change detection. In *2019 IEEE Intelligent Transportation Systems Conference (ITSC)*, pages 3405–3411. IEEE, 2019.

[19] Briti Gangopadhyay, Harshit Soora, and Pallab Dasgupta. Hierarchical program-triggered reinforcement learning agents for automated driving. *IEEE Transactions on Intelligent Transportation Systems*, 23(8):10902–10911, 2021.

[20] Vilde B Gjaerum, Inga Strumke, Ole Andreas Alsos, and Anastasios M Lekkas. Explaining a deep reinforcement learning docking agent using linear model trees with user adapted visualization. *Journal of Marine Science and Engineering*, 9(11):1178, 2021.

[21] Sim Kuan Goh, Zhi Jun Lim, Sameer Alam, and Narendra Pratap Singh. Tunnel gaussian process model for learning interpretable flight's landing parameters. *Journal of Guidance, Control, and Dynamics*, 44(12):2263–2275, 2021.

[22] Jingjing Gu, Qiang Zhou, Jingyuan Yang, Yanchi Liu, Fuzhen Zhuang, Yanchao Zhao, and Hui Xiong. Exploiting interpretable patterns for flow prediction in dockless bike sharing systems. *IEEE Transactions on Knowledge and Data Engineering*, 34(2):640–652, 2020.

[23] Balint Gyevnar. Cars that explain: Building trust in autonomous vehicles through explanations and conversations.

[24] Balint Gyevnar, Massimiliano Tamborski, Cheng Wang, Christopher G Lucas, Shay B Cohen, and Stefano V Albrecht. A human-centric method for generating causal explanations in natural language for autonomous vehicle motion planning. *arXiv preprint arXiv:2206.08783*, 2022.

[25] Eun Hak Lee, Kyoungtae Kim, Seung-Young Kho, Dong-Kyu Kim, and Shin-Hyung Cho. Estimating express train preference of urban railway passengers based on extreme gradient boosting (xgboost) using smart card data. *Transportation Research Record*, 2675(11):64–76, 2021.

[26] Jian He, Yong Hao, and Xiaoqiong Wang. An interpretable aid decision-making model for flag state control ship detention based on smote and xgboost. *Journal of Marine Science and Engineering*, 9(2):156, 2021.

[27] Lei He, Nabil Aouf, and Bifeng Song. Explainable deep reinforcement learning for uav autonomous path planning. *Aerospace science and technology*, 118:107052, 2021.

[28] Afroza Hossain. *Application of Interpretable Machine Learning in Flight Delay Detection*. The University of Texas at Arlington, 2021.

[29] Sakif Hossain, Fatema T Johora, Jorg P Muller, Sven Hartmann, and Andreas Reinhardt. Sfmgnet: A physics-based neural network to predict pedestrian trajectories. *arXiv preprint arXiv:2202.02791*, 2022.

[30] Yamin Huang, Linying Chen, Rudy R Negenborn, and PHAJM Van Gelder. A ship collision avoidance system for human-machine cooperation during collision avoidance. *Ocean Engineering*, 217:107913, 2020.

[31] Mir Riyanul Islam, Mobyen Uddin Ahmed, Shaibal Barua, and Shahina Begum. A systematic review of explainable artificial intelligence in terms of different application domains and tasks. *Applied Sciences*, 12(3):1353, 2022.

[32] Sida Jiang, Christer Persson, and Joel Akesson. Punctuality prediction: combined probability approach and random forest modelling with railway delay statistics in sweden. In *2019 IEEE Intelligent Transportation Systems Conference (ITSC)*, pages 2797–2802. IEEE, 2019.

[33] Xia Jiang, Jian Zhang, and Bo Wang. Energy-efficient driving for adaptive traffic signal control environment via explainable reinforcement learning. *Applied Sciences*, 12(11):5380, 2022.

[34] Arash Kalatian and Bilal Farooq. Decoding pedestrian and automated vehicle interactions using immersive virtual reality and interpretable deep learning. *Transportation research part C: emerging technologies*, 124:102962, 2021.

[35] Blen M Keneni, Devinder Kaur, Ali Al Bataineh, Vijaya K Devabhaktuni, Ahmad Y Javaid, Jack D Zaientz, and Robert P Marinier. Evolving rule-based explainable artificial intelligence for unmanned aerial vehicles. *IEEE Access*, 7:17001–17016, 2019.

[36] Gyeongho Kim and Sunghoon Lim. Development of an interpretable maritime accident prediction system using machine learning techniques. *IEEE Access*, 10:41313–41329, 2022.

[37] Jinkyu Kim, Anna Rohrbach, Trevor Darrell, John Canny, and Zeynep Akata. Textual explanations for self-driving vehicles. In *Proceedings of the European conference on computer vision (ECCV)*, pages 563–578, 2018, CVF Open Access.

[38] Trevor Kistan, Alessandro Gardi, and Roberto Sabatini. Machine learning and cognitive ergonomics in air traffic management: Recent developments and considerations for certification. *Aerospace*, 5(4):103, 2018.

[39] Friedrich Kruber, Jonas Wurst, and Michael Botsch. An unsupervised random forest clustering technique for automatic traffic scenario categorization. In *2018 21st International conference on intelligent transportation systems (ITSC)*, pages 2811–2818. IEEE, 2018.

[40] Sebastian Lapuschkin, Stephan Waldchen, Alexander Binder, Gregoire Montavon, Wojciech Samek, and Klaus-Robert Muller. Unmasking clever hans predictors and assessing what machines really learn. *Nature Communications*, 10(1):1096, 2019.

[41] Runmei Li, Yongchao Hu, and Qiuhong Liang. T2f-lstm method for long-term traffic volume prediction. *IEEE Transactions on Fuzzy Systems*, 28(12):3256–3264, 2020.

[42] Yanfen Li, Hanxiang Wang, L Minh Dang, Tan N Nguyen, Dongil Han, Ahyun Lee, Insung Jang, and Hyeonjoon Moon. A deep learning-based hybrid framework for object detection and recognition in autonomous driving. *IEEE Access*, 8:194228–194239, 2020.

[43] Xiao Lu, Yaonan Wang, Xuanyu Zhou, Zhenjun Zhang, and Zhigang Ling. Traffic sign recognition via multi-modal tree-structure embedded multi-task learning. *IEEE Transactions on Intelligent Transportation Systems*, 18(4):960–972, 2016.

[44] Xiaolei Ma, Chuan Ding, Sen Luan, Yong Wang, and Yunpeng Wang. Prioritizing influential factors for freeway incident clearance time prediction using the gradient boosting decision trees method. *IEEE Transactions on Intelligent Transportation Systems*, 18(9):2303–2310, 2017.

[45] Harsh Mankodiya, Dhairya Jadav, Rajesh Gupta, Sudeep Tanwar, Wei-Chiang Hong, and Ravi Sharma. Od-xai: Explainable ai-based semantic object detection for autonomous vehicles. *Applied Sciences*, 12(11):5310, 2022.

[46] Harsh Mankodiya, Mohammad S Obaidat, Rajesh Gupta, and Sudeep Tanwar. Xai-av: Explainable artificial intelligence for trust management in autonomous vehicles. In *2021 International Conference on Communications, Computing, Cybersecurity, and Informatics (CCCI)*, pages 1–5. IEEE, 2021.

[47] Aboli Marathe, Pushkar Jain, Rahee Walambe, and Ketan Kotecha. Restorex-ai: A contrastive approach towards guiding image restoration via explainable ai systems. In *Proceedings of the IEEE/CVF Conference on Computer Vision and Pattern Recognition*, pages 3030–3039, 2022.

[48] Tim Miller. Explanation in artificial intelligence: Insights from the social sciences. *Artificial Intelligence*, 267:1–38, 2019.

[49] Leticia Monje, Ramon A Carrasco, Carlos Rosado, and Manuel Sanchez-Montanes. Deep learning xai for bus passenger forecasting: A use case in spain. *Mathematics*, 10(9):1428, 2022.

[50] Artur Movsessian, David Garcia Cava, and Dmitri Tcherniak. Interpretable machine learning in damage detection using shapley additive explanations. *ASCE-ASME Journal of Risk and Uncertainty in Engineering Systems, Part B: Mechanical Engineering*, 8(2):021101, 2022.

[51] W James Murdoch, Chandan Singh, Karl Kumbier, Reza Abbasi-Asl, and Bin Yu. Definitions, methods, and applications in interpretable machine learning. *Proceedings of the National Academy of Sciences*, 116(44):22071–22080, 2019.

[52] Marion Neumeier, Michael Botsch, Andreas Tollkuhn, and Thomas Berberich. Variational autoencoder-based vehicle trajectory prediction with an interpretable latent space. In *2021 IEEE International Intelligent Transportation Systems Conference (ITSC)*, pages 820–827. IEEE, 2021.

[53] Tomasz Nowak, Michal R Nowicki, Krzysztof Cwian, and Piotr Skrzypczynski. How to improve object detection in a driver assistance system applying explainable deep learning. In *2019 IEEE Intelligent Vehicles Symposium (IV)*, pages 226–231. IEEE, 2019.

[54] Daniel Omeiza, Helena Webb, Marina Jirotka, and Lars Kunze. Explanations in autonomous driving: A survey. *IEEE Transactions on Intelligent Transportation Systems*, 23(8):10142–10162, 2021.

[55] Luca Oneto, Irene Buselli, Alessandro Lulli, Renzo Canepa, Simone Petralli, and Davide Anguita. A dynamic, interpretable, and robust hybrid data analytics system for train movements in large-scale railway networks. *International Journal of Data Science and Analytics*, 9:95–111, 2020.

[56] International Civil Aviation Organization. Global aviation security plan. `https://www.icao.int/Security/Documents/GLOBAL\%20AVIATION\%20SECURITY\%20PLAN\%20EN.pdf`.

[57] Janak Parmar, Pritikana Das, and Sanjaykumar M Dave. A machine learning approach for modelling parking duration in urban land-use. *Physica A: Statistical Mechanics and its Applications*, 572:125873, 2021.

[58] Alun Preece, Dan Harborne, Dave Braines, Richard Tomsett, and Supriyo Chakraborty. Stakeholders in explainable ai. *arXiv preprint arXiv:1810.00184*, 2018.

[59] Stefano Giovanni Rizzo, Giovanna Vantini, and Sanjay Chawla. Reinforcement learning with explainability for traffic signal control. In *2019 IEEE Intelligent Transportation Systems Conference (ITSC)*, pages 3567–3572. IEEE, 2019.

[60] Gaith Rjoub, Jamal Bentahar, and Omar Abdel Wahab. Explainable ai-based federated deep reinforcement learning for trusted autonomous driving. In *2022 International Wireless Communications and Mobile Computing (IWCMC)*, pages 318–323. IEEE, 2022.

[61] Abbas Sadat, Sergio Casas, Mengye Ren, Xinyu Wu, Pranaab Dhawan, and Raquel Urtasun. Perceive, predict, and plan: Safe motion planning through interpretable semantic representations. In *Computer Vision–ECCV 2020: 16th European Conference, Glasgow, UK, August 23–28, 2020, Proceedings, Part XXIII 16*, pages 414–430. Springer, 2020.

[62] Bibhudhendu Shukla, Ip-Shing Fan, and Ian Jennions. Opportunities for explainable artificial intelligence in aerospace predictive maintenance. *PHM Society European Conference*, volume 5, page 11, 2020.

[63] Ruoyu Sun, Shaochi Hu, Huijing Zhao, Mathieu Moze, Francois Aioun, and Franck Guillemard. Human-like highway trajectory modeling based on inverse reinforcement learning. In *2019 IEEE Intelligent Transportation Systems Conference (ITSC)*, pages 1482–1489. IEEE, 2019.

[64] W Unctad. United nations conference on trade and development. *Review of Maritime Transport*, 2014.

[65] Manjunatha Veerappa, Mathias Anneken, and Nadia Burkart. Evaluation of interpretable association rule mining methods on time-series in the maritime domain. In *Pattern Recognition. ICPR International Workshops and Challenges: Virtual Event, January 10–15, 2021, Proceedings, Part III*, pages 204–218. Springer, 2021.

[66] Manjunatha Veerappa, Mathias Anneken, Nadia Burkart, and Marco F Huber. Validation of xai explanations for multivariate time series classification in the maritime domain. *Journal of Computational Science*, 58:101539, 2022.

[67] Erik Veitch and Ole Andreas Alsos. Human-centered explainable artificial intelligence for marine autonomous surface vehicles. *Journal of Marine Science and Engineering*, 9(11):1227, 2021.

[68] Kai Wang and Fugee Tsung. Sparse and robust multivariate functional principal component analysis for passenger flow pattern discovery in metro systems. *IEEE Transactions on Intelligent Transportation Systems*, 23(7):8367–8379, 2021.

[69] Pengyue Wang, Yan Li, Shashi Shekhar, and William F Northrop. Uncertainty estimation with distributional reinforcement learning for applications in intelligent transportation systems: A case study. In *2019 IEEE Intelligent Transportation Systems Conference (ITSC)*, pages 3822–3827. IEEE, 2019.

[70] Wei Wang, Hanyu Zhang, Tong Li, Jianhua Guo, Wei Huang, Yun Wei, and Jinde Cao. An interpretable model for short term traffic flow prediction. *Mathematics and Computers in Simulation*, 171:264–278, 2020.

[71] Yibing Xie, Nichakorn Pongsakornsathien, Alessandro Gardi, and Roberto Sabatini. Explanation of machine-learning solutions in air-traffic management. *Aerospace*, 8(8):224, 2021.

[72] Wei Xiong, Zhenyu Xiong, and Yaqi Cui. An explainable attention network for fine-grained ship classification using remote-sensing images. *IEEE Transactions on Geoscience and Remote Sensing*, 60:1–14, 2022.

[73] Yanyan Xu, Qing-Jie Kong, Reinhard Klette, and Yuncai Liu. Accurate and interpretable Bayesian mars for traffic flow prediction. *IEEE Transactions on Intelligent Transportation Systems*, 15(6):2457–2469, 2014.

[74] Gopikrishna Yadam, Akshata Kishore Moharir, and Ishita Srivastava. Explainable and visually interpretable machine learning for flight sciences. In *2020 IEEE International Conference on Electronics, Computing and Communication Technologies (CONECCT)*, pages 1–6. IEEE, 2020.

[75] Ran Yan and Shuaian Wang. Ship detention prediction using anomaly detection in port state control: Model and explanation. *Electronic Research Archive*, 30:3679–3691, 2022.

[76] Ran Yan, Shuaian Wang, Lu Zhen, and Gilbert Laporte. Emerging approaches applied to maritime transport research: Past and future. *Communications in Transportation Research*, 1:100011, 2021.

[77] Wenyuan Zeng, Wenjie Luo, Simon Suo, Abbas Sadat, Bin Yang, Sergio Casas, and Raquel Urtasun. End-to-end interpretable neural motion planner. In *Proceedings of the IEEE/CVF Conference on Computer Vision and Pattern Recognition*, pages 8660–8669. IEEE, 2019.

[78] Xuehao Zhai, Fangce Guo, and Aruna Sivakumar. Opening the black box: An interpretability evaluation framework using land-use probing tasks to understand deep learning traffic prediction models. *Available at SSRN 4109859.*

[79] Dalin Zhang, Yi Xu, Yunjuan Peng, Chenyue Du, Nan Wang, Mincong Tang, Lingyun Lu, and Jiqiang Liu. An interpretable station delay prediction model based on graph community neural network and time-series fuzzy decision tree. *IEEE Transactions on Fuzzy Systems*, 31(2):421–433, 2022.

[80] Yanru Zhang and Ali Haghani. A gradient boosting method to improve travel time prediction. *Transportation Research Part C: Emerging Technologies*, 58:308–324, 2015.

[81] Feng Zhou, X Jessie Yang, and Joost CF de Winter. Using eye-tracking data to predict situation awareness in real time during takeover transitions in conditionally automated driving. *IEEE Transactions on Intelligent Transportation Systems*, 23(3):2284–2295, 2021.

[82] Ruiqi Zhu and Huiyu Zhou. Railway passenger flow forecast based on hybrid pvar-nn model. In *2020 IEEE 5th International Conference on Intelligent Transportation Engineering (ICITE)*, pages 190–194. IEEE, 2020.

II

Interpretable Methods for ITS Applications

CHAPTER 2

Towards Safe, Explainable, and Regulated Autonomous Driving

Shahin Atakishiyev

Department of Computing Science, University of Alberta, Edmonton, Canada

Mohammad Salameh

Huawei Technologies Canada Co., Ltd, Edmonton, Canada

Hengshuai Yao and Randy Goebel

Department of Computing Science, University of Alberta, Edmonton, Canada

CONTENTS

DOI: 10.1201/9781003324140-2

THERE HAS BEEN recent and growing interest in the development and deployment of autonomous vehicles, encouraged by the empirical successes of powerful artificial intelligence techniques (AI), especially in the applications of deep learning and reinforcement learning. However, as demonstrated by recent traffic accidents, autonomous driving technology is not fully reliable for safe deployment. As AI is the main technology behind the intelligent navigation systems of self-driving vehicles, both the stakeholders and transportation regulators require their AI-driven software architecture to be safe, explainable, and regulatory compliant. In this paper, we propose a design framework that integrates autonomous control, explainable AI (XAI), and regulatory compliance to address this issue, and then provide an initial validation of the framework with a critical analysis in a case study. Moreover, we describe relevant XAI approaches that can help achieve the goals of the framework.

2.1 INTRODUCTION

Autonomous driving is a rapidly growing field that has attracted increasing attention over the last decade. According to a recent report by Intel, the deployment of autonomous cars will reduce on-road travel by approximately 250 million hours and save about 585,000 lives per year between the years 2035 and 2045, just in the USA [24]. While these advantages certainly encourage the use of autonomous vehicles, there is also major public concern about the safety of this technology. This concern arises mainly from reports of recent accidents [28, 39, 55] with the involvement of autonomous or semi-autonomous cars, primarily attributed to improper use of semi-autonomous functions. This issue is a major drawback, impeding self-operating vehicles from being acceptable by road users and society at a wider level. As artificial intelligence techniques power autonomous vehicles' real-time decisions and actions, a malfunction of the vehicle's intelligent control system is considered the main focus of analysis in such mishaps. Hence, both road users and regulators require that the AI systems of autonomous vehicles should be "explainable," meaning that real-time decisions of such cars, particularly in critical traffic scenarios, should be intelligible in addition to being robust and safe (e.g., see Figure 2.1). In this case explainability needs not only to provide transparency on an individual failure but also broadly inform the process of public transportation regulatory compliance [11, 38]. Urging the right to an explanation, several regulators have established safety and compliance standards for Intelligent Transportation Systems. In this context, we are developing a general framework that can make autonomous vehicle manufacturers and involved users improve safety and regulatory compliance, and help autonomous vehicles become more publicly trustable and acceptable. With such a focus, our paper makes the following contributions:

- We propose a general design framework for explainable autonomous driving systems and validate the framework with a use case;

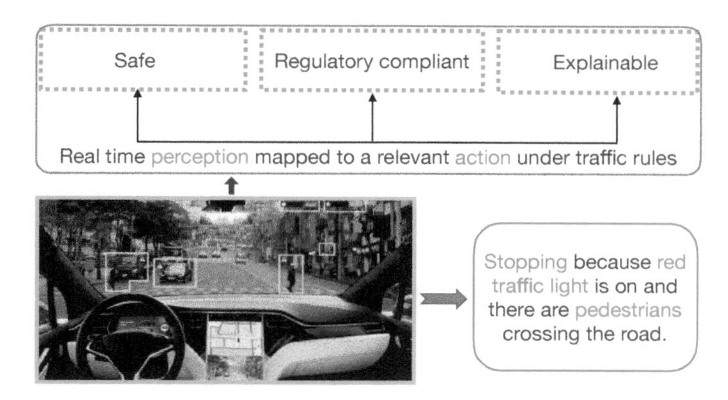

Figure 2.1 A graphical illustration of an autonomous car on its perception-action mapping that is safe, explainable, and regulatory compliant. *Safe* because the vehicle drives under traffic rules and does not hit people and touch other objects. *Explainable* because the vehicle provides a rationale for the taken action. *Regulatory compliant* because the vehicle follows all the traffic rules and guidelines. The red-colored text implies perception and the green-colored text is the corresponding action. An image in the left corner from [25].

- We present AI approaches from an algorithmic point of view that can help achieve explainability in autonomous driving and provide rationales behind an automated vehicle's real-time decisions.

The rest of the article is structured as follows. In Section 2.2, we provide a brief overview of modern autonomous driving and show the need for explainability in this technology. We introduce a relevant design framework in Section 2.3 and substantiate it with a case study in Section 2.4. Finally, in Section 2.5, we provide potential AI approaches that can help attain explainable autonomous driving systems and sum up the article with conclusions.

2.2 A GLANCE AT STATE-OF-THE-ART AUTONOMOUS DRIVING

Autonomous cars, also known as self-driving or driverless cars, are capable of perceiving their environment and making real-time driving decisions with the help of intelligent driving control systems. To capture the operational environment, autonomous vehicles leverage a variety of passive sensors (e.g., collecting information from the surrounding without emitting a wave, such as visible spectrum cameras) and active sensors (e.g., sending a wave and receiving a reflected signal from objects, such as lidar). Sensor devices detect changes in the environment and enable the driving system of the car to make real-time decisions on the captured information [8, 54]. Current autonomous vehicles deployed on real roads are classified as having different levels of automation. SAE International has defined six different levels of autonomous driving based on the expected in-vehicle technologies and level of intelligent system, namely Level 0 – No automation, Level 1 – Driving assistance, Level 2 – Partial automation, Level 3 – Conditional automation, Level 4 – High automation, and Level 5 – Full automation [37]. The anticipated increase of automation levels escalates reliance on an intelligent driving system rather than a human driver, particularly in Level 3 and above. However, such vehicles, even at Level 3, have recently caused several road accidents, cited above, that have led to severe injuries or even loss of human lives. Why did the accident happen? What malfunction of the driving system led to the crash? These questions naturally raise serious ethical and safety issues and provide the motivation for *explainable* AI (XAI) systems. In this context, the General Data Protection Regulation (GDPR) of the European Union (EU) established guidelines to promote a "right to explanation" for users, enacted in 2016 and carried into effect in May 2018 [13].

In another example, The National Highway Traffic Safety Administration (NHTSA) of the US Department of Transportation has issued a federal guideline on automated vehicle policy to attain enhanced road safety [27]. Current and future generation autonomous vehicles must comply with these emerging regulations, and their intelligent driving system should be explainable, transparent, and acceptably safe. In this regard, we propose a straightforward framework that considers the motivation for these requirements and then identify computational and legislative components that we believe are necessary for safe and transparent autonomous driving.

2.3 AN XAI FRAMEWORK FOR AUTONOMOUS DRIVING

We present a general design framework in which methods for developing end-to-end autonomous driving, XAI, and regulatory compliance are connected. In this approach, the framework consists of three main components: (1) an end-to-end autonomous systems component, (2) a safety-regulatory compliance component, and (3) an XAI component. Explainability in the context of autonomous driving can be thought of as the ability of an intelligent driving system to provide transparency with comprehensible explanations (1) for any action, (2) to support failure investigation, and (3) in support of the process of public transportation regulatory compliance. We describe the role of the aforementioned components individually in the following subsections.

2.3.1 An End-to-End Autonomous Systems Component

We need a simple but precise description of what we mean by "end-to-end autonomous systems." To start, we need to be able to refer to the set of actions that *any* autonomous vehicle is capable of executing. We consider the set of possible autonomous actions of automated vehicles as

$$A = \{a_1, a_2, \ldots a_n\}.$$

Notice we consider the list of executable actions as a finite repertoire of actions that can be selected by a predictive model, whether that model is constructed by hand or by a machine learning algorithm or by some combination. Example actions are things like "turn right," or "accelerate." For now, we refrain from considering complex actions like "decelerate and turn right" but acknowledge such actions will be possible in any given set A of actions, depending on the vehicle.

We use the notation C to denote an autonomous system controller and E to denote the set of all possible autonomous system operating environments. The overall function of an end-to-end controller is to map an environment E to an action A; we have an informal description of the role of the autonomous system controller as

$$C : E \mapsto A.$$

This mapping is intended to denote how *every* controller's responsibility is to map a perceived or sensed environment to an autonomous system function. We can provide a descriptive definition of an end-to-end autonomous controller as follows: A control system C is an *end-to-end control system* or $\boldsymbol{eeC}$, if C is a total function that maps every instance of an environment

$$e \in E$$

to a relevant action

$$a \in A.$$

Even with such a high-level description, we can note that the most essential attribute of an eeC is that it provides a complete mapping from any sensed environment to an action selection. We want this simple definition because we are not directly interested in sub-controllers for sub-actions; we are rather interested in autonomous controllers that provide complete control for high-level actions of any particular autonomous system, and which can be scrutinized for safety compliance for the whole scope of autonomous operation. Therefore, the eeC component is primarily responsible for perceiving the operational scene accurately using sensors such as video cameras, ultrasonics, lidar, radar, and other sensors, and enabling the car to take relevant actions.

2.3.2 A Safety-Regulatory Compliance Component

The role of our framework's safety-regulatory component, $\boldsymbol{srC}$, is to represent the function of a regulatory agency, whose primary role is to verify the safety of any configuration of eeC for an autonomous vehicle's repertoire of actions A.

We first note that safety compliance is a function that confirms the safety and reliability of an eeC system. This could be as pragmatic as an inspection of individual vehicle safety (e.g., some jurisdictions require processes that certify essential safety functions of individual vehicles for re-licensing), but more and more is associated with sophisticated compliance testing of eeC components from manufacturers, to establish their public safety approval (e.g., [48]). The srC components will vary widely from jurisdiction to jurisdiction (e.g., [11]). In addition, AI and machine learning techniques have largely emphasized the construction of predictive models for eeC systems. We can also note that all existing methods for software testing apply to this srC task, and we expect the evolution of srC processes will increasingly rely on the automation of compliance testing against all eeC systems. The complexity of srC systems lies within the scope of certified testing methods to confirm a threshold of safety. For example, from a compliance repertoire of N safety tests, a regulator may require a

90% performance of any particular eeC. With these expectations, the srC component should guarantee the following safety concepts as defined by [29]'s work:

- An autonomous car does not hit any static or dynamic objects on the road
- An autonomous car understands its performance abilities on the road
- An autonomous car does not block the traffic whether it is in motion or in a parking state
- An autonomous car does not exceed the speed limit specified in a road segment
- An autonomous car does not block the routes of an emergency vehicle
- An autonomous car watches out for cars driving at a red traffic light
- An autonomous car gets ready for a takeover by a backup driver in case automation capabilities are lost or malfunction

Thus, a proper combination of the eeC and srC components is a key to the correctness of the autonomous driving software system.

2.3.3 An XAI Component

The third component of our framework is a general XAI mechanism. At the highest level, the XAI component of our framework is responsible for providing transparency on how an eeC makes a selection from a repertoire of possible actions A. While the eeC and srC components have always been part of traditional autonomous driving systems, the explainability ability in real-time decisions remains unclear in the current state of the art. In particular, providing explanations in critical driving decisions is an essential factor to establish trust in the automation capability of autonomous driving technology. The difficulty of the challenge is directly related to those methods used to construct the control mechanism C of the eeC. For example, a machine learning technique will need to address the selection of actions based on the interpretation of the sensed environment as accurately as possible. This challenge is coupled with the demands created for the "explainee," which may range from natural language interaction with a human driver (e.g., why did the system suggest having the driver take over steering?) to the potentially very complex setting of compliance safety parameters for regulatory compliance testing software. In terms of delivery, the explanations of the XAI component could be communicated in two ways as specified in Figure 2.2:

- **Textual Explanations:** Natural language-based explanations can justify decisions of an automated vehicle linguistically to the stakeholders. For example, if a vehicle changes its route at some time step unexpectedly, it could produce such an explanation: "Changing predefined route *because* there is a traffic gridlock ahead."
- **Visual Explanations:** In this case, explanations are delivered to end-users in a visual format. For instance, a passenger can see the visual perception of

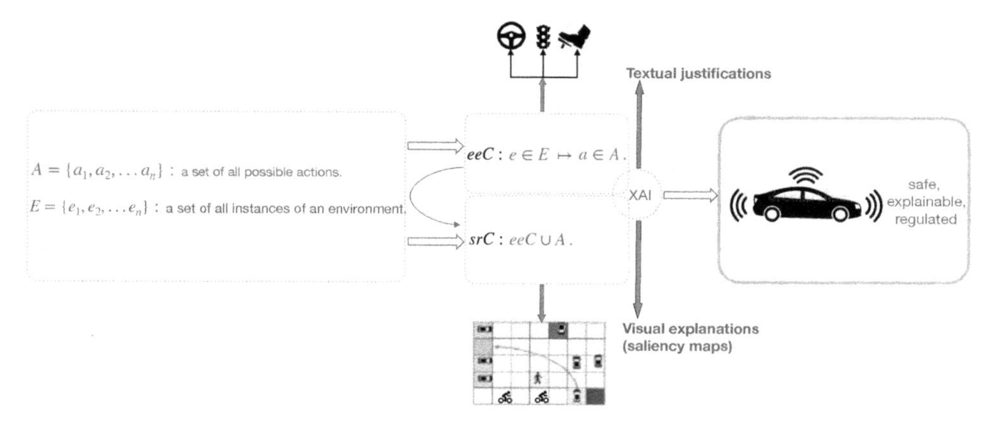

Figure 2.2 A graphical description of the proposed XAI framework for autonomous driving.

a car in an interface provided and judge appropriateness of real-time actions with respect to the visual information. Recently, [16]s' work, using saliency maps, shows how agents make their real-time decisions in Atari games. So, it would be worthwhile to apply the concept of saliency maps in the form of visual explanations to understand why a vehicle makes particular decisions in specific time steps.

The graphical description of the proposed framework is shown in Figure 2.2. As indicated, the three components of the framework are interrelated, and each component has a concrete role in the framework.

2.4 CASE STUDY: THE ROLE OF THE FRAMEWORK IN A POST-ACCIDENT INVESTIGATION

Transportation regulators within their specific jurisdictions initiate a set of protocols to analyze the cause of traffic accidents with the involvement of autonomous vehicles. In such cases, the primarily investigated questions are: Who caused the accident? Is this a fault of a vehicle's autonomous system, or did other road conditions and participants trigger the mishap? It is expected that autonomous cars can perform safe actions in all potential traffic scenarios that human drivers have commonly handled. ISO 26262 has established international safety standards for road automobiles that define safety rules and principles for passenger vehicles [18]. This standard requires that a self-driving car provide intelligible and evidence-based rationales for the safety of its decisions in the operational environment. Moreover, a vehicle should also provide information on potential residual risks of taking particular actions. As an example, we show a traffic scenario with an accident and describe the role of our framework to help identify the factors leading to this accident and to help further handle the situation appropriately according to traffic regulations.

For our simple case study, we first assume a simple traffic scenario in an uncontrolled four-way intersection where an accident with the presence of an autonomous

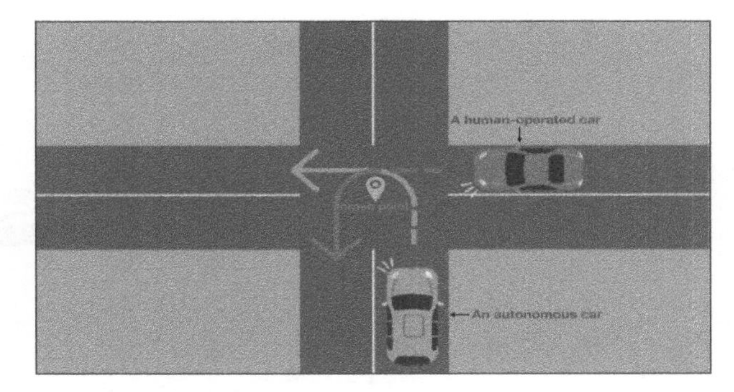

Figure 2.3 A traffic accident in an uncontrolled four-way intersection with autonomous and human-operated cars. Both vehicles try to turn left and incorrect coordination of time and maneuvers leads to a crash. As there may not be witnesses or other helpful arguments, an explainable AI-based driving system, storing a history of explanations alongside the relevant actions of an autonomous car can help understand which party made the wrong decision that caused the mishap. Graphics adapted from [7].

car (i.e., ego car) and another vehicle happens while both are attempting to make left turns in a given time step t_n (Figure 2.3). Accident investigators analyze the decisive actions of both cars at that particular time step. Suppose an autonomous vehicle records its action at t_n and provides an *explanation* as a justification for a taken action and the quantitative description of a residual risk associated with the performed action. Such functionality can help the investigators understand the main cause of the accident, i.e., whether an autonomous or other car took the wrong action. It immediately turns out that providing human interpretable justifications on a history of performed actions is a powerful tool both from a regulatory perspective in a post-accident analysis and from the lens of debugging and improving the existing automated system.

The other benefit of our framework is that it can help diminish the *responsibility*, *liability*, and *semantic* gaps in autonomous driving, as identified by [6]. In terms of the responsibility gap, the framework, with its ability to explain a history of temporal actions, can help regulators and inspectors conclude whom to blame for the mishap. Once the leading actor of the accident is identified, financial responsibility can also be determined, and the liability gap can be resolved. Finally, the debuggability of intelligent driving can reduce the semantic gap: By inspecting the history of previously taken actions, AI engineers can take the opportunity to improve an existing system by deploying improved AI techniques. Therefore, we see that an intelligent driving system with explainability features and safety verification within the predefined standards has manifold benefits both from legal and consumer perspectives.

2.5 XAI HEREAFTER: WHAT TO EXPECT?

We infer that end-to-end learning and motion trajectory of autonomous vehicles is based on precisely mapping perceived observations to corresponding actions. To date, perception has been mainly carried out through convolutional neural network (CNN) architectures and their augmented variations, and reinforcement learning (RL)-based approaches have proven computationally robust and adequate to map such real-time perception of states to relevant actions. Based on this intuition, we can note that explainable autonomous driving can be achieved by explaining the decisions of some combination of CNN/RL learning architecture. To conform to the framework presented, we provide directions for (1) explainable vision, (2) explainable RL, (3) representation of knowledge acquired from the operational environment, and (4) a question-driven software hierarchy for comprehensible action selection.

2.5.1 Use of Trusted Explainable Vision: Understanding Limitations and Improving CNNs

In recent studies, autonomous driving researchers and practitioners have already attempted to leverage CNN-based end-to-end learning. For example, Bojarski et al. [5] used convolutional neural networks to map camera inputs to steering actions and obtained impressive results with minimal training data. Their CNN architecture learned a complete set of driving operations for driving in situations with and without lane markings in sunny, rainy, and cloudy weather conditions. In a further work, [50] combined a dilated CNN with a long-short term memory (LSTM) network to predict a self-driving vehicle's future motions. Another empirically successful end-to-end learning example from vision is [47]s' study, where the authors used reinforcement learning along with a convolutional encoder to drive an autonomous car safely while considering road users (e.g., pedestrians) and respecting driving rules (e.g., lane keeping, collision prevention, and traffic light detection). Their ablation study of the method proved effectiveness of the proposed approach by winning the "Camera Only" track of the CARLA competition challenge [9]. So, the end-to-end learning approach has demonstrated its effectiveness with an appropriate choice of real-time computational decision-making methods. Consequently, explainable vision-directed actions for autonomous vehicles are based on how high-level features are used to detect and identify objects. CNNs, as the standard deep learning architectures, are "black-box" models, so there is a need to develop *explainable CNN* (XCNN) architectures to comply with the requirements of the proposed framework. In this direction, there has been recent research on explaining predictions of CNNs: DeepLift [36], Class Activation Maps (CAM) [58] and its extended versions such as Grad-CAM [33], Grad-CAM ++ [10], Guided Grad-CAM [46], as well as heuristics-based Deep Visual Explanations [1,2]. Motivated by such vision-based explainability methods, some recent studies have attempted to generate visual explanations for autonomous driving tasks. In a related project, [4] developed a method, called VisualBackProp, that shows which set of pixels are primarily and the most influential in triggering the predictions of convolutional networks. The authors Kim and Canny [20] introduced a vision-based framework using *causal attention* that shows what parts of an image control the steering

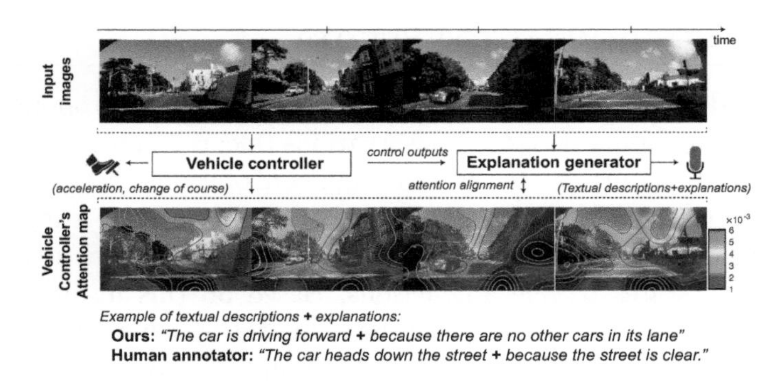

Figure 2.4 The intelligent driving system produces a textual explanation in a natural language, and a visual explanation with an attention mechanism on a taken action. Original source [21].

wheel for appropriate actions. The authors extended their work in their further study and produce intelligible textual explanations on the vehicle's actions [21] (e.g., see Figure 2.4). Such post hoc explanations are a promising step towards transparent autonomous driving and can be helpful to understand critical decisions of an automated vehicle from legislative and stakeholders' perspectives, as mentioned in our case study.

Taking a further strategic step, in addition to these post-hoc explanations, it is strongly argued that autonomous driving control systems also need to provide *intrinsic* explanations, where the developed system already becomes interpretable by design [30]. While post hoc explanations are useful for root cause identification in accident analysis, intrinsic explanations can be helpful to prevent such accidents. For instance, a back-up driver or an in-vehicle passenger can see textual explanations of critical decisions of a car with a suitable XAI interface while driving: in case these explanations do not reflect the actual decisions, the back-up driver can control the car with their input (if available) or end the trip and prevent potential accidents in more complicated situations, foreseeable ahead.

There are also at least two additional limitations of CNNs. Firstly, they need an enormous volume of training data to perform well in image recognition and object classification tasks. Moreover, conventional CNNs require all possible orientations of a sample image during training, in order to accurately identify and recognize unseen images which have various orientations and poses. They do not learn pose-invariant objects and this property hampers CNN 's modeling abilities. To overcome this downside of CNNs, [17] proposed the concept of *capsules*, a set of artificial neurons that capture orientation and represent pose parameters such as size, position, and skewing of an object. In their subsequent work, [31] they used a *dynamic routing* algorithm to train information passage between capsules at two successive layers. This feature enables the neural network to detect segments of an image and represent spatial representations differences between them although it is not yet clear how effective this can be. The hope is that the capsule neural network can detect an object in different shapes and poses even if a new sample of the same image has

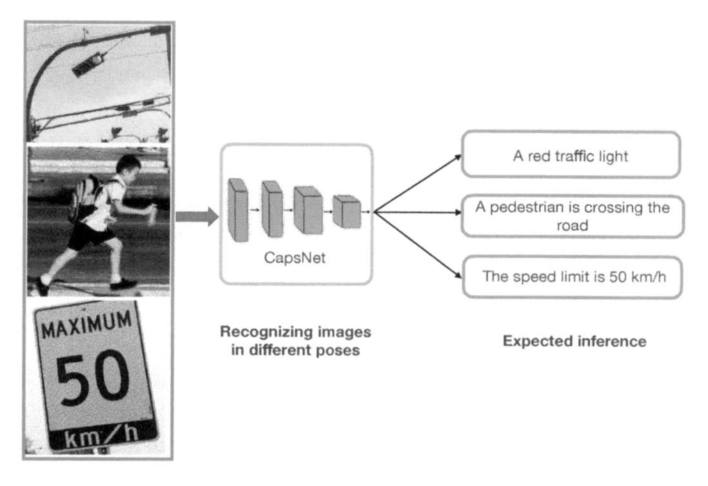

Figure 2.5 Three exemplary images in different poses: A traffic light sign affected by wind, a pedestrian leaning forward, and a speed limit sign with a changed vertical orientation. While CNNs work well with properly aligned images, changed shapes and poses may result in reduced accuracy in object recognition and classification tasks leading to further high-stakes consequences. CapsNet architectures, on the other hand, are able to detect objects with different poses. Similar to the provided images, we can see many other static and dynamic objects in everyday traffic that can be in various poses and shapes in specific times. CapsNet with its ability to detect objects in any poses can improve accuracy in object recognition and classification tasks in addition to its explainability advantage in perception tasks of autonomous driving.

not previously been used in the training phase. Sabour et al.'s most recent work [32] further improves the concept of dynamic routing and learns primary capsule encoders, which detect very tiny pieces of an image.

From the perspective of explainability, the learning mechanism of a capsule network makes this architecture intrinsically interpretable. For example, in the medical domain, Sharoudnejad et al. have empirically shown that likelihood and instantiation parameter vector values provide rational explanations in classification tasks on the 28×28 MNIST dataset and MRI images of patients with brain cancer [34]. Similarly, [23] have shown that their explainable capsule network, called X-Caps, carries human-interpretable features in the vectors of its capsules encoded as high-level visual attributes.

Overall, both explainability and the ability to recognize objects in different poses make CapsNets promising in vision tasks of autonomous driving. In particular, static road objects such as speed signs and traffic lights often undergo the impact of adverse weather conditions (e.g., snow and wind) and collisions by vehicles that change their stance angle and form (e.g., see Figure 2.5). As CNNs ultimately require regular shape and orientations of objects for real-time perception, CapsNet can improve their accuracy and consider the likelihood of the aforementioned pose, shape, and position changes along the traffic. Therefore, a CapsNet architecture's superior abilities with explainability and improved accuracy on object recognition tasks are promising research directions for the vision problems of autonomous driving.

2.5.2 Explanation Opportunity Using Model-Based Reinforcement Learning

Autonomous vehicles make sequential decisions throughout their motion trajectory within a setting formally characterized as a Markov Decision Process (MDP). The field of reinforcement learning (e.g., [41]) provides an MDP implementation architecture whose high-level goal is to estimate the differential reward of any possible action and to incrementally adjust the priority of making decisions by computing a policy that produces ranking preferred actions. An RL agent's interaction with the environment as an MDP can be implemented either as model-free or model-based RL. In a model-free setting (such as Q-learning), the algorithm does not have access to the dynamics of the environment (i.e., transition or reward function). It estimates the value function directly from a sensed experience [41]. So, model-free RL lacks explainability for learned policies. On the other hand, in the case of a model-based RL approach, an agent firstly tries to understand the world as prior knowledge, then develops a model to represent this world [22,26,41,52,53]. This approach in model-based RL is known as *planning*. The idea of planning in RL is essential for understanding the explicit components of decision-making and has powered further model-based architectures (i.e., the Dyna architectures [40, 44, 51]). While in a model-free setting, an RL agent *directly* learns a policy with the environment through interaction that produces a reward, in the Dyna and linear-Dyna style architectures an agent simultaneously learns a world model whilst learning an optimal policy through interactions with the world. Such a structure of planning makes it naturally *explainable*. Whether based on approximations of state descriptions from the world or an imaginary state defined by the model, the planning process uses a model representation to generate a predicted future trajectory. According to the model projection, an "optimal" action is decided at each planning step, which provides a predicted state and a predicted reward. The predicted states and rewards can be analyzed and visualized for the planned trajectory, thus providing an explanation of why the agent favors the choice of a particular action at a specific time step. Within that context, the potential benefits and perspectives of using Dyna architectures and model-based RL are huge for XAI and autonomous driving.

2.5.3 Leveraging Predictive Knowledge

It is important to specify how an agent could both represent and use the knowledge collected through interaction with the environment. This knowledge can be considered a compendium of predictions that the agent makes in anticipation of a selection from possible actions. Within the RL literature, such an approach to knowledge representation is called *predictive knowledge*, and has received significant attention in RL studies [12,42,45,49]. An agent regularly expects a response from the environment by making many predictions about the dynamics of the environment in accordance with the autonomous system's behavior. Therefore, predictive knowledge, as one of the essential notions of reinforcement learning, can be considered as prior knowledge about the possible worlds in which the corresponding actions might be taken. In order to consider a prediction as knowledge, a prediction should comply with the requirements of knowledge; first, a prediction should carry fundamental elements of epistemology – justification and truth, in itself [19]. Some, for example, [49] and [35] have shown that General Value Functions (GVFs) [42], an architecture to learn numerous value functions from experience, are a promising proposal for the robust representation of predictive knowledge. In particular, [35] has empirically validated that GVFs, as a scalable representation, form a predictive model of an agent's interaction with its environment and can represent an operational environment. Recent works on GVFs have proven their value in perception [15] and policy learning problems of real-world autonomous driving [14]. Motivated by these results, it is beneficial to further investigate the concept of predictive knowledge and its representation with GVFs, to evaluate the robustness of such a formalism for autonomous driving. However, it is noteworthy to point out that GVFs commit to a particular encoding of predictive knowledge, which leaves their explanatory value undetermined.

2.5.4 Temporal Questions and Question Hierarchies

If there is one common design principle that is shared by the autonomous driving industry, it is the commitment to arrange the control software in a hierarchical structure. To have a self-explainable decision procedure for autonomous driving, it is important that this hierarchy becomes *question driven*. In particular, the questions frequently arising while driving are those like "Am I going to see the traffic lights turning yellow shortly?", " Why is the car in front braking?", "Will the car in the front right of me cut to my lane?", etc. All these questions seem to come to us naturally when we drive. However, this concept of asking questions receives very little attention from current research on AI methods for autonomous driving. Understanding why human beings have this ability is vital to advancing the safety of autonomous machines. These questions don't emerge randomly but subconsciously; considering answers to these questions prepares us for a safe drive.

To better understand the usefulness of the question-answering concept, assuming an autonomous car's intelligent driving system produces such a question-answer pair in a natural language in a particular time step: Q: " *Why* did the car change its lane?" – A: "*Because* an obstruction was observed ahead in the current lane." Such a question-answering pair would mainly be helpful in case of accident investigation, and also could help inspectors to ask further follow-up questions.

From the RL perspective, one potential approach to generate temporal questions based on the ongoing actions would be to use the concept of *options* [43]. Options are the generalized concept of actions that has a policy for taking action with *terminal* conditions. The *option-critic architecture* was recently proposed by [3]. Both internal policies and terminal conditions of options have been experimentally successful in end-to-end learning of options in the Arcade Learning Environment (ALE).

The option-critic architecture is useful for temporal questions in autonomous driving; because driving-related questions are often temporal, new questions can be generated for the subsequent actions after just several seconds. Hence, it is important to study the formulation of questioning, with research focused on the generation, organization, and evaluation of questions for transparent autonomous driving [56]. Note also that the sensitivity of driving decisions varies dynamically in real time, creating different levels of risk for the vehicle. In general, low-risk actions and options are preferred. However, in safety stringent situations, we need to explore efficiently (possibly in real-time) to manage possible dangers. This step requires a representation of the risks in decision making with a principled way of evaluating the risks to decide which risk level the vehicle is to undertake. Experiments by [57] showed that considering no risks but only acting according to the maximum reward principle as in traditional RL is not always the best decision and can fail to discover the optimal policy. In contrast, acting according to a variety of levels of risk can find an optimal policy in environments with a diverse range of different transition and reward dynamics. We can thus infer and conclude that decision-making for autonomous vehicles is time-sensitive, often in sub-seconds, and a well-composed question hierarchy may not only help to select the present action but also determine subsequent actions that help sustain safe driving. The hierarchical structure can also provide temporal, informative, and reliable explanations in critical traffic situations with appropriate benefits.

2.6 CONCLUSIONS

As a result of our ongoing study, we presented a general design framework for XAI -based autonomous driving. We validated the framework with a case study of a post-accident analysis and showed that the proposed framework could have multi-faceted benefits to autonomous driving systems. First, the concept of the intrinsic and post-hoc explanations is not only limited to providing transparency on real-time decisions but also provides further opportunities to debug, fix, and enhance existing intelligent driving systems. Moreover, we showed that the principles of the proposed framework could address three actual issues, namely, responsibility, liability, and semantic gaps in the realm of autonomous driving. Finally, we presented some XAI approaches as future work and elucidated their potential in the explainability of autonomous driving.

While the presented propositions are promising directions to follow, whether these concepts work effectively in a real driving environment remains unclear for now and is a limitation of our preliminary study. As a next step, we are performing empirical research conforming to the principles of the presented framework. Currently, we are trying to incorporate visual question answering with model-based reinforcement learning to achieve explainable vision mapped to explainable actions. Two interesting areas for deeper investigation are (1) a comparative analysis of model-free and model-based reinforcement learning approaches on the same task, and (2) whether the interpretability of an AI architecture for intelligent driving results in reduced accuracy of the interpretable algorithm compared to its original version on the same task. Therefore the contributions of this paper are mainly theoretical and further empirical studies are needed for proof of concept. We anticipate that further practical work based on these propositions can be helpful towards public acceptance of autonomous driving technology.

2.7 ACKNOWLEDGMENTS

We acknowledge support from the Alberta Machine Intelligence Institute (Amii), from the Computing Science Department of the University of Alberta, and the Natural Sciences and Engineering Research Council of Canada (NSERC). Shahin Atakishiyev also acknowledges support from the Ministry of Science and Education of the Republic of Azerbaijan.

Bibliography

[1] Housam Babiker and Randy Goebel. Using KL-divergence to focus Deep Visual Explanation. *31st Neural Information Processing Systems Conference (NIPS), Interpretable ML Symposium. Long Beach, CA, USA*, 2017.

[2] Housam Khalifa Bashier Babiker and Randy Goebel. An Introduction to Deep Visual Explanation. *31st Neural Information Processing Systems Conference (NIPS), Long Beach, CA, USA*, 2017.

[3] Pierre-Luc Bacon, Jean Harb, and Doina Precup. The option-critic architecture. In *Proceedings of the AAAI Conference on Artificial Intelligence*, volume 31, 2017, AAAI.

[4] Mariusz Bojarski, Anna Choromanska, Krzysztof Choromanski, Bernhard Firner, Larry Jackel, Urs Muller, and Karol Zieba. Visualbackprop: visualizing CNNs for autonomous driving. *arXiv preprint arXiv:1611.05418*, 2, 2016.

[5] Mariusz Bojarski, Davide Del Testa, Daniel Dworakowski, Bernhard Firner, Beat Flepp, Prasoon Goyal, Lawrence D Jackel, Mathew Monfort, Urs Muller, Jiakai Zhang, et al. End to end learning for self-driving cars. *arXiv preprint arXiv:1604.07316*, 2016.

[6] Simon Burton, Ibrahim Habli, Tom Lawton, John McDermid, Phillip Morgan, and Zoe Porter. Mind the gaps: Assuring the safety of autonomous systems from an engineering, ethical, and legal perspective. *Artificial Intelligence*, 279:103201, 2020.

[7] Cameron McKay. Uncontrolled intersection rules in Alberta, 2021. Accessed on April 10, 2022.

[8] Sean Campbell, Niall O'Mahony, Lenka Krpalcova, Daniel Riordan, Joseph Walsh, Aidan Murphy, and Conor Ryan. Sensor technology in autonomous vehicles: A review. In *2018 29th Irish Signals and Systems Conference (ISSC)*, pages 1–4. IEEE, 2018.

[9] CARLA's blog. The CARLA Autonomous Driving Challenge, 2019. (Accessed on November 1, 2021).

[10] Aditya Chattopadhay, Anirban Sarkar, Prantik Howlader, and Vineeth N Balasubramanian. Grad-cam++: Generalized gradient-based visual explanations for deep convolutional networks. In *2018 IEEE Winter Conference on Applications of Computer Vision (WACV)*, pages 839–847. IEEE, 2018.

[11] Dentons. Global Guide to Autonomous Vehicles 2021. 2021.

[12] Gary L Drescher. *Made-up minds: A constructivist approach to artificial intelligence.* MIT Press, 1991.

[13] GDPR. Regulation EU 2016/679 of the European Parliament and of the Council of 27 April 2016. *Official Journal of the European Union*, 2016.

[14] Daniel Graves, Nhat M Nguyen, Kimia Hassanzadeh, and Jun Jin. Learning predictive representations in autonomous driving to improve deep reinforcement learning. *arXiv preprint arXiv:2006.15110*, 2020.

[15] Daniel Graves, Kasra Rezaee, and Sean Scheideman. Perception as prediction using general value functions in autonomous driving applications. In *2019 IEEE/RSJ International Conference on Intelligent Robots and Systems (IROS)*, pages 1202–1209. IEEE, 2019.

[16] Samuel Greydanus, Anurag Koul, Jonathan Dodge, and Alan Fern. Visualizing and understanding atari agents. In *International Conference on Machine Learning*, pages 1792–1801. PMLR, 2018.

[17] Geoffrey E Hinton, Alex Krizhevsky, and Sida D Wang. Transforming autoencoders. In *International Conference on Artificial Neural Networks*, pages 44–51. Springer, 2011.

[18] ISO 26262. Road vehicles-functional safety. *International Standard ISO/FDIS*, 26262, 2011.

[19] Alex Kearney and Patrick M Pilarski. When is a prediction knowledge? *arXiv preprint arXiv:1904.09024*, 2019.

[20] Jinkyu Kim and John Canny. Interpretable learning for self-driving cars by visualizing causal attention. In *Proceedings of the IEEE International Conference on Computer Vision*, pages 2942–2950. IEEE, 2017.

[21] Jinkyu Kim, Anna Rohrbach, Trevor Darrell, John Canny, and Zeynep Akata. Textual explanations for self-driving vehicles. In *Proceedings of the European Conference on Computer Vision (ECCV)*, pages 563–578. CVF, 2018.

[22] B Ravi Kiran, Ibrahim Sobh, Victor Talpaert, Patrick Mannion, Ahmad A Al Sallab, Senthil Yogamani, and Patrick Perez. Deep reinforcement learning for autonomous driving: A survey. *IEEE Transactions on Intelligent Transportation Systems*, 23(6):4909–4926, 2021.

[23] Rodney LaLonde, Drew Torigian, and Ulas Bagci. Encoding visual attributes in capsules for explainable medical diagnoses. In *International Conference on Medical Image Computing and Computer-Assisted Intervention*, pages 294–304. Springer, 2020.

[24] Roger Lanctot et al. Accelerating the future: The economic impact of the emerging passenger economy. *Strategy Analytics*, 5, 2017.

[25] Todd Litman. Autonomous vehicle implementation predictions: Implications for transport planning. 2021.

[26] Thomas M Moerland, Joost Broekens, and Catholijn M Jonker. Model-based reinforcement learning: A survey. *arXiv preprint arXiv:2006.16712*, 2020.

[27] NHTSA. *Federal automated vehicles policy: Accelerating the next revolution in roadway safety.* US Department of Transportation, 2016.

[28] NTSB. Collision between a sport utility vehicle operating with partial driving automation and a crash attenuator Mountain View, California. 2020.

[29] Andreas Reschka. Safety concept for autonomous vehicles. In *Autonomous Driving*, pages 473–496. Springer, 2016.

[30] Cynthia Rudin. Stop explaining black box machine learning models for high stakes decisions and use interpretable models instead. *Nature Machine Intelligence*, 1(5):206–215, 2019.

[31] Sara Sabour, Nicholas Frosst, and Geoffrey E Hinton. Dynamic routing between capsules. In *Advances in Neural Information Processing Systems*, volume 30, 2017.

[32] Sara Sabour, Andrea Tagliasacchi, Soroosh Yazdani, Geoffrey Hinton, and David J Fleet. Unsupervised part representation by flow capsules. In *International Conference on Machine Learning*, pages 9213–9223. PMLR, 2021.

[33] Ramprasaath R Selvaraju, Michael Cogswell, Abhishek Das, Ramakrishna Vedantam, Devi Parikh, and Dhruv Batra. Grad-CAM: Visual explanations from deep networks via gradient-based localization. In *Proceedings of the IEEE International Conference on Computer Vision*, pages 618–626. IEEE, 2017.

[34] Atefeh Shahroudnejad, Parnian Afshar, Konstantinos N Plataniotis, and Arash Mohammadi. Improved explainability of capsule networks: Relevance path by agreement. In *2018 IEEE Global Conference on Signal and Information Processing (GlobalSIP)*, pages 549–553. IEEE, 2018.

[35] Craig Sherstan. Representation and general value functions. *PhD thesis, University of Alberta*, 2020.

[36] Avanti Shrikumar, Peyton Greenside, and Anshul Kundaje. Learning important features through propagating activation differences. In *International Conference on Machine Learning*, pages 3145–3153. PMLR, 2017.

[37] Jennifer Shuttleworth. SAE standard news: J3016 automated-driving graphic update, 2019. Accessed online on August 16, 2021.

[38] R. Slagter and R. Voster. Autonomous Compliance Standing on the shoulders of RegTech! *Compact*, 2017.

[39] Neville A Stanton, Paul M Salmon, Guy H Walker, and Maggie Stanton. Models and methods for collision analysis: a comparison study based on the Uber collision with a pedestrian. *Safety Science*, 120:117–128, 2019.

[40] Richard S Sutton. Dyna, an integrated architecture for learning, planning, and reacting. *ACM Sigart Bulletin*, 2(4):160–163, 1991.

[41] Richard S Sutton and Andrew G Barto. *Reinforcement learning: An introduction.* MIT Press, Second edition, 2018.

[42] Richard S Sutton, Joseph Modayil, Michael Delp, Thomas Degris, Patrick M Pilarski, Adam White, and Doina Precup. Horde: A scalable real-time architecture for learning knowledge from unsupervised sensorimotor interaction. In *The 10th International Conference on Autonomous Agents and Multiagent Systems-Volume 2*, pages 761–768. IFAAMAS, 2011.

[43] Richard S Sutton, Doina Precup, and Satinder Singh. Between MDPs and semi-MDPs: A framework for temporal abstraction in reinforcement learning. *Artificial Intelligence*, 112(1–2):181–211, 1999.

[44] Richard S. Sutton, Csaba Szepesvari, Alborz Geramifard, and Michael H. Bowling. Dyna-style planning with linear function approximation and prioritized sweeping. In David A. McAllester and Petri Myllymäki, editors, *UAI 2008, Proceedings of the 24th Conference in Uncertainty in Artificial Intelligence, Helsinki, Finland, July 9-12, 2008*, pages 528–536. AUAI Press, 2008.

[45] Richard S Sutton and Brian Tanner. Temporal-difference networks. In *Advances in Neural Information Processing Systems*, pages 1377–1384, 2005.

[46] Ziqi Tang, Kangway V Chuang, Charles DeCarli, Lee-Way Jin, Laurel Beckett, Michael J Keiser, and Brittany N Dugger. Interpretable classification of Alzheimer's disease pathologies with a convolutional neural network pipeline. *Nature Communications*, 10(1):1–14, 2019.

[47] Marin Toromanoff, Emilie Wirbel, and Fabien Moutarde. End-to-end model-free reinforcement learning for urban driving using implicit affordances. In *Proceedings of the IEEE/CVF Conference on Computer Vision and Pattern Recognition*, pages 7153–7162. IEEE/CVF, 2020.

[48] Transport Canada. Guidelines for testing automated driving systems in Canada. *Ministry of Transportation of Canada*, 2021.

[49] Adam White. Developing a predictive approach to knowledge. *PhD thesis, University of Alberta*, 2015.

[50] Huazhe Xu, Yang Gao, Fisher Yu, and Trevor Darrell. End-to-end learning of driving models from large-scale video datasets. In *Proceedings of the IEEE Conference on Computer Vision and Pattern Recognition*, pages 2174–2182. IEEE, 2017.

[51] Hengshuai Yao, Shalabh Bhatnagar, Dongcui Diao, Richard S Sutton, and Csaba Szepesvari. Multi-step dyna planning for policy evaluation and control. In Y. Bengio, D. Schuurmans, J. Lafferty, C. Williams, and A. Culotta, editors, *Advances in Neural Information Processing Systems*, volume 22. Curran Associates, Inc., 2009.

[52] Hengshuai Yao and Csaba Szepesvari. Approximate policy iteration with linear action models. In *Proceedings of the AAAI Conference on Artificial Intelligence*, volume 26, pages 1212–1218. AAAI, 2012.

[53] Hengshuai Yao, Csaba Szepesvari, Bernardo Avila Pires, and Xinhua Zhang. Pseudo-mdps and factored linear action models. In *2014 IEEE Symposium on Adaptive Dynamic Programming and Reinforcement Learning (ADPRL)*, pages 1–9. IEEE, 2014.

[54] De Jong Yeong, Gustavo Velasco-Hernandez, John Barry, Joseph Walsh, et al. Sensor and sensor fusion technology in autonomous vehicles: A review. *Sensors*, 21(6):2140, 2021.

[55] Ekim Yurtsever, Jacob Lambert, Alexander Carballo, and Kazuya Takeda. A survey of autonomous driving: Common practices and emerging technologies. *IEEE Access*, 8:58443–58469, 2020.

[56] Eloi Zablocki, Hedi Ben-Younes, Patrick Perez, and Matthieu Cord. Explainability of deep vision-based autonomous driving systems: Review and challenges. *International Journal of Computer Vision*, pages 1–28, 2022.

[57] Shangtong Zhang, Borislav Mavrin, Linglong Kong, Bo Liu, and Hengshuai Yao. QUOTA: the quantile option architecture for reinforcement learning. *CoRR*, abs/1811.02073, 2018.

[58] Bolei Zhou, Aditya Khosla, Agata Lapedriza, Aude Oliva, and Antonio Torralba. Learning deep features for discriminative localization. In *Proceedings of the IEEE Conference on Computer Vision and Pattern Recognition*, pages 2921–2929. IEEE, 2016.

CHAPTER 3

Explainable Machine Learning Method for Predicting Road-Traffic Accident Injury Severity in Addis Ababa City Based on a New Graph Feature Selection Technique

Yassine Akhiat

Department of Informatics, LPAIS Laboratory, USMBA, Fez Morocco

Younes Bouchlaghem

Department of Informatics, UAE, Tetouan Morocco

Ahmed Zinedine

Department of Informatics, LPAIS Laboratory, USMBA, Fez Morocco

Mohamed Chahhou

Department of Informatics, UAE, Tetouan Morocco

CONTENTS

DOI: 10.1201/9781003324140-3

ROAD TRAFFIC ACCIDENTS (RTAs) are the leading cause of injuries and fatalities worldwide, especially in developing countries. Addis Ababa, with a demographic population estimated with at least 4 million dwellers, is the largest urban center in Ethiopia has the high road crash rate in the world. Recently, RTAs have become a challenging public health concern within the city as the trend of RTAs injuries and death is increasing drastically. To date, few studies have been proposed to identify the crucial factors contributing to road crashes and to classify fatalities. In addition, most AI applications lack explainability/transparency. In this paper, we developed an explainable method capable of classifying accident severity accurately and providing interpretations and explanations about the pivotal features and factors that may lead to road accidents based on the proposed graph feature selection for the sake of improving road safety.

The recorded results on Addis Ababa RTAs dataset confirm the validity of our proposed method and its superiority using five evaluation metrics and three well-known classification algorithms.

3.1 INTRODUCTION

Nowadays, the rapid growth of the population and the number of vehicles on the road has made road traffic accident one of the concerns of the modern times [2,19]. World Health Organization (WHO) has reported top ten disastrous reasons for taking human's life, and unfortunately road accidents come at ninth place, and by 2030 road traffic injuries are estimated to become the fifth leading cause of mortality globally [18,19]. The Global status report on road safety 2018, launched by WHO in December 2018, highlights that the number of annual road traffic deaths has attained 1.35 million. The highest number of RTA deaths are reported in particular in developing and low-income countries and its number is increasing from time to time [30,39]. This is resulting in negative consequences for the environment, health and the economy sector.

In Addis Ababa the capital of Ethiopia with a population of more than 4 million people, the situation has been worsened as the number of vehicles has increased consequently due to increased traffic flow and conflicts between vehicles and pedestrians [2,38]. The capital is encountering the highest rate TRAs resulting in fatalities and various levels of injuries. According to Addis Ababa Traffic Police Commission data, around 300 people are killed and 12,000 are seriously injured due to RTAs each

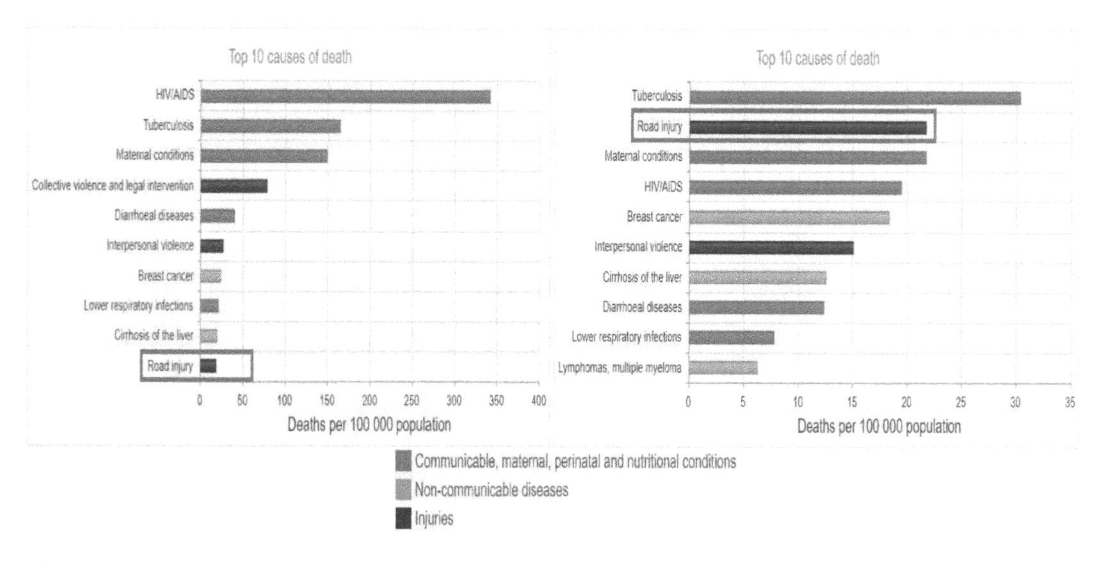

Figure 3.1 Top ten causes of death in Ethiopia per 100,000 population in 2000 and 2019.

year in Addis Ababa [11]. Furthermore, the numbers of road traffic injuries and deaths have been increasing from time to time in Ethiopia as it had the highest death rate due to road traffic injuries worldwide at 28 per 100,000 population in 2019 according to WHO (see Figures 3.1 and 3.2).

The cost of RTAs has brought unprecedented challenges that influence on the community in various levels [11,27,38]. Understanding circumstances leading to road traffic accidents is crucial to improve road safety via developing Intelligent Transportation Systems (ITS) counting on collected RTAs datasets. Machine learning is now a trending and powerful technique that facilitates the analyzing and processing

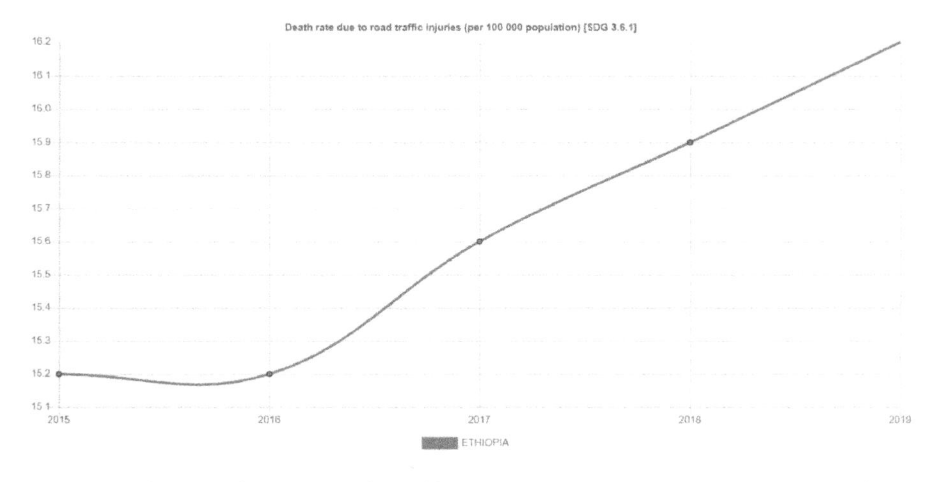

Figure 3.2 Death rate due to road traffic injuries per 100,000 population from 2015 and 2019.

huge amounts of data to quantify the hidden patterns among the data characteristics [10, 34]. This can contribute a lot to the decision-makers to propose improved road traffic safety and traffic crashes controlling policies, rules, and systems. The massive amount of RTA data generated through many resources and different features of governmental and non-governmental organizations can be used for early accident prediction and identification of the main factors of the accident using a plethora of machine learning tools. In literature, there are few researches that have been carried out regarding road traffic accidents in Ethiopia to address burden of RTA, types of vehicles involved in injury, severity of injury, situation of the victims during accident, body region injured by the accident, and other associated factors [10]. In this study, we put more emphasize of classifying accident severity and identifying their major impacting variables.

The costs of fatalities and injuries due to traffic accidents have a great impact on society [14, 26]. Recently, many researchers have paid increasing attention to detecting crucial factors that significantly affect injury severity due to car crashes [26]. A plethora of studies have been conducted to tackle this problem including neural networks, log-linear models, and fuzzy systems [28].

The authors of [25] have proposed a new system for injury severity levels classification based on real data gathered from Addis Ababa traffic office. The main purpose of the study was the development of a machine learning model able to analyze road accident severity for the sake of enhancing road safety in Addis Ababa. The limitation of this study is that the authors have selected the most informative features and factors counting on the importance of Decision Tree which in not the best choice as it can overfit on noisy features [9]. Therefore, training machine learning models on noisy features may lead to unreliable and unstable results [26, 28]. Another study is carried out in [15] to evaluate the crucial characteristics (factors) affecting injury severity levels.

In this paper, a new intelligent transportation system based on graph feature selection and ensemble learning is introduced to identify the major causes of road traffic accidents. Therefore, this research study addresses the following thorough key points:

- Identifying the major contributing features that can cause a road traffic accident and design mitigation.
- Which machine learning classification technique performs well in identifying the major causes of a road traffic accident?
- Providing accident severity classification.
- Offering an explainable and interpretable ITS.
- Providing some recommendations to enhance road safety.

The reminder of this paper is structured as follows: first, Section 3.2 and subsections are devoted to the proposed RTA method for accident severity classification ending up with the role of expandability in transportation sector. Section 3.3

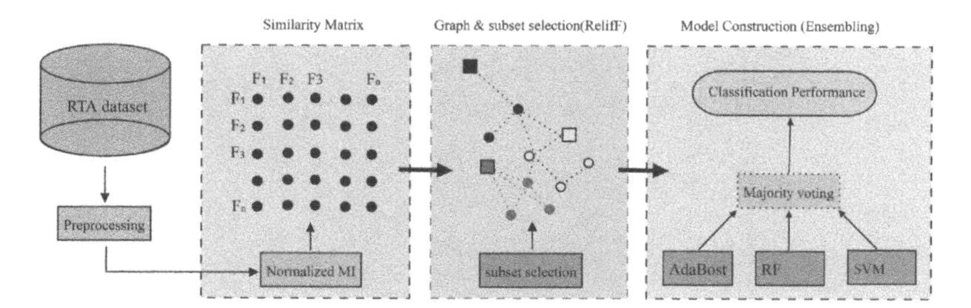

Figure 3.3 Flowchart of the proposed method. The squares denote the representative feature using ReliefF technique.

introduced the experimental setups, dataset description, and findings. Section 3.4 concluded this work by providing some recommendations.

3.2 METHODOLOGY

In this paper, we have proposed a novel graph feature selection method to identify and select the most relevant features of RTA dataset. The method consists of five main steps (see Figure 3.3). First, based on RTA data, we create a $N \times N$ symmetric similarity matrix (N is the number of features in RTA data) of pairwise feature correlation where the diagonal is initialized by the individual correlation of each feature with itself. Second, we construct a weighted graph using the similarity matrix. Third, relying on the well-known community detection technique [3], we cluster the graph into sub-graphs (communities) of the same characteristics. Fourth, detecting the representative feature from each sub-graph counting on ReliefF feature importance method [4,23,35,36]. The selected optimal subset which includes relevant and informative features is capable of classifying accident severity accurately. Fifth, the selected subset is meant to build an ensemble learning classifier to enhance stability, reliability, and accuracy.

3.2.1 Similarity Matrix

After processing RTA dataset by cleaning it and increasing its quality by removing noisy features handling outliers and missing values as well as converting categorical variables into the most suitable format for machine learning methods, we create an $N \times N$ matrix (N is the number of features) counting on some similarity measures such as Normalized Mutual Information [3]. The weight W_{ij} between two variables is computed using the following formula:

$$W_{ij} = NMI(F_i,\ F_j) = \frac{I(F_i,\ F_j)}{\sqrt{H(F_i)H(F_j)}} \tag{3.1}$$

where $I(F_i, F_j)$ is the Mutual Information between F_i and F_j, and $H(.)$ denotes their Entropy.

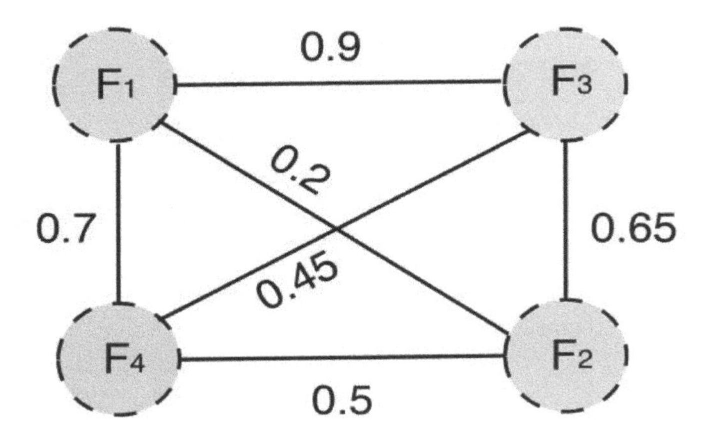

Figure 3.4 Example of weighted graph.

3.2.2 Graph Representation and Community Detection

A graph, or a network, G is simply a couple of collections $(N; E)$, where N is a set of nodes and E is a set of edges, so that each edge being a pair of nodes in N. Nodes are also referred to as vertices. Edges are also referred to as links or connections. If each edge is an unordered couple of vertices, the edge is undirected. Thus, the graph is an undirected (Figure 3.4). Otherwise, if each edge is an ordered couple of vertices, the edge is directed from one vertex to the other. So, the graph is a directed graph. In this case, an ordered couple of vertices $(v_1; v_2)$ is an edge directed from vertex v_1 to vertex v_2. If each edge has an associated numeric value called weight, the edge is weighted, and the graph is a weighted graph. In this study, since the correlation between j^{th} feature and i^{th} feature is the same as the correlation between the ith feature and j^{th} feature, we only used the weighted graph where each vertex represents one feature and the edges are used to describe the correlation between the corresponding pair of features [5,7,8]. See Figure 3.5, where the initial graph is partitioned to communities using the entire features of RTA dataset.

3.2.3 Feature Selection

Feature selection is an important pre-processing technique used for the purpose of selecting the relevant features and discarding the irrelevant, redundant and noisy ones, aiming to obtain the best performing subset of original features without any transformation [6,9,13]. As a result, the constructed models counting on the identified feature subset are more interpretable, explainable and readable. The main reasons for applying feature selection are the following:

- The facilitation of model's interpretation
- The reduction of resources requirement (short training time, small storage capacity, etc.)

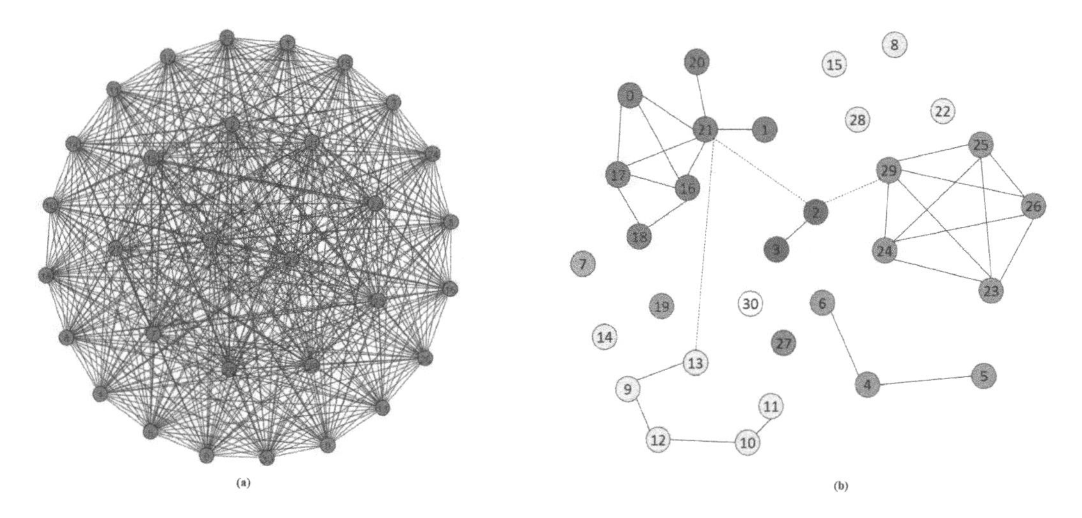

Figure 3.5 Using community detection algorithm to cluster the initial graph into homogenous groups of features (a) Initial graph and (b) partitioned into communities.

- Avoiding the curse of dimensionality
- Avoiding over-fitting problem, automatically leading to a better model
- generalization.
- Improving accuracy: less noise in data which means an improved modeling accuracy.

Feature selection is an active research filed in machine learning as it is an important pre-processing, finding success in different real problem applications [16]. In general, feature selection algorithms are categorized into supervised, Semi-supervised and Unsupervised feature selection according to the availability of label information. Supervised feature selection methods usually come in three flavors: Filter, Wrapper and Embedded approach [5, 7–9]. The suggested graph feature selection method in this paper is counting on community detection in order to reduce the search space and select the best performing feature subset that could help to develop the accident prediction model. Whereas those irrelevant variables to this research objective were removed from the gathered RTA dataset. The choice of the graph representation is made relying on the fact that graphs are powerful and many-sided data structure which allow us to easily represents real-world datasets and the interactions between different type of features [29].

3.2.3.1 ReliefF-Based Feature Selection

The main idea behind ReliefF, which is an extension of relief, is to estimate feature importance according to how well feature values discriminate concept (endpoint) values among instances that are similar (near) to each other. In contrast to relief, the variant ReliefF method can handle the classification of multi-class tasks by selecting a sample randomly from training. Then it tries to find the k nearest neighbors from the

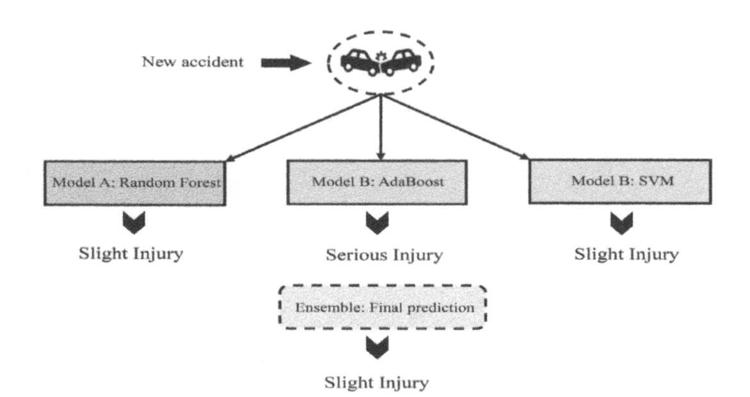

Figure 3.6 Using community detection algorithm to cluster the initial graph into homogenous groups of features.

same class (nearest hit) and from the opposite classes (nearest misses). Therefore, the quality of features is updated according to their powerful efficiency in making proper separation of the examples from different classes.

After clustering the initial graph into groups of homogeneous features in such a way that attributes in the same cluster (group) have more same characteristics than those in other clusters. The process of selecting the best features to be in the final ensemble consists of choosing just one feature from each cluster to avoid the selection of correlated and redundant features to be included in the final subset. The quantification and the selection of the important features have performed relying on RelifF's feature importance Therefore, the proposed algorithm is advantageous and unique in many ways. As the majority of attributes inside each cluster are similar, the introduced method can efficiently handle missing values problem by replacing the outliers and missing values counting on other features' values within the same group. In addition, the feature generation enhancement is another merit of our method where new features can be created using the best characteristics that better relate to the target.

3.2.4 Ensemble Learning

Ensemble methods or multiple classifier systems are considered the state-of-the-art solution for various machine learning problems [31]. These methods improve the predictive performance of a single classifier by training multiple models and combining their predictions via voting schemes (see Figure 3.6). Relying on the fact that a set of experts is better than a single one, we have combined three classifiers together including SVM, RF, and AdaBoost. Consequently, this employed ensemble technique could produce accurate, reliable and more stable results [20, 31].

3.2.5 Explainable Artificial Intelligence

Recently, artificial intelligence and machine learning methods have proved their dramatic success in many sensitive areas of daily life (for example, criminal justice, medicine, transportation, security, and finance). However, the black-box design of the

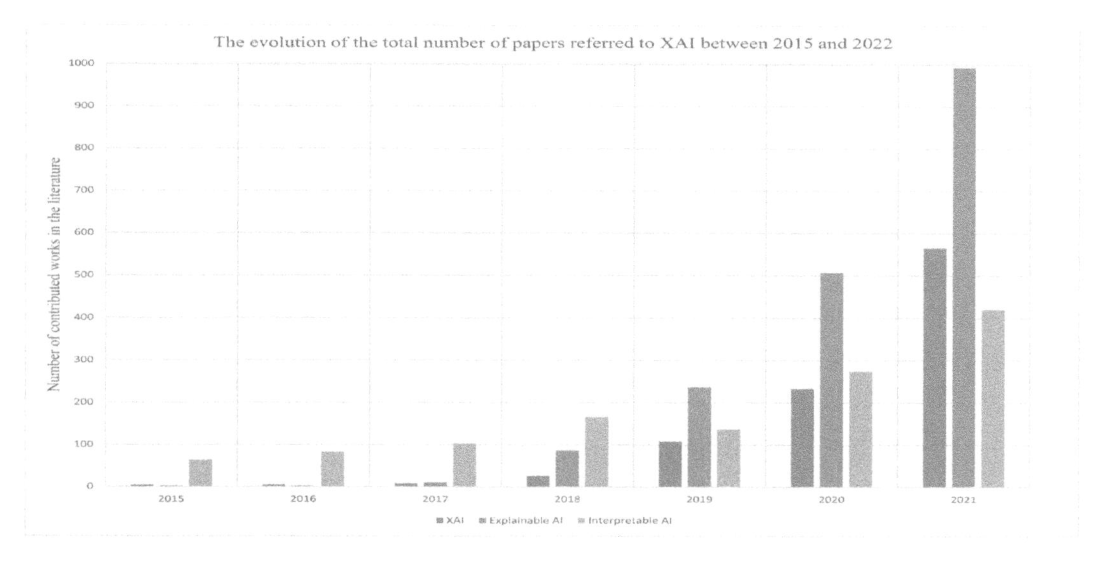

Figure 3.7 The evolution of the number of total published papers whose title, abstract or keywords refer to the domain of XAI between 2015 and 2021. The statistics are retrieved from Scopus database using the terms (XAI, Interpretable AI and Explainable AI).

most machine learning and Deep Learning techniques make it difficult to understand the underlying and the inner behaviors of each trained models [12]. Whereas in many different domains such as medical and transportation sectors, providing explanation and transparency is as crucial as providing good results [21]. Thus, explanations for machine decisions and predictions are needed to justify their reliability. To adjust the lack of interpretability of artificial intelligence based-systems, their inner processes, and their generated outputs, explainable AI (XAI) has been emerged lately [32]. In order to justify that XAI is the the research trend in AI, a literature search using the keywords "XAI," "Interpretable AI," and "Explainable AI" has been conducted of publications listed at Scopus during the last 7 years (from 2015 to 2021). The obtained results are presented in Figures 3.7 which clearly justifies that fact.

A good machine learning method is not the one that achieves only good results but also the one that could provide and offer transparent explanations whenever it is crucial and important (see Figure 3.8). Lately, A plethora of method-based interpretability are introduced in the literature [22]. In this study, we have put more emphasize on feature selection-based interpretability/expandability relying on the importance generated using ReliefF feature selection technique so as to enhance transparency and clarity. In addition, it could be helpful for experts to take good decisions when it is necessary by understanding how the model makes predictions by assessing which features are deemed important during model training.

3.3 EXPERIMENTAL RESULTS AND DISCUSSION

This section describes and discusses the practical implementation and experimental results of this work. It provides detailed information and description about the

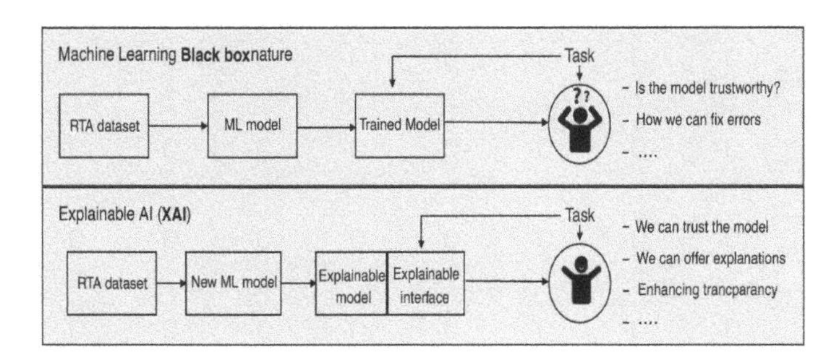

Figure 3.8 Machine Learning vs Explainable AI.

employed road traffic dataset, the conducted experiments and state-of-the-art machine learning used for accident severity classification. In addition, it discusses the reported results by evaluating the performance of classification models and reporting their results.

3.3.1 Dataset Description

The RTA dataset [1] employed in this work is gathered from Addis Ababa Sub city police departments. It contains manual records of road traffic accident of three years flow traffic from 2017 to 2020. Personal and sensitive information have been excluded from RTA dataset for privacy reasons. RTA is a multi-class classification dataset with 32 features and 12,316 examples (accident). Road traffic injuries in Ethiopia are classified into the following three groups:

- **Fatal Injury**: if a road user involved in the accident died within 30 days of the crash date.
- **Serious Injury**: if at least one person is hospitalized for a period of 24 hours or more
- **Slight injury**: if at least one person suffers from a road traffic injury that requires outpatient medical treatment or hospitalization for less than 24 hours.

Table 3.1 provides a description of the dataset. the selected attributes with their description and data type are presented in Table 3.2.

TABLE 3.1 Descriptive Characteristics of RTA Dataset

Dataset	Features	Instances	Accident severity (class label)	
			Class	Distribution (%)
			Slight Injury(class 0)	10,415 (**84.56**)
RTA dataset	32	12,316	Serious Injury(class 1)	1743(**14.15**)
			Fatal Injury(class 2)	158(**1.28**)

The bold values indicate the percentage of imbalances for each class.

TABLE 3.2 RTA Dataset Characteristics

No	Variable name	Type	Description
1	Accident_ID	Numerical	The identifier of an accident
2	Age_band_of_driver	Categorical	The age group of the driver causing the accident
3	Sex_of_driver	Categorical	The sex of the driver causing the accident
4	Educational_level	Categorical	The educational level of the driver causing the accident
5	Vehicle_driver_relation	Categorical	Shows whether the driver is the owner, employed driver, or other.
6	Driving_experience	Categorical	The driving experience of the driver
7	Lanes_or_Median	Categorical	The Lane type in which the vehicle was moving
8	Types_of_junction	Categorical	The type of road junction
9	Road_surface_type	Categorical	Shows whether the road is asphalt, earth road, or other
10	Light_condition	Categorical	The light condition at the time of the accident
11	Weather_condition	Categorical	The weather condition at the time of the accident
12	Type_of_collision	Categorical	Shows the collision type of the vehicles
13	Vehicle_movement	Categorical	The driver action before the accident
15	Pedestrian_movement	Categorical	The pedestrian action before the accident
15	Cause_of_accident	Categorical	The cause of the accident
16	Accident_severity	Categorical	The severity of the accident
17	Time	Categorical	The time of accident
18	Day_of_week	Categorical	Day of accident
19	Type_of_vehicle	Categorical	Type of vehicle involved in the accident
20	Owner_of_vehicle	Categorical	The owner of the vehicle
21	Service_year_of_vehicle	Categorical	The service year of the vehicle
22	Defect_of_vehicle	Categorical	The failure of the vehicle
23	Area_accident_occured	Categorical	Gives information about the location of the accident
24	Road_allignment	Categorical	The category of the road
25	Number_of_vehicles_involved	Numerical	Total number of involved vehicles
26	Number_of_casualties	Numerical	Total number of involved causalities
27	Casualty_class	Categorical	The causality class
28	Sex_of_casualty	Categorical	The sex of the causality
29	Age_band_of_casualty	Categorical	Age band of involved causality
30	Casualty_severity	Categorical	The injury severity of causality
31	Work_of_casuality	Categorical	The work of the causality
32	Fitness_of_casuality	Categorical	The fitness of causality
33	Accident_severity	Categorical	The accident severity

TABLE 3.3 Handling Imbalanced RTA Dataset via SMOTE Technique

	Class	Samples	Distribution
Without SMOTE	**Class** = slight injury	**n**=10,415	(**84.56**%)
	Class = serious injury	**n**=1743	(**14.15**%)
	Class = Fatal injury	**n**=158	(**1.28**%)
With SMOTE	**Class** = slight injury	**n**=10,415	(**33.33**%)
	Class = serious injury	**n**=10,415	(**33.33**%)
	Class = Fatal injury	**n**=10,415	(**33.33**%)
Up-sampled data shape		31,245	(**100**)%

The bold values indicate the percentage of imbalances for each class.

3.3.1.1 *Data Transformation*

Data transformation is used to convert the data instances into the most suitable format for the machine learning algorithms. RTA dataset used in this study includes categorical features only. Since the most machine learning algorithm cannot be applied directly on categorical data, data transformation is mandatory before training phase. In this study, we have mapped any categorical data into numerical format via one hot encoding technique [33].

3.3.1.2 *Handling Imbalance Data*

Imbalanced data consists of a skewed class proportion (unequal class distribution) where the target class has an uneven distribution of samples. Some classes have a small number of observations (called minority classes) and the other have a very high number of observations (called majority classes). As shown in Table 3.1, the RTA dataset is visibly unbalanced dataset as the slight injury class is the majority class with a distribution of 84.56%. This means that RTA dataset is biased towards the majority class (slight injury) in the dataset. Which may lead the classification algorithm trained on imbalance dataset to be biased toward the same class. For this reason, we have prevented the impact of imbalanced data issue by increasing the number of rare samples of minority classes using the well-known technique called Synthetic Minority Over-sampling Technique (SMOTE) [24]. The results of applying SMOTE technique on RTA dataset is presented in Table 3.3.

3.3.2 Classification Algorithms and Evaluation Metrics

Through this paper, we have evaluated the performance of the proposed system when used in conjunction with three classification algorithms (AdaBoost, Random Forest, and Support Vector Machine). In this section, we will provide a brief overview of each classifier as well as model evaluation metrics.

- **Random Forest (RF)**: Random Forest is a robust algorithm in different applications. Based on the aggregation technique, RF combines several individual classifications or regression trees. Several bootstrap samples are drawn from the

TABLE 3.4 Confusion Matrix

		Predicted class		
		Slight Injury	Serious Injury	Fatal Injury
Actual class	Slight Injury	**TP**		
	Serious Injury		**TP**	
	Fatal Injury			**TP**

training data; then, a set of un-pruned decision trees are constructed on each bootstrap samples, so all trees of the forest are maximal trees [9].

- **AdaBoost**: It is an ensemble learning algorithm which was initially built to enhance the efficiency of binary classifiers. It is an iterative approach to learn from the mistakes of weak classifiers, and strengthen them to become strong ones [37].
- **Support Vector Machines (SVM)**: is a powerful classification and regression algorithm. However, it is widely used as it produces excellent performance. SVM has the ability to solve linear and non-linear. The main idea behind SVM is that it tries to find a hyperplane that perfectly discriminates between data points of different classes [17].

Model evaluation metrics are indispensable to assess model performance. The choice of evaluation metrics depends on the machine learning task at hand (such as classification, regression, clustering, community detection, etc.) as well as the type of dataset. The performance of classifiers can also be evaluated on those metrics calculated using the so-called Confusion Matrix of 3×3 (3 is the number of classes) as is shown in Table 3.4. The diagonal values represent the correct predicted classes whereas, the rest values are the wrong predicted classes.

- **TN (True Negative)**: when the actual class was negative and predicted as negative
- **TP (True Positive)**: when the actual class case was positive and predicted as positive
- **FN (False Negative)**: when the actual class was positive but predicted as negative
- **FP (False Positive)**: when the actual class was negative but predicted as positive

Using the confusion matrix, several standard evaluation metrics can be defined and computed for multi-class classification tasks as we have handled the imbalanced RTA dataset previously in Section 3.3.1.2. In the following, we defined the used evaluation metrics in this study:

$$Accuracy = \frac{TP + TN}{TP + TN + FP + FN} \tag{3.2}$$

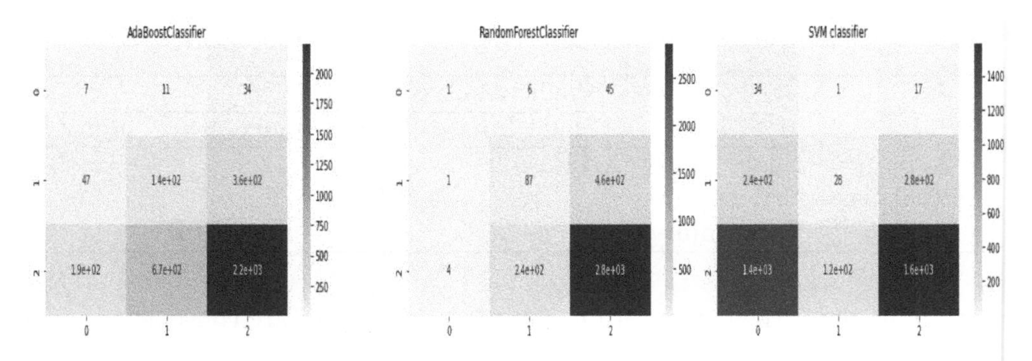

Figure 3.9 Confusion Matrix of injury severity using three models (AdaBoost, SVM, and RF).

$$Precision = \frac{TP}{TP + FP} \tag{3.3}$$

$$Recall/TPR/Sensitivity = \frac{TP}{TP + FN} \tag{3.4}$$

$$F_1 - score = \frac{Precision \times Recall}{Precision + Recall} \tag{3.5}$$

AUC-ROC: The area under the ROC curve measures the performance of the classifier at various thresholds settings. It plots the true positive rate (TPR) on the y-axis versus the false positive rate (FPR) on the x-axis.

3.3.3 Experiment Settings

The experimental results for the four experiments conducted in this study to evaluate the applicability of the proposed RTA system are presented below.

Experiment 1: Baseline model without feature selection

In this first experiment, a baseline lunching ground model is developed to be used as a benchmarking model for later model comparison and used for model assessment by using the three selected classifiers (SVM, AdaBoost, and RF). The performance results obtained after the experiment has been carried out and presented in Figure 3.9 and Table 3.5.

TABLE 3.5 Performance Results of Experiment One

Algorithms	Performance Metrics				
	Accuracy (%)	Precision (%)	Recall (%)	F1-Score (%)	ROC-AUC (%)
SVM	44	35	**40**	37	53
AdaBoost	65	**44**	37	**40**	53
RF	**80**	43	**37**	**40**	**54**

Bold values represent the best performance results on the data set.

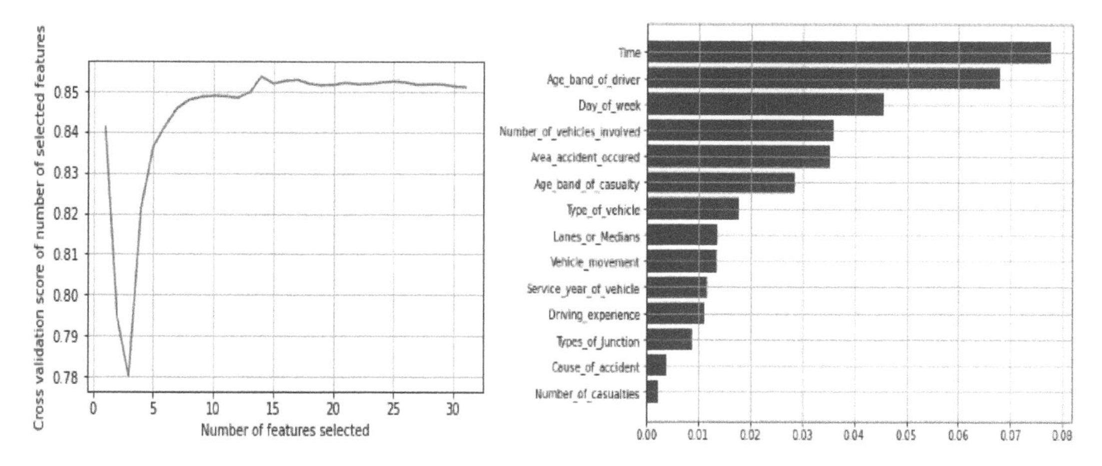

Figure 3.10 AUC score versus the number of selected features and the selected features with their importances using ReliefF-based feature selection.

The results report the performance of the first experiment using the previously discussed classification algorithms. The following can be observed: although the whole multi-class RTA classification dataset is used to train the three classifiers, the models did not perform well almost across all metrics except for RF that showed its superiority as it achieved 80% accuracy when compared with other models. This resulting from the high number of mistakes in prediction especially for serious and fatal injury classes as it is clearly illustrated through the confusion matrices (Figure 3.9).

Experiment 2: Model building with feature selection

In this second experiment, we have trained the selected models using only the selected features generated counting on the proposed graph feature selection methods. The selected variables using the proposed framework processes S_{best} undergo the second step of the algorithm are meant to be fed to the final ensemble learning constituted of three well-known classification algorithms (SVM, RF, and AdaBoost). The number of the most effective features is found to be 14 as demonstrated in Figure 3.10. Thus, the selected features S_{best} are highlighted in Figure 3.10.

To scrutinize the impact exercised on model performance before and after using the proposed feature selection refers to Table 3.6).

TABLE 3.6 Model Performance before and after Feature Selection (Experiment Two)

Algorithms	Performance Metrics				
	Accuracy (%)	Precision (%)	Recall (%)	F1-Score (%)	ROC-AUC (%)
SVM	60	35	40	37	54
AdaBoost	69	61	69	65	65
RF	**86**	**69**	**69**	**69**	**70**

Bold values represent the best performance results on the data set.

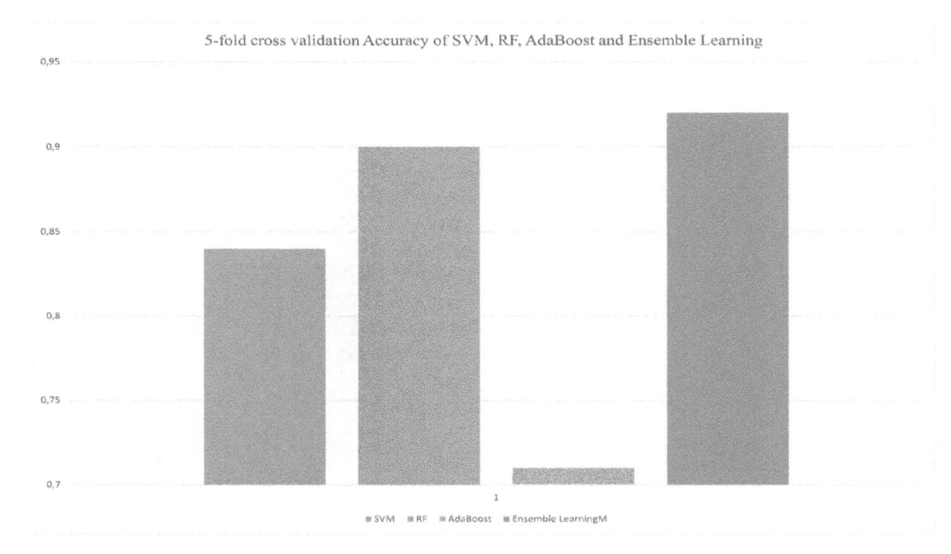

Figure 3.11 Experiment three performance.

It is clear that there's a significant improvement in the model performance using SVM, AdaBoost, and RF model for classification when feature selection is used as the uninformative and noisy features are removed. Additionally, regarding the manner how features are clustered into communities including features of similar characteristics, the selected features are not correlated as they are selected from different communities.

Experiment 3: Training accident severity classifier

In this experiment an ensemble learning technique is developed to combine the predictions of the three models used previously in the experiment one for the sake of helping to improve machine learning results by combining SVM, RF, and AdaBoost models to ameliorate the predictive power compared to a single model (see Figure 3.6). The max voting technique is applied after training each model on the whole RTA dataset. This strategy relies on several model prediction for each accident (instance) which are considered as the voting phase. At the end, the predictions with the majority voting are used as the final prediction (see Figure 3.11). All models are assessed using repeated five fold cross validation, and they have 100 estimators (n-estimators $= 100$) and default parameters are used. The obtained results of this experiment are summarized in figure 200.

It is notably conspicuous that using ensemble learning technique rather than single models is a good alternative as the AUC score is in a cumulative growth reaching the percentage of 92%. Whereas, the RF maximum performance is restrictively reduced to the following percentage of 90, AdaBoost to 71, and SVM to 84.

Experiment 4: Explainability

The main objective of this study is to develop an accident severity classification model to predict the injury severity and spotlight major factors of RTA in Addis

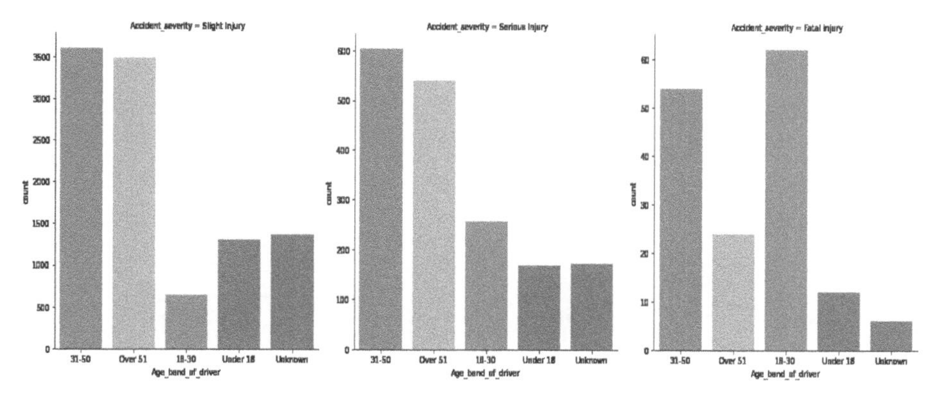

Figure 3.12 The impact of *Age band of driver* factor on accident severity.

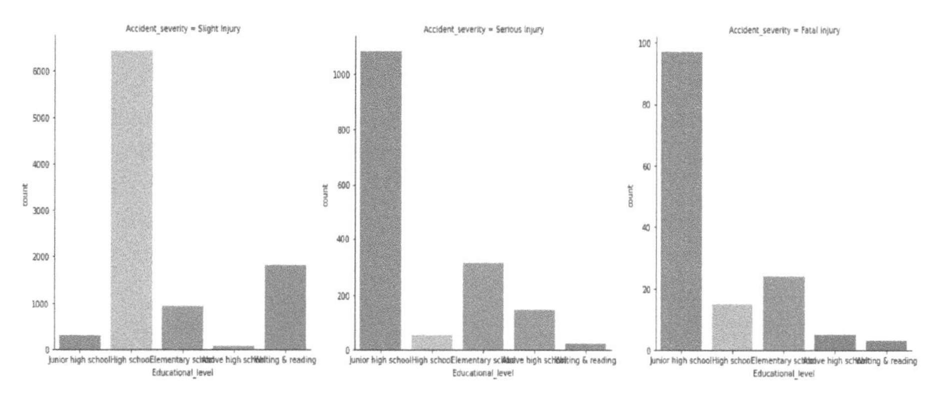

Figure 3.13 The impact of *Education level* factor on accident severity.

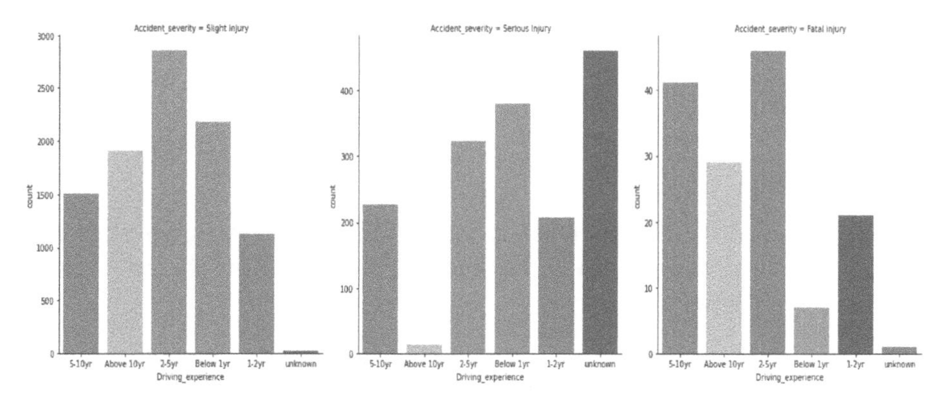

Figure 3.14 The impact of *Driving experience* factor on accident severity.

Ababa city by applying machine learning techniques. As the result shown in the above and current sections (see Figures 3.12–3.15), the final findings of the study help answering the pre-stated research questions which have been formerly posed at the introductory outset of the study.

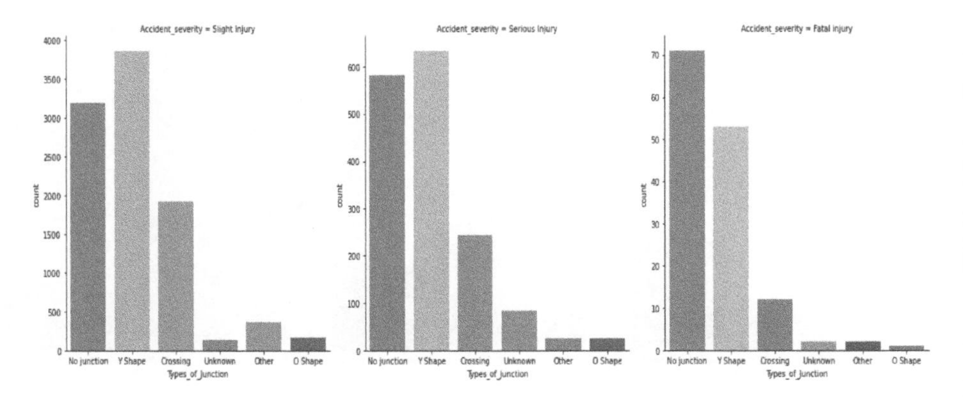

Figure 3.15 The impact of *Junction type* factor on accident severity.

- **What are the major features that may lead to a traffic accident?**

There is a variegated array of influential variables which impact the traffic accident of injury severity. The RTA dataset encompasses information about the drivers, roads, whether, and causality. Drawing from this applicable data, it is inferentially deducible that the most relevant features were identified and used for model building (14 relevant features). Building on the result of the best model as a preliminary groundwork, we have found out the following findings:

- Male Youthful youngsters whose age scale vacillates between 18 and 30 are more likely to be liable to the accident injury (especially fatal injury as illustrated in Figure 3.12).
- light condition is also a highly influential determinant playing a crucial role in determining the likelihood of being subject to nocturnal severity injury.
- Drivers practitioners whose practical knowledge based driving experience wavers between 2 and 5 years are among the vulnerable groups susceptible to accident injury (see Figure 3.14).
- The asymmetrical discrepancies or unequal imbalances in the drivers' educational level is an indispensable key-factor responsible for the rate of accidental injuries (see Figure 3.13).
- A two-way lane which is diametrically divergent in the form of Y-shaped structure are also among the impactful impetuses which may increasingly escalate the probability of RTA in Addis Ababa (see Figure 3.15).
- Etc.

3.4 CONCLUSION AND RECOMMENDATIONS

Road Traffic Accident causalities are extremely high in Addis Ababa, the capital of Ethiopia. Male young adults and pedestrians are found to be significantly associated with fatal injuries and at higher risk from RTAs. urgently expeditious precautions and preventive countermeasures are to be thoroughly taken to extenuate the exponential upsurge of accident fatalities, an enormous responsibility which rests firmly on the healthcare sector with its professional crew of emergency paramedics and police officers who should strictly regulate or systematize the traffic light navigation. Hence, understanding circumstances and factors affecting a crash severity is of high priority to improve road safety.

In this study, we have proposed a new graph feature selection-based method for accident severity prediction. The method is developed based on RTA dataset collected from Addis Ababa sub city police department to manually record RTA from 2017 to 2020. Fourteen revelatory variables are detected as crucial factors leading to efficiently predicting injury severity of accident causalities and then identifying the main causes of accidents.

Regularly delivered Pedagogic driving lessons, know-how logistics of introspecting the mechanical status of the driven vehicles and conscious-raising campaigns with slogans stressing road safety and responsible commitment should be disseminated by the local authorities to sidestep the life-threatening risk of accidental car collisions. In correlative harmony with what is articulately pinpointed above, it is preferably advisable to incorporate the advantageous serviceability provided by Intelligent Transportation Systems brought forth by progressive developments in artificial intelligence and technological industry.

Bibliography

[1] Road traffic accident dataset of addis ababa city, 2020.

[2] Teferi Abegaz and Samson Gebremedhin. Magnitude of road traffic accident related injuries and fatalities in ethiopia. *PloS one*, 14(1):e0202240, 2019.

[3] Elyazid Akachar, Brahim Ouhbi, and Bouchra Frikh. Acsimcd: A 2-phase framework for detecting meaningful communities in dynamic social networks. *Future Generation Computer Systems*, 125:399–420, 2021.

[4] Yassine Akhiat. Feature selection methods for high dimensional data. 2021.

[5] Yassine Akhiat, Youssef Asnaoui, Mohamed Chahhou, and Ahmed Zinedine. A new graph feature selection approach. In *2020 6th IEEE Congress on Information Science and Technology (CiSt)*, pages 156–161. IEEE, 2021.

[6] Yassine Akhiat, Mohamed Chahhou, and Ahmed Zinedine. Feature selection based on pairwise evalution. In *2017 Intelligent Systems and Computer Vision (ISCV)*, pages 1–6. IEEE, 2017.

[7] Yassine Akhiat, Mohamed Chahhou, and Ahmed Zinedine. Feature selection based on graph representation. In *2018 IEEE 5th International Congress on Information Science and Technology (CiSt)*, pages 232–237. IEEE, 2018.

[8] Yassine Akhiat, Mohamed Chahhou, and Ahmed Zinedine. Ensemble feature selection algorithm. *International Journal of Intelligent Systems and Applications*, 11(1):24, 2019.

[9] Yassine Akhiat, Youness Manzali, Mohamed Chahhou, and Ahmed Zinedine. A new noisy random forest based method for feature selection. *Cybernetics and Information Technologies*, 21(2):10–28, 2021.

[10] Saravanan Alagarsamy, M Malathi, M Manonmani, T Sanathani, and A Senthil Kumar. Prediction of road accidents using machine learning technique. In *2021 5th International Conference on Electronics, Communication and Aerospace Technology (ICECA)*, pages 1695–1701. IEEE, 2021.

[11] Aderaw Anteneh and Bilal Shikur Endris. Injury related adult deaths in addis ababa, ethiopia: analysis of data from verbal autopsy. *BMC public health*, 20(1):1–8, 2020.

[12] Alejandro Barredo Arrieta, Natalia Diaz-Rodriguez, Javier Del Ser, Adrien Bennetot, Siham Tabik, Alberto Barbado, Salvador Garcia, Sergio Gil-Lopez, Daniel Molina, Richard Benjamins, et al. Explainable artificial intelligence (xai): Concepts, taxonomies, opportunities and challenges toward responsible ai. *Information fusion*, 58:82–115, 2020.

[13] Youssef Asnaoui, Yassine Akhiat, and Ahmed Zinedine. Feature selection based on attributes clustering. *2021 Fifth International Conference On Intelligent Computing in Data Sciences (ICDS)*, pages 1–5, IEEE, 2021.

[14] Ayesha Ata, Muhammad Adnan Khan, Sagheer Abbas, Gulzar Ahmad, and Areej Fatima. Modelling smart road traffic congestion control system using machine learning techniques. *Neural Network World*, 29(2):99–110, 2019.

[15] Ararso Baru, Aklilu Azazh, and Lemlem Beza. Injury severity levels and associated factors among road traffic collision victims referred to emergency departments of selected public hospitals in addis ababa, ethiopia: the study based on the haddon matrix. *BMC emergency medicine*, 19(1):1–10, 2019.

[16] Younes Bouchlaghem, Yassine Akhiat, and Souad Amjad. Feature selection: A review and comparative study. In *E3S Web of Conferences*, volume 351, page 01046. EDP Sciences, 2022.

[17] Jair Cervantes, Farid Garcia-Lamont, Lisbeth Rodriguez-Mazahua, and Asdrubal Lopez. A comprehensive survey on support vector machine classification: Applications, challenges and trends. *Neurocomputing*, 408:189–215, 2020.

[18] Arun Chand, S Jayesh, and AB Bhasi. Road traffic accidents: An overview of data sources, analysis techniques and contributing factors. *Materials Today: Proceedings*, 47:5135–5141, 2021.

[19] Sharon Chekijian, Melinda Paul, Vanessa P Kohl, David M Walker, Anthony J Tomassoni, David C Cone, and Federico E Vaca. The global burden of road injury: its relevance to the emergency physician. *Emergency Medicine International*, 2014.

[20] Xibin Dong, Zhiwen Yu, Wenming Cao, Yifan Shi, and Qianli Ma. A survey on ensemble learning. *Frontiers of Computer Science*, 14(2):241–258, 2020.

[21] Filip Karlo Dosilovic, Mario Brcic, and Nikica Hlupic. Explainable artificial intelligence: A survey. In *2018 41st International convention on information and communication technology, electronics and microelectronics (MIPRO)*, pages 0210–0215. IEEE, 2018.

[22] Jack Dunn, Luca Mingardi, and Ying Daisy Zhuo. Comparing interpretability and explainability for feature selection. *arXiv preprint arXiv:2105.05328*, 2021.

[23] Carlos Eiras-Franco, Bertha Guijarro-Berdinas, Amparo Alonso-Betanzos, and Antonio Bahamonde. Scalable feature selection using relieff aided by locality-sensitive hashing. *International Journal of Intelligent Systems*, 36(11):6161–6179, 2021.

[24] Alberto Fernandez, Salvador Garcia, Francisco Herrera, and Nitesh V Chawla. Smote for learning from imbalanced data: progress and challenges, marking the 15-year anniversary. *Journal of Artificial Intelligence Research*, 61:863–905, 2018.

[25] Rediet Fikru Gebresenbet and Anteneh Dirar Aliyu. Injury severity level and associated factors among road traffic accident victims attending emergency department of tirunesh beijing hospital, addis ababa, ethiopia: A cross sectional hospital-based study. *PLoS One*, 14(9):e0222793, 2019.

[26] Ali J Ghandour, Huda Hammoud, and Samar Al-Hajj. Analyzing factors associated with fatal road crashes: a machine learning approach. *International Journal of Environmental Research and Public Health*, 17(11):4111, 2020.

[27] Girma Gemechu Hordofa, Sahilu Assegid, Abiot Girma, and Tesfaye Dagne Weldemarium. Prevalence of fatality and associated factors of road traffic accidents among victims reported to burayu town police stations, between 2010 and 2015, ethiopia. *Journal of Transport & Health*, 10:186–193, 2018.

[28] Bulbula Kumeda, Fengli Zhang, Fan Zhou, Sadiq Hussain, Ammar Almasri, and Maregu Assefa. Classification of road traffic accident data using machine learning algorithms. In *2019 IEEE 11th International Conference on Communication Software and Networks (ICCSN)*, pages 682–687. IEEE, 2019.

[29] Jundong Li, Kewei Cheng, Suhang Wang, Fred Morstatter, Robert P Trevino, Jiliang Tang, and Huan Liu. Feature selection: A data perspective. *ACM computing surveys (CSUR)*, 50(6):1–45, 2017.

[30] Alfonso Navarro-Espinoza, Oscar Roberto Lopez-Bonilla, Enrique Efren Garcia-Guerrero, Esteban Tlelo-Cuautle, Didier Lopez-Mancilla, Carlos Hernandez-Mejia, and Everardo Inzunza-Gonzalez. Traffic flow prediction for smart traffic lights using machine learning algorithms. *Technologies*, 10(1):5, 2022.

[31] Omer Sagi and Lior Rokach. Ensemble learning: A survey. *Wiley Interdisciplinary Reviews: Data Mining and Knowledge Discovery*, 8(4):e1249, 2018.

[32] Wojciech Samek, Gregoire Montavon, Andrea Vedaldi, Lars Kai Hansen, and Klaus-Robert Muller. *Explainable AI: interpreting, explaining and visualizing deep learning*, volume 11700. Springer Nature, 2019. XI, 439.

[33] Cedric Seger. An investigation of categorical variable encoding techniques in machine learning: binary versus one-hot and feature hashing, 2018.

[34] Jaspreet Singh, Gurvinder Singh, Prithvipal Singh, and Mandeep Kaur. Evaluation and classification of road accidents using machine learning techniques. In *Emerging Research in Computing, Information, Communication and Applications*, pages 193–204. Springer, 2019.

[35] Hongwei Tan, Guodong Wang, Wendong Wang, and Zili Zhang. Feature selection based on distance correlation: a filter algorithm. *Journal of Applied Statistics*, 49(2):411–426, 2022.

[36] Ryan J Urbanowicz, Melissa Meeker, William La Cava, Randal S Olson, and Jason H Moore. Relief-based feature selection: Introduction and review. *Journal of biomedical informatics*, 85:189–203, 2018.

[37] Wenyang Wang and Dongchu Sun. The improved adaboost algorithms for imbalanced data classification. *Information Sciences*, 563:358–374, 2021.

[38] Ashenafi Habte Woyessa, Worku Dechasa Heyi, Nesru Hiko Ture, and Burtukan Kebede Moti. Patterns of road traffic accident, nature of related injuries, and post-crash outcome determinants in western ethiopia-a hospital based study. *African journal of emergency medicine*, 11(1):123–131, 2021.

[39] George Yannis, Anastasios Dragomanovits, Alexandra Laiou, Francesca La Torre, Lorenzo Domenichini, Thomas Richter, Stephan Ruhl, Daniel Graham, and Niovi Karathodorou. Road traffic accident prediction modelling: a literature review. In *Proceedings of the institution of civil engineers-transport*, volume 170, pages 245–254. Thomas Telford Ltd, 2017.

CHAPTER 4

COVID-19 Pandemic Effects on Traffic Crash Patterns and Injuries in Barcelona, Spain: An Interpretable Approach

Ahmad Aiash and Francesc Robusté
Civil Engineering School, UPC–BarcelonaTech, Barcelona, Spain

CONTENTS

THE COVID-19 PANDEMIC has certainly affected our lives in many different aspects. Traffic, in general, and traffic crashes, in particular, have both been affected by the pandemic. This study, therefore, is conducted to examine the influence of the pandemic on traffic crashes different levels of injuries in Barcelona, Spain. Time-series analysis and forecast are both exploited to achieve the objective of conducting this study. The results show a dramatic drop in traffic crashes, especially during the state of alarm duration that was imposed in March 2020. All levels of injuries that resulted from traffic crashes have shown approximately 39%, 30%, and 36% reduction for slight, severe, and fatal injuries, respectively. Additionally, the impact of certain

DOI: 10.1201/9781003324140-4

chosen risk factors is found to be the same after employing two exhaustive chi-square automatic interaction (CHAID) models. Male injured persons and the time of the day represented by evening and night timing are found to have higher probabilities of having severe or fatal injuries both before and during the pandemic periods. The two applied exhaustive CHAID models have proven their ability to provide interpretable results.

4.1 INTRODUCTION

Corona Virus Disease (COVID-19) which is linked to severe acute respiratory syndrome (SARS) and some types of the common cold is a new strain of coronavirus that is spread in late 2019 [43]. This virus can be transmitted from one person to another when respiratory droplets of an infected person are disseminated or being contacted by another person. Eventually, the COVID-19 outbreak has changed into a COVID-19 pandemic to change the implemented strategies that deal with the virus [40]. One of the measures that have been implemented is imposing a curfew in different areas and regions to contain the number of infected cases (Koh, 2020). Until now, the unprecedented number of confirmed cases is exceeding 548 million cases globally, while the number of deaths is exceeding 6 million deaths [45]. Based on the European Centre for Disease Prevention and Control [16], more than 230 million cases are confirmed in Europe, with more than 2 million deaths.

This pandemic has, for sure, influenced our life from different perspectives alongside the different measures that were imposed and implemented that may also have changed our way of living. A research study [39] found that 25% of people who answered a conducted survey stopped working. This category had shown worse mental and physical health compared to other categories. Consistently, a 31.7% increase in suicide was captured in Hong Kong among people who were 65 years and older during the pandemic [22]. Conforming to the previously mentioned situation, a study [32] found that the increasing number of COVID-19 cases had increased the levels of dissatisfaction and anxiety.

Moving from the mental and physical health aspect, traffic has also been influenced by the pandemic globally due to the measures that were imposed and the cautions from people to avoid unnecessary social connections. The impact of COVID-19 on traffic was found in South Korea related to the number of vehicles that were utilized during the pandemic. The study [21] found that the number of vehicles was 9.7% lower than in 2019. However, the results also showed that the number of vehicles started to increase again as the number of patients decreased. For the environmental aspect related to traffic, a notable decrease was recognized for seven harmful gases during the pandemic in a Northwestern United States city [29]. In Kazakhstan, benzene and toluene concentrations were found to be two to three times higher during the same seasons in 2015–2019 compared to the COVID-19 pandemic period as an impact of the traffic-free urban condition [5]. Another COVID-19 restriction effect is that vehicle miles and driving days were both reduced among adolescents [13].

Traffic crashes are part of the other aspects of the other categories that have been intensively and passively affected by the pandemic due to the series of different occurred events worldwide. Curfew is one of these events that has been implemented as a measure for restricting unnecessary movement and lessening the number of confirmed cases. Considering the bright side of the pandemic, 17,600 injuries and 200 fatalities as a result of traffic crashes were avoided in Turkey when people stayed at their homes during past March and April in 2020 compared to the previous same periods [42]. In Tarragona, Spain, the number of traffic crashes drastically decreased during the period between March and April during the pandemic [31]. Another study [25] examined the Emergency Traumatology Service data at a tertiary hospital within the Spanish National Health System. The results revealed that there was a reduction in the number of traffic crash case visits besides other types of crashes during the pandemic. A huge drop in the number of traffic crash fatalities was noticed in Peru as the number of fatalities decreased by 12.22 million men per month [34]. In Louisiana, the United States, a study [38] found a large reduction in the number of traffic crashes, injuries resulting from traffic crashes, and distracted drivers. The findings also found that the age category for persons who were involved in incidents during the pandemic was between 25–64. During the lockdown in Telangana, India, a drop of 77.9% in the cases of trauma that is related to traffic crashes was noticed [44]. However, there was no detected change in fragility fracture incidence cases for the elderly during the lockdown. A reduction in traffic crashes by 68% was found for the category that represented people who were older than 60 years old [28].

Despite the reduction in crashes [14] and traffic volumes during the pandemic in Greece, there was an increase in speeding, harsh acceleration and braking, and using the mobile phone [7]. Another research [37] detected a minor impact of stay-at-home restrictions on human mobility, whereas the same research found an increase in mobility after re-opening. Similarly, drivers were found to have a higher tendency to commit speed traffic violations that were involved in fatal crashes during the lockdown second month [20]. Following the same trend, the number of traffic crash injuries, motor vehicle fatalities, and speeding violations increased during the pandemic [30]. A similar case was found in Malaysia as the number of traffic crashes was high despite the decrease in the volume of traffic [27]. In Missouri, United States, the number of severe or fatal injuries that resulted from traffic crashes had no significant change [4]. In contradiction, a research study [11] found a decrease in traffic violations during complete movement restriction in Qatar with more than 70%. In addition, the study showed there was a reduction in traffic crashes that accompanied the imposed preventive measures.

The main objective of this study is to examine the impact of the COVID-19 pandemic period on traffic crash injuries and patterns in Barcelona, Spain. For this reason, different time intervals are exploited alongside different techniques. Time series analysis is carried out to reveal traffic crash injuries during the selected time interval including the pandemic and then compare it with previous traffic crash injury occurrences. Additionally, a forecast model represented by a simple seasonal exponential smoothing technique is employed to compare the actual injuries with predictions besides detecting abnormal cases. Lastly, two identical supervised learning techniques represented by two exhaustive chi-square automatic interaction detector (CHAID) models are employed to compare the impact of chosen risk factors on traffic crash injuries during the pandemic and previous years. Using models with inherent interpretability is a crucial factor to gain experts' trust in such algorithms. The results can help road authorities in detecting and observing injuries during the pandemic in order to decide whether new rules should be implemented to confront unusual traffic crash cases besides determining the correlations between several risk factors and traffic crash injuries. The remainder of this chapter is organized as follows; the methodology section is set to be Section 4.4. The results are presented in the third section. The conclusions are presented in the last section of this chapter.

4.2 METHODOLOGY

4.2.1 Data Description

Injuries that are resulted from traffic crashes are, indeed, representing a challenge worldwide. During the COVID-19 pandemic, several changes are noticed regarding traffic crashes, in general, and the injuries that are resulted from traffic crashes, as explained before. In Barcelona, Spain, similar to the rest of the world, this phenomenon, is still under examination. The Open Data Service of Barcelona's City Hall [6] is providing the service of public access to the data related to traffic crashes alongside other types of data. Eventually, the spotted data that is required for carrying out the analysis to conduct this study is gathered from this open-access database. Mainly the data utilized in this study is the same for all different types of analysis that are carried out, however, the differences in choosing which factors are studied.

As shown in Table 4.1, the gender predictor has two categories male and female. The age variable has five different classifications including 0–15 years old group, 16–24 years old group, 25–40 years old group, 41–64 years old group, and lastly 65 years old and older injured persons. The time of the day has three categories including morning (6 a.m. to 1 p.m.), evening (2 p.m. to 9 p.m.), and night timing (9 p.m. to 5 a.m.). This time classification is based on the collected data when the crash occurred. The injured person variable has pedestrian, passenger, and driver categories. The reasons for choosing these factors are based on previously conducted studies related to traffic crashes as these risk factors have already shown a significant impact on traffic crash injuries (Aiash and Robusté, 2021). For the first exhaustive CHAID tree, data from 2016 to 2019 is utilized. 2020 data is used to represent the COVID-19 pandemic. All blank categories are eliminated from both datasets. The data shown in Table 4.1 are only utilized for the CHAID tree models. For the simple seasonal exponential smoothing, all datasets without eliminating blank categories for four years that occurred before 2020 including the years 2016, 2017, 2018, and 2019 are utilized as this part of the paper is only focusing on the date with only focusing on severe injury cases. Eventually, 870 severe injury cases are included for this purpose.

4.2.2 Time Series Forecasting

In this study, simple seasonal exponential smoothing is exploited to predict the pandemic period with considering the traffic crashes severe injuries. Exponential

TABLE 4.1 Injuries during Different Years Related to Chosen Risk Factors

Risk factor	Category	2020 Severe or fatal injury	2020 Slight injury	2016-2019 Severe or fatal injury	2016-2019 Slight injury
Time of day	Morning	27	482	488	23,274
	Evening	36	1616	160	5215
	Night	94	2319	291	17,725
Gender	Male	111	4270	634	28,058
	Female	46	147	305	18,156
Age	0–15	1	121	25	1513
	16–24	18	644	121	6893
	25–40	53	1806	296	19,160
	41–64	65	1609	359	15,377
	65+	20	237	138	3271
Injured	Driver	37	299	269	4573
	Passenger	15	415	86	8989
	Pedestrian	105	3703	584	32,652

smoothing is considered one of the most popular modules used in time series. This popularity can be referred to its simplicity, efficiency, and reasonable accuracy [9]. Due to the nature of traffic crashes data, the seasonal type of exponential smoothing is selected. For the observation and time interval, four years are included to conduct the prediction estimations for 2020 including the years 2016, 2017, 2018, and 2019. IBM Watson Studio Cloud is exploited to apply the time series model that provides the aimed forecast model. While applying the forecast model, several settings are set to achieve accurate results. Firstly, the time interval is estimated per day assuming the weekday first day is Monday with an increment value of 1. For the missing value, the linear interpolation replacement method is chosen with the lowest data quality of 5%. The forecast option is utilized to extend the prediction estimation for the 2020 period based on the previously occurred severe injuries during the four years period. The number of extended future days is chosen as 365 days to cover the entire 2020 year and conduct the aimed comparison between the predictions and the actual number of severe injuries that occurred during the 2020 year. The IBM time series algorithm is designed by IBM with the help of Ruey Tsay at The University of Chicago. More details can be found regarding time series algorithms [2, 3, 12, 15, 17, 18, 33, 36].

After the previous setups are chosen, the model is conducted, and results are displayed and determined. A simple seasonal exponential smoothing method is applied. This method has level and season parameters. $Y(t = 1, 2, ..., n)$ which is the univariate time series under investigation. The total number of observations is n. $L(t)$ is the level parameters for the applied model. The seasonal length is $S.S(t)$ is the season parameter. $\hat{Y}_t(k)$ is the model estimated k-step ahead forecast at time t for Y series. α is the weight of level smoothing. γ is the trend smoothing weight. Formulas 4.1–4.3 are presented to grasp the overall simple seasonal exponential smoothing algorithm estimations.

$$L(T) = \begin{cases} (Y(t) - S(t-s)) + (1-\alpha)L(t-1), & \text{if } Y(t) \text{ is not missing} \\ L(t-1), & \text{else} \end{cases} \tag{4.1}$$

$$S(T) = \begin{cases} \gamma(Y(t) - L(t)) + (1-\gamma)S(t-s), & \text{if } Y(t) \text{ is not missing} \\ S(t-s), & \text{else} \end{cases} \tag{4.2}$$

$$\hat{Y}_t(k) = L(t) + S(t+k-s) \tag{4.3}$$

Twelve measures are exploited and depicted to present the forecast model goodness of fit for severe injury category including equations from 4 to 15. Mean squared error (MSE), root mean squared deviation (RMSE), root mean squared prediction error (RMSPE), mean absolute error (MAE), mean absolute percentage error (MAPE), maximum absolute error (MaxAE), maximum absolute percentage error (MaxAPE), Akaike information criteria (AIC), Bayesian information criteria (BIC), R-squared (R^2), stationary R-squared ((R_s^2), and Ljung-Box (Q) are all determined to estimate the goodness of fit for the applied time-series forecast model. $Z(t)$ is $Y(t)$ or transformed $Y(t)$. ($\overline{\Delta Z}$) is the simple mean model for the differenced transformed series

$$MSE = \frac{\sum(Y(t) - \hat{Y}(t))^2}{n-k} \tag{4.4}$$

$$RMSE = \sqrt{MSE} \tag{4.5}$$

$$MAE = \frac{1}{n}\sum|Y(t) - \hat{Y}(t)| \tag{4.6}$$

$$MAPE = \frac{1}{n}\sum\left|\frac{Y(t) - \hat{Y}(t)}{Y(t)}\right| \tag{4.7}$$

$$RMSPE = \sqrt{\frac{100}{n}\sum\left(\left|\frac{Y(t) - \hat{Y}(t)}{Y(t)}\right|\right)^2} \tag{4.8}$$

$$MaxAE = \max(|Y(t) - \hat{Y}(t)|) \tag{4.9}$$

$$MaxAPE = 100\max\left(\left|\frac{Y(t) - \hat{Y}(t)}{Y(t)}\right|\right) \tag{4.10}$$

$$AIC = n \times \ln(\frac{\sum(Y(t) - \hat{Y}(t)^2)}{n}) + 2k \tag{4.11}$$

$$BIC = n \times \ln(\frac{\sum(Y(t) - \hat{Y}(t)^2)}{n}) + k \times \ln(n) \tag{4.12}$$

$$R^2 = 1 - \frac{\sum(Y(t) - \hat{Y}(t))^2}{\sum(Y(t) - \bar{Y})^2} \tag{4.13}$$

$$R_s^2 = 1 - \frac{\sum_t(Z(t) - \hat{Z}(t))^2}{\sum_t(Z(t) - \overline{\Delta Y})} \tag{4.14}$$

$$Q(K) = n(n+2)\sum_{k=1}^{K}\frac{r_k^2}{n-k} \tag{4.15}$$

4.2.3 Risk Factor Analysis

Chi-square automatic interaction detector (CHAID) which lies under supervised learning techniques that belong to the decision tree (DT) technique that is normally a non-parametric technique is exploited in this study. This technique does not have any dependency on any functional form and no prior probabilistic knowledge requirement [24] and is developed by [19]. One of the main reasons for applying the CHAID tree is its ability to develop a simple chart to detect the correlations between the independent variables and the dependent variable compared to other techniques that cannot provide this easiness in interpretation. It simply acts as a decision tree with the exception of using chi-squared based criterion instead of using the information gain or gain ratio criteria. Additionally, it is considered an efficient statistical method for segmentation or tree growing. The outputs of the CHAID tree are highly meaningful and easy to interpret. This process keeps repeating till the fully grown tree is formed. Part of the advantages of applying the CHAID tree is its ability in producing more than two categories as the CHAID tree is not a binary tree technique. Therefore, it can work with different types of variables and can create a wider tree than binary tree techniques. However, in this study, exhaustive CHAID is employed as it is identical to normal CHAID in the statistical tests but much safer in finding useful splits as it can continue to merge categories of independent variables until two super categories are left. Thus, the best split for each independent variable is found and then compare based on the p-values to choose which independent variable to split. Exhaustive CHAID is basically a modification of CHAID that is developed by [8].

By following a previously conducted study that employed the CHAID tree [1], exhaustive CHAID [35] and utilized IBM Watson with using SPSS modeler set, 70% is set for the training set and 30% for is set for the testing set. Pearson chi-square statistic and its corresponding p-values are calculated because the injury level is a categorical variable as shown in equations 4.16 and 4.17, respectively. X_d^2 follows $d = (J-1)(I-1)$ a chi-square distribution degree of freedom. The independent variable is X, and the dependent variable is Y. n_{ij} is the observed cell frequency. $\hat{m}_{ij}$ is the expected cell frequency ($x_n = i$, $y_n = j$). During the merge process, categories are compared to consider only the records that belong to the comparison categories. During the split step, all the current nodes are used as the categories are considered in p-value calculations. The likelihood ratio chi-square and its p-value is calculated in equations 4.18–4.20, respectively based on the observed and expected frequencies. The expected frequencies are calculated based on two cases. The expected frequencies with no case weights are calculated as shown in equations 4.21 and 4.22. The expected frequencies with specified weights are calculated in equations 4.23 and 4.24. The estimated parameters are α_i and β_i. Further details related to CHAID algorithms are mentioned in the IBM guide [23]

$$n_{ij} = \sum_n f_n I(x_n = i \wedge y_n = j) \tag{4.16}$$

$$X^2 = 2\sum_{j=1}^{J}\sum_{i=1}^{I}\frac{n_{ij} - \hat{m}_{ij}^2}{\hat{m}_{ij}} \tag{4.17}$$

$$p = Pr(X_d^2 > X^2) \tag{4.18}$$

$$G^2 = 2\sum_{j=1}^{J}\sum_{i=1}^{I} n_{ij}\ln(\frac{n_{ij}}{\hat{m}_{ij}}) \tag{4.19}$$

$$p = Pr(X_d^2 > G^2) \tag{4.20}$$

$$\hat{m}_{ij} = \frac{n_{i.}n_{.j}}{n_{..}} \tag{4.21}$$

Where

$$n_{i.} = \sum_{j=1}^{J} n_{ij}, \quad n_{.j} = \sum_{i=1}^{I} n_{ij}, \quad n_{..} = \sum_{j=1}^{J}\sum_{i=1}^{I} n_{ij} \tag{4.22}$$

$$\hat{m}_{ij} = \overline{w}_{ij}^{-1}\alpha_i\beta_i \tag{4.23}$$

Where

$$\overline{w}_{ij} = \frac{w_{ij}}{n_{ij}}, \quad w_{ij} = \sum_{n\in D} w_n f_n I(x_n = 1 \wedge y_n = j) \tag{4.24}$$

4.3 RESULTS AND DISCUSSION

4.3.1 Injuries Analysis

To understand the traffic crash injury occurrences during the pandemic in Barcelona, a radar chart is prepared to depict the variations that happened as shown in Figure 4.1. As the pandemic happened in 2020, both years 2019 and 2020 are included in this chart for comparison. The first confirmed case related to COVID-19 was identified in Spain in late January 2020, followed by declaring the state of alarm on the 14th of March, and then ended on the June 21 [26]. This can be obviously noticed in Figure 4.1. The total number of traffic crash injuries that consists of the three levels of injuries is included. During January and February, both months have similar total numbers for both years. Then, the number of traffic crash injuries is drastically dropped during the following months.

For time series, time decomposition analysis is exploited to examine the differences between the two periods before and after the pandemic. The times series analysis is separated based on the resulting level of injury. The slight injury is presented in Figure 4.2. As shown, the number of traffic crashes is almost similar till the beginning of March 2020, when the numbers started to drastically drop as the state of alarm is set on the March 14, 2020. Then, the numbers started to increase again in July of the same year. For the total slight injuries, 2019 had 11,620 injuries, and 2020 had only 7059. For severe injuries that are shown in Figure 4.3, the trend was different. Indeed, fewer numbers of severe injuries can be noticeably detected during the same period when the state of alarm was set. The total number of severe injuries was

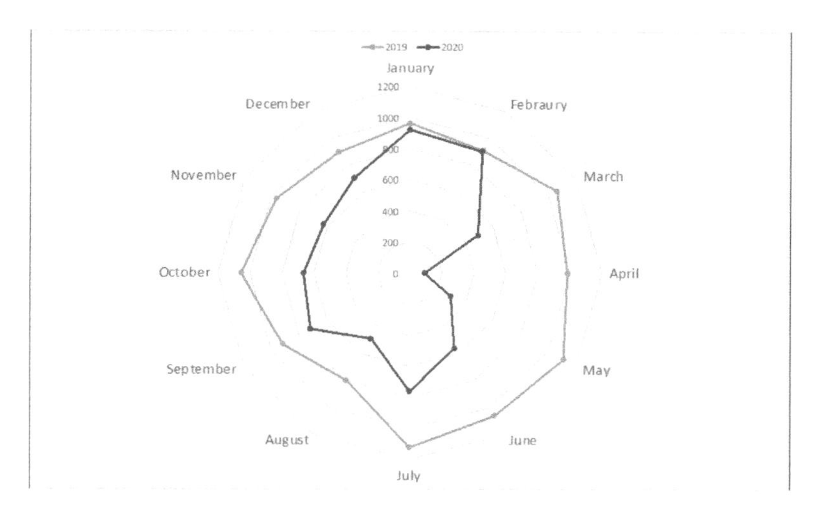

Figure 4.1 Traffic crashes injuries during 2019 and 2020.

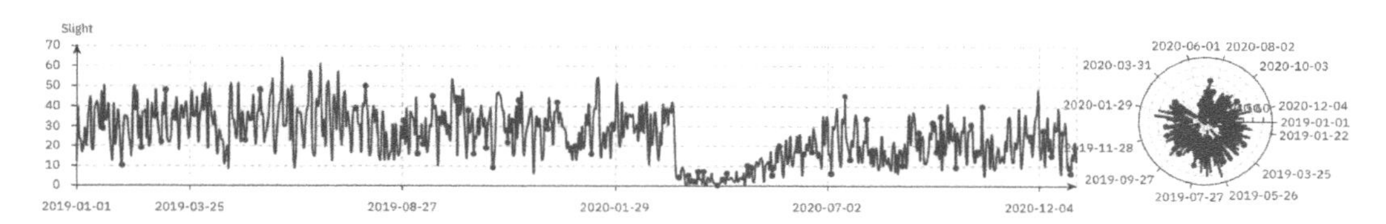

Figure 4.2 Slight injury time series analysis during 2019 and 2020.

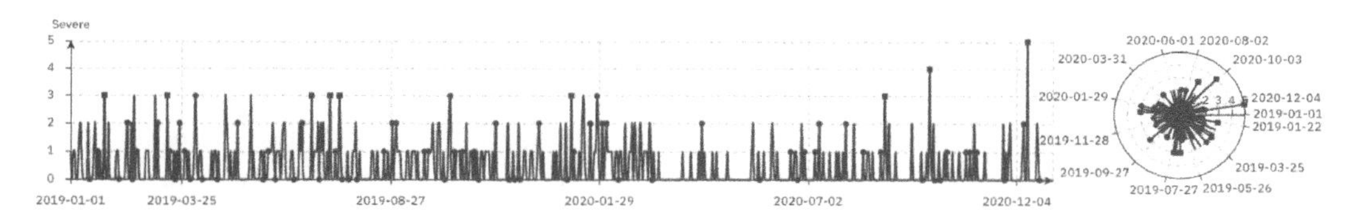

Figure 4.3 Severe injury time series analysis during 2019 and 2020.

202 in 2019, while in 2020, severe injuries number was 147. However, in September 2020, there was an abnormal increase in severe injury cases per day compared to 2019. There were four severe injuries that occurred on the September 30, 2020. This number was not found on any day in September 2019. This abnormal trend is also detected in October with more reported severe cases. Similarly, in December 2020, there were five severe injuries in only one day which was on the December 12, 2020, which is higher than any day compared to 2019 for the same month. Figure 4.4 is depicting the number of fatal injuries that are reported during 2019 and 2020. One of the noticeable observations is that the number of fatalities was highly affected by the state of alarm and the imposed restrictions as these fatalities cases were almost zero during that entire period. The total number of fatal injuries in 2019 was 22 deaths, while in 2020, this number was 14.

The number of severe injuries that are found in unusual cases compared to previous years is recalling the examination of traffic volume during those specific durations

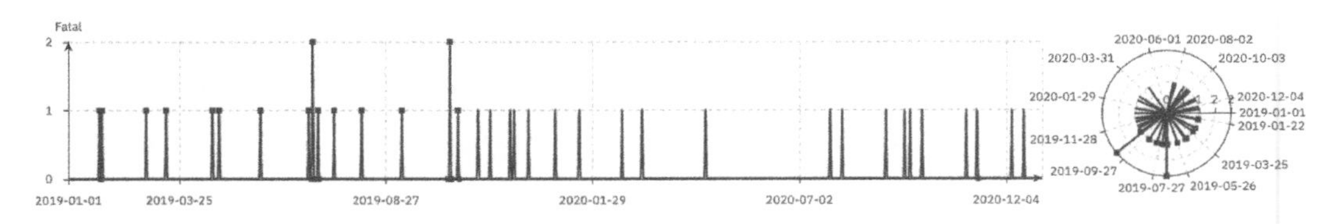

Figure 4.4 Fatal injury time series analysis during 2019 and 2020.

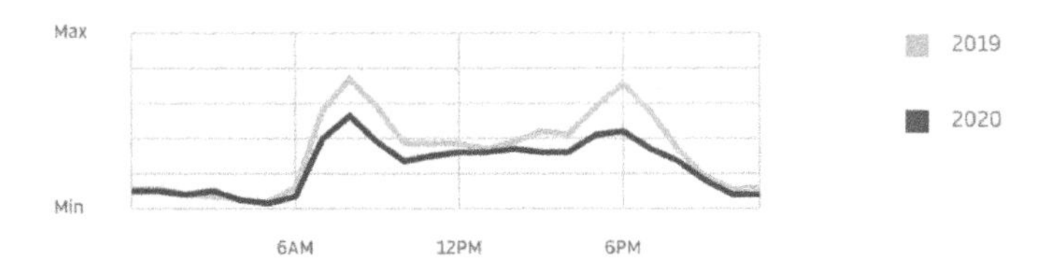

Figure 4.5 Average daily traffic during work days in September based on [41].

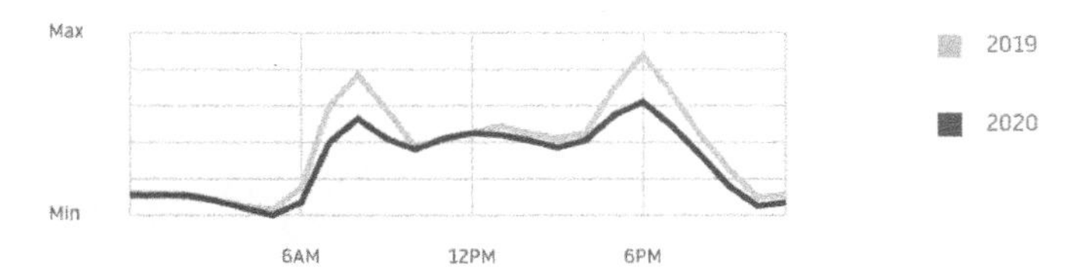

Figure 4.6 Average daily traffic during work days in December based on [41].

to determine whether traffic volume can be the reason for these cases. Therefore, traffic volumes for both September and December are both included and depicted in Figures 4.5 and 4.6. The data and figures are both gathered and brought from the TomTom data report for Barcelona [41] based on data availability. The average traffic volume on working days for both 2019 and 2020 is shown. It can be concluded that the traffic volume has indeed decreased in 2020 compared to 2019 during both months. However, after the peak hours that have the highest traffic volume in both years, the average traffic volume gap between 2019 and 2020 is almost straightened.

For the same reason of unusual severe injury cases, the overall statistics for several characteristics are presented to grasp the correlation between the pandemic and different aspects. The data for these aspects are gathered from the above-mentioned Open Data Service of Barcelona's City Hall. Part of these characteristics is the total number of registered vehicles. Figure 4.7 is depicting the total number of vehicles that are registered in the city of Barcelona. The total number of vehicles for the 2019 year is compared with the 2020 year. As shown, during the state of alarm the number of registered vehicles is, obviously, dramatically dropped especially in April. However, this trend has changed as the number started to increase again reaching a higher total number of vehicles compared to the previous year. Then, the spike decreased again to almost equalize the previous year's numbers. Another aspect that is examined is the reasons for traffic crash injuries for both 2019 and 2020 which are presented in Figure 4.8. Alcohol, a road in poor condition, drugs or medicine, overspeeding, meteorological factors, and objects or animals on the road are all included and shown with

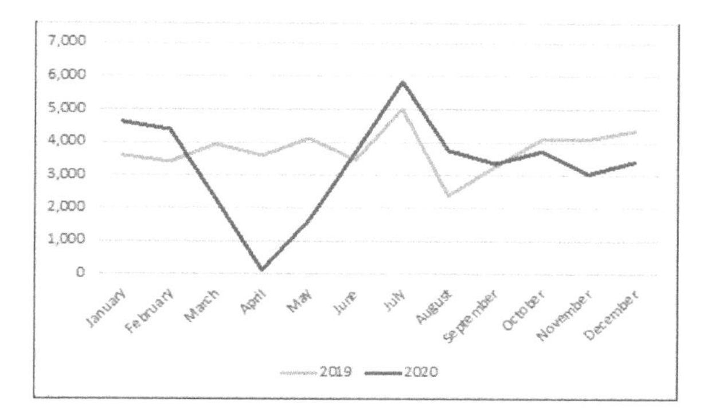

Figure 4.7 Number of vehicle registrations in Barcelona.

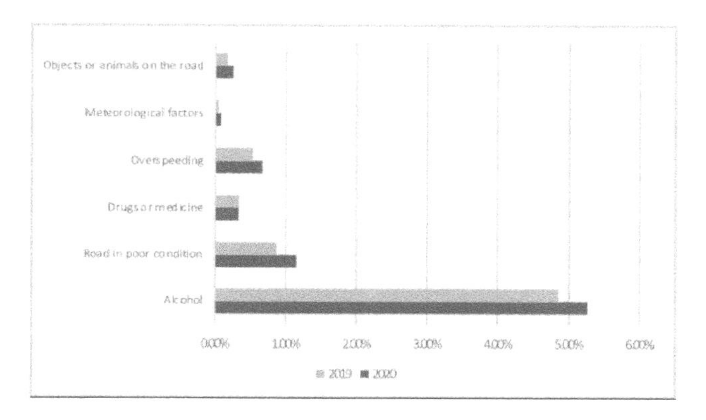

Figure 4.8 Percentages of several reasons for traffic crash injuries in Barcelona.

their percentages that are determined based on the total number of reported reasons on the total number of crashes for both 2019 and 2020, individually. It can be shown that alcohol, a road in poor condition, overspeeding, and objects or animals on the road all have slightly increased in 2020 compared to 2019. It is worth mentioning that the low percentages for all included reasons are based on the reported reasons from the data source, as other factors are reported as "There is no mediated cause." It can be concluded from those different overall statistics that the pandemic has influenced these aspects which may in turn affect the traffic crash injury cases and lead to this number of cases.

4.3.2 Forecast Model and Goodness of Fit

Table 4.2 is showing the results of the model goodness of fit measures. The results show an acceptable result for the carried measures regarding the goodness of fit for the applied model. Table 4.3 is showing the results for the Ljung-Box. Df is the degree of freedom for the number of parameters that are free to vary when considering the severe injury target variable. Q(#) is the Ljung-Box statistics. The significance value is larger than 0.05 indicating that the model has random residual errors, which is a good indication for the model. In Figure 4.9, the actual and upper forecasts are all presented alongside the actual crash reports that included only severe injuries.

TABLE 4.2 Forecast Model Fit for Severe Injury

MSE	0.729
RMSE	0.854
RMSPE	51.400
MAE	0.696
MAPE	48.338
MAXAE	5.377
MAXAPE	90.257
AIC	−459.223
BIC	−448.649
R-squared	0.006
Stationary R-squared	0.741

TABLE 4.3 Ljung-Box Q for the Forecast Model of the Severe Injury

Q(#)	15.803
df	16.000
Significance	0.467

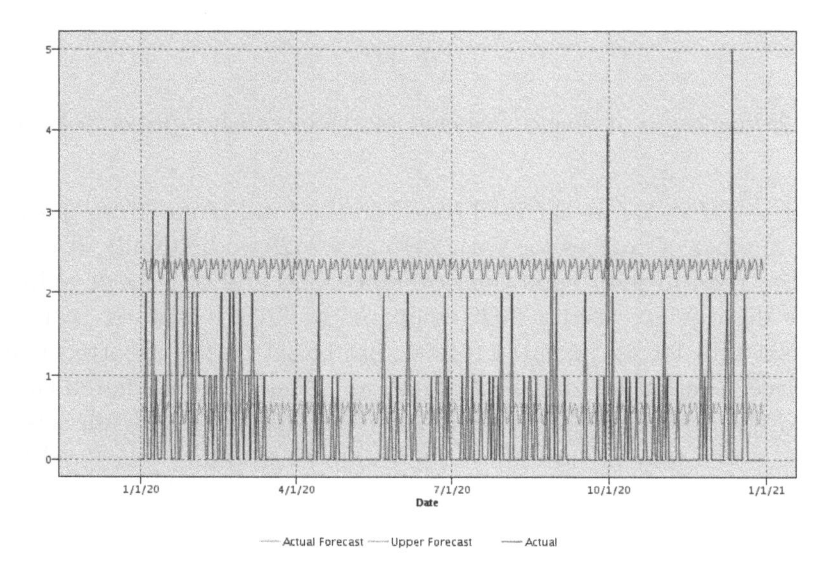

Figure 4.9 Forecast results vs actual observations for severe injuries in 2020.

As observed previously, the number of severe injuries that occurred per day during September and December are also captured by the forecast as abnormal cases because their results exceeded the upper limits for the forecast.

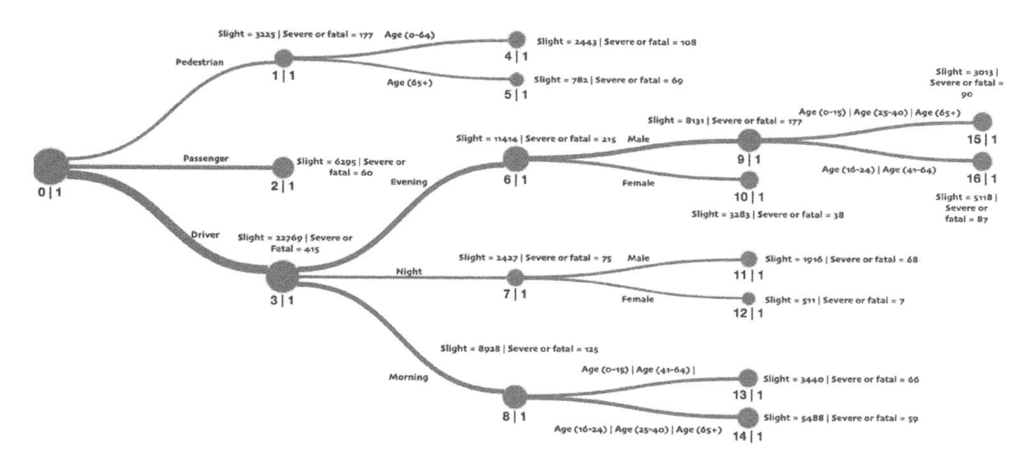

Figure 4.10 The structure of CHAID for traffic crash injuries between 2016–2019.

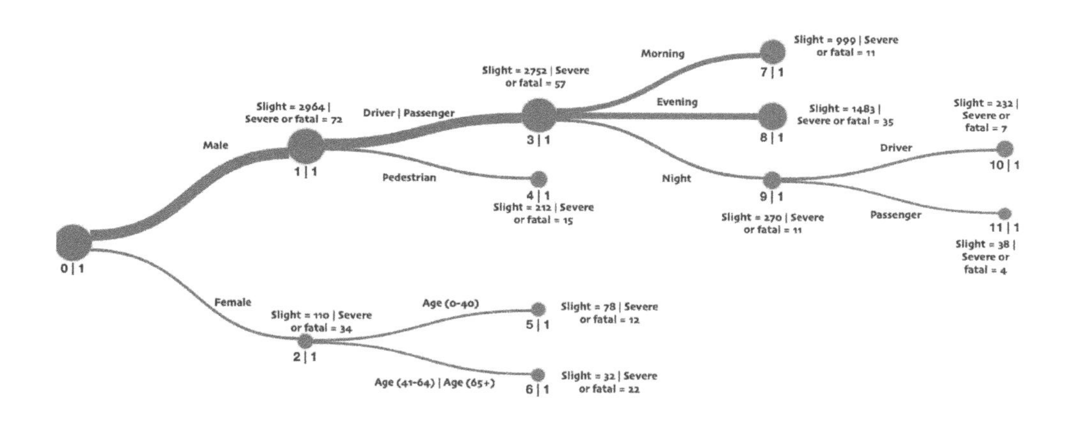

Figure 4.11 The structure of CHAID for traffic crash injuries in 2020.

4.3.3 Risk Factor Analysis

To compare the correlations between four selected independent variables, two identical exhaustive CHAID models are employed. The four selected predictors are gender, age, time of the day, and injured person The training and testing prediction accuracy values for the first exhaustive CHAID are 98.02% and 97.98%, respectively. The training and testing prediction accuracy values for the second model are 96.67% and 96.34%, respectively. Figures 4.10 and 4.11 depict the results of both employed models. As shown, for traffic crashes that happened during 2016–2019, the injured person category is the most important variable, while for 2020, gender is the most important variable based on the applied exhaustive CHAID. For the first model, pedestrians and drivers are at a higher risk of having severe or fatal injuries compared to passengers. The branch of the pedestrian category is, then split based on the age category. Elderly pedestrians are shown to be more involved in severe or fatal injuries compared to other age categories. Driver branch is split based on the time of the day predictor.

The evening period has the highest number of severe and fatal injuries compared to morning and night timing. Both evening and night timing are split based on gender, while the morning category is split based on age. Males in all branches have a higher number of severe or fatal injuries compared to females. For the second model, the male category is split based on the injured type, while the female category is split based on the age group. Drivers and passengers branch has a higher number of both slight, severe, and fatal injuries. Both morning and night timing have the same number of severe and fatal injuries which is less than the number of injuries during the evening timing. The females who are in age groups between 41 and older are showing the highest odds of having severe or fatal injuries compared to all other age categories in 2020.

Pedestrians in both years intervals are vulnerable road users when comparing the total severe and fatal injuries to total injuries that occurred for all injured persons. However, drivers have the highest number of severe or fatal injuries in total during the 2016–2019-time interval. On other branches of CHAID for 2016–2019, the age category has more complex variations compared to 2020 due to having different sets from different groups at the same branch. The time of the day risk factor shows that evening and night timing for both time intervals have a higher number of severe or fatal injuries compared to morning timing. Male injured persons are still considered a risk category in 2020 similar to the 2016–2019-time interval.

4.4 CONCLUSIONS

The COVID-19 pandemic is certainly affected our lives and many other different aspects globally. Traffic crashes are one of those aspects that are affected by this pandemic. However, the influence of COVID-19 varied in different regions related to traffic crashes, but generally, the number of traffic crashes decreased drastically worldwide due to the imposed restrictions to restrain the spread of this virus and to lessen the number of deaths and infected cases. Several studies in different countries have confirmed the impact of this pandemic. However, some studies have noticed an uncommon increase in certain traffic violations or traffic crashes after the restrictions are eased across some regions. In Barcelona, until now, no similar studies are conducted to examine the impact of COVID-19 on traffic crash injuries. Therefore, this study is conducted to provide the overall status before and during the pandemic regarding traffic crashes. The main analysis approach that is carried out in this study is time-series analysis and forecast to understand this potential impact on traffic crashes different levels of injuries that occurred in 2020 and then compare it with previous years. Exhaustive CHAID tree models are, then, employed to prepare a comparison between the pandemic period and four previous years.

The results have shown a dramatic drop in the number of traffic crashes during the pandemic, especially when the state of alarm started in March. This pattern continued till July as the number increased to reach a similar total number of the previous year. Slight injuries have shown a total percentage decrease of 39%. For the total decrease of severe injuries, a total percentage of 30% reduction during 2020 compared to 2019. Similarly, fatal injuries have dropped by 36%. However, despite the total reduction in all the levels of injuries that occurred in 2020, there was an abnormal increase during some days in severe injuries cases that happened in 2020 during September and December. Therefore, a forecast model is applied based on the four previous years including 2016, 2017, 2018, and 2019. These anomalous cases have also exceeded the upper limit of the forecast model to confirm this captured increase.

For the risk factors comparison, the results show that male injured person is still more involved in severe and fatal injuries compared to females before and during the pandemic. Evening and night timing has also a similar trend in 2020 when compared to 2016–2019 in having a higher number of severe and fatal injuries. Pedestrians during 2016–2019 have a much higher number of severe or fatal injuries compared to 2020. However, the number of severe or fatal injuries for pedestrians during 2020 is still high compared to other categories when focusing on the overall number of injuries. In 2020, elderly female injured persons are showing higher probabilities of having severe or fatal injuries compared to other age categories. Although the pandemic has affected the number of injuries, the impact of certain risk factors is found to be the same as before and during the pandemic. It can be shown that both CHAID models that are employed to examine the correlations have proven their ability to provide interpretable results.

Bibliography

[1] A. Aiash and F. Robuste, "Traffic accident severity analysis in Barcelona using a binary probit and CHAID tree," International Journal of Injury Control and Safety Promotion, vol. 29, no. 2, pp. 256–264, 2021.

[2] A. Alfons, C. Croux and S. Gelper, "Sparse least trimmed squares regression for analyzing high-dimensional large data sets," The Annals of Applied Statistics, vol. 7, no. 1, pp. 226–248, 2013.

[3] A. C. Harvey, Forecasting, structural time series models and the Kalman filter, Cambridge: Cambridge University Press, 1989.

[4] A. I. Qureshi, W. Huang, S. Khan, I. Lobanova, F. Siddiq, C. R. Gomez and M. F. K. Suri, "Mandated societal lockdown and road traffic accidents," Accident Analysis & Prevention, vol. 146, p. 105747, 2020.

[5] A. Kerimray, N. Baimatova, O. P. Ibragimova, B. Bukenov, B. Kenessov, P. Plotitsyn and F. Karaca, "Assessing air quality changes in large cities during COVID-19 lockdowns: The impacts of traffic-free urban conditions in Almaty, Kazakhstan," Science of The Total Environment, vol. 730, p. 139179, 2020.

[6] Barcelona's City Hall Open Data Service, "Open Data BCN," Barcelona's City Hall Open Data Service, 2020. [Online]. Available: https://opendata-ajuntament.barcelona.cat/en/.

[7] C. Katrakazas, E. Michelaraki, M. Sekadakis and G. Yannis, "A descriptive analysis of the effect of the COVID-19 pandemic on driving behavior and road safety," Transportation Research Interdisciplinary Perspectives, vol. 7, p. 100186, 2020.

[8] D. Biggs, B. De Ville and E. Suen, "A method of choosing multiway partitions for classification and decision trees," Journal of Applied Statistics, vol. 18, no. 1, pp. 49–62, 1991.

[9] D. C. Montgomery, L. A. Johnson and J. S. Gardiner, Forecasting and Time Series Analysis, McGraw-Hill, 1990.

[10] D. Koh, "COVID-19 lockdowns throughout the world," Occupational Medicine, vol. 70, no. 5, p. 322, 2020.

[11] D. Muley, M. S. Ghanim, A. Mohammad and M. Kharbeche, "Quantifying the impact of COVID-19 preventive measures on traffic in the State of Qatar," Transport Policy, vol. 103, pp. 45–59, 2021.

[12] D. Pena, G. C. Tiao and R. S. Tsay, A course in time series analysis, New York: John Wiley and Sons, 2001.

[13] D. Stavrinos, B. McManus, S. Mrug, H. He, B. Gresham, M. G. Albright, A. M. Svancara, C. Whittington, A. Underhill and D. M. White, "Adolescent driving behavior before and during restrictions related to COVID-19," Accident Analysis & Prevention, vol. 144, p. 105686, 2020.

[14] E. I. Sakelliadis, K. D. Katsos, E. I. Zouzia, C. A. Spiliopoulou and S. Tsiodras, "Impact of Covid-19 lockdown on characteristics of autopsy cases in Greece. Comparison between 2019 and 2020," Forensic Science International, vol. 313, p. 110365, 2020.

[15] E. S. Gardner, "Exponential smoothing: The state of the art," Journal of Forecasting, vol. 4, pp. 1–28, 1985.

[16] ECDC, 2022. [Online]. Available: https://www.ecdc.europa.eu/en/geographical-distribution-2019-ncov-cases.

[17] G. E. P. Box, G. M. Jenkins, G. C. Reinsel and G. M. Ljung, Time Series Analysis: Forecasting and Control, 3rd Edition, Englewood Cliffs, N.J.: Prentice Hall, 1994.

[18] G. Melard, "A fast algorithm for the exact likelihood of autoregressive-moving average models," Applied Statistics, vol. 33, no. 1, pp. 104–119, 1984.

[19] G. V. Kass, "An exploratory technique for investigating large quantities of categorical data," Journal of the Royal Statistical Society: Series C (Applied Statistics), vol. 29, no. 2, pp. 119–127, 1980.

[20] H. Inada, L. Ashraf and S. Campbell, "COVID-19 lockdown and fatal motor vehicle collisions due to speed-related traffic violations in Japan: a time-series study," Injury Prevention, vol. 27, pp. 98–100, 2021.

[21] H. Lee, S. J. Park, G. R. Lee, J. E. Kim, J. H. Lee, Y. Jung and E. W. Nam, "The relationship between trends in COVID-19 prevalence and traffic levels in South Korea," International Journal of Infectious Diseases, vol. 96, pp. 399–407, 2020.

[22] H. V. Solomon, "COVID-19 checklist: Mask, gloves, and video chatting with grandpa," Psychiatry Research, vol. 288, p. 112986, 2020.

[23] IBM, "IBM SPSS Modeler 18.0 Algorithms Guide," IBM Corporation, 2016.

[24] J. Abellan, G. Lopez and J. de Ona, "Analysis of traffic accident severity using Decision Rules via Decision Trees," Expert Systems with Applications, vol. 40, no. 15, pp. 6047–6054, 2013.

[25] J. H. Nunez, A. Sallent, K. Lakhani, E. Guerra-Farfan, N. Vidal, S. Ekhtiari and J. Minguell, "Impact of the COVID-19 pandemic on an Emergency Traumatology Service: Experience at a Tertiary Trauma Centre in Spain," Injury, vol. 51, no. 7, pp. 1414–1418, 2020.

[26] J. Henriquez, E. Gonzalo-Almorox, M. Garcia-Goni and F. Paolucci, "The first months of the COVID-19 pandemic in Spain," Health Policy Technol, vol. 9, no. 4, p. 560–574, 2020.

[27] J. Prasetijo, G. Zhang, Z. M. Jawi, M. E. Mahyeddin, Z. F. Zainal, M. Isradi and N. Muthukrishnan, "Crash model based on integrated design consistency with low traffic volumes (due to health disaster (COVID-19)/movement control order)," Innovative Infrastructure Solutions, vol. 6, p. 22, 2021.

[28] J. Shen, C. Wang, C. Dong, Z. Tang and H. Sun, "Reductions in mortality resulting from COVID-19 quarantine measures in China," Journal of Public Health, p. 1–7, 2021.

[29] J. Xiang, E. Austin, T. Gould, T. Larson, J. Shirai, Y. Liu, J. Marshall and E. Seto, "Impacts of the COVID-19 responses on traffic-related air pollution in a Northwestern US city," Science of the Total Environment, vol. 747, p. 141325, 2020.

[30] M. W. Meyer, "COVID Lockdowns, Social Distancing, and Fatal Car Crashes: More Deaths on Hobbesian Highways?," Cambridge Journal of Evidence-Based Policing, vol. 4, pp. 238–259, 2020.

[31] O. Saladie, E. Bustamante and A. Gutierrez, "COVID-19 lockdown and reduction of traffic accidents in Tarragona province, Spain," Transportation Research Interdisciplinary Perspectives, vol. 8, p. 100218, 2020.

[32] P. de Pedraza and M. Guzi, "Life Dissatisfaction and Anxiety in COVID-19 pandemic," Global Labor Organization (GLO), Essen, 2020.

[33] P. J. Brockwell and R. A. Davis, Time Series: Theory and Methods 2nd edition, Springer-Verlag, 1991.

[34] R. J. C. Calderon-Anyosa and J. S. Kaufman, "Impact of COVID-19 lockdown policy on homicide, suicide, and motor vehicle deaths in Peru," Preventive Medicine, vol. 143, p. 106331, 2021.

[35] S. AlKheder, F. AlRukaibi and A. Aiash, "Risk analysis of traffic accidents' severities: An application of three data mining models," ISA Transactions, vol. 106, pp. 213–220, 2020.

[36] S. G. Makridakis, S. C. Wheelwright and R. J. Hyndman, Forecasting: Methods and applications 3rd edition, New York: John Wiley and Sons, 1997.

[37] S. Hu, C. Xiong, M. Yang, H. Younes, W. Luo and L. Zhang, "A big-data driven approach to analyzing and modeling human mobility trend under non-pharmaceutical interventions during COVID-19 pandemic," Transportation Research Part C: Emerging Technologies, vol. 124, p. 102955, 2021.

[38] S. R. Barnes, L.-P. Beland, J. Huh and D. Kim, "The Effect of COVID-19 Lockdown on Mobility and Traffic Accidents: Evidence from Louisiana," Global Labor Organization (GLO), Essen, 2020.

[39] S. X. Zhang, Y. Wang, A. Rauch and F. Wei, "Unprecedented disruption of lives and work: Health, distress and life satisfaction of working adults in China one month into the COVID-19 outbreak," Psychiatry Research, vol. 288, p. 112958, 2020.

[40] The Lancet Infectious Diseases, "COVID-19, a pandemic or not?," The Lancet Infectious Diseases, vol. 20, no. 4, p. 383, 2020.

[41] TomTom, "Barcelona traffic," 2020. [Online]. Available: https://www.tomtom.com/en_gb/traffic-index/barcelona-traffic/.

[42] U. Oguzoglu, "COVID-19 Lockdowns and Decline in Traffic Related Deaths and Injuries," IZA, p. 13278, 2020.

[43] UNICEF; WHO; IFRC, "Key Messages and Actions for COVID-19 Prevention and Control in Schools March 2020," World Health Organization, 2020.

[44] V. R. Maryada, P. Mulpur, A. V. Guravareddy, S. K. Pedamallu and B. V. Bhasker, "Impact of COVID-19 pandemic on Orthopaedic Trauma Volumes: a Multi-Centre Perspective From the State of Telangana," Indian Journal of Orthopaedics, vol. 54, pp. 368–373, 2020.

[45] World Health Organization, 2022. [Online]. Available: https://covid19.who.int.

CHAPTER 5

Advances in Explainable Reinforcement Learning: An Intelligent Transportation Systems Perspective

Rudy Milani, Maximilian Moll, and Stefan Pickl

Universitaet der Bundeswehr Muenchen, Werner-Heisenberg-Weg 39, Neubiberg 85577, Germany

CONTENTS

IN THE LAST DECADES, Reinforcement Learning (RL) started entering every-days life since its super-human performances. Recently, RL has been applied to problems related to transportation systems, e.g., self-driving cars and traffic control, bringing contributions to the new field of Intelligent Transportation Systems. In these cases, each action is of vital importance, and for this reason, understanding why the RL

DOI: 10.1201/9781003324140-5

agent is choosing a particular plan instead of a different one can help the human user find the right decision. Therefore, the focus of research in this field was recently centred on Explainable Reinforcement Learning (XRL) methods, i.e., algorithms that are interpretable and understandable to users. Consequently, an enormous amount of papers have been published. In this chapter, we summarize the most important advances in the XRL area over the last six years, focusing on outlining the pros and cons of each approach, and proposing a new classification based on how the explanations are generated and presented. Finally, a discussion highlighting open questions and possible future work are presented.

5.1 INTRODUCTION

In the last years, Machine Learning (ML) started becoming central in the solution of problems relative to Intelligent Transportation Systems (ITS) [33]. In particular, Reinforcement Learning (RL) could be widely applied to find outstanding results, e.g., self-driving cars [15]. Despite the best performances being quite recent, the idea to apply RL in ITS problems exists since the early 2000s [1], where the problem was the control of traffic lights using a simple RL algorithm. The main complications in ITS are modelling and optimization. In the majority of cases, the need for an expert for creating an ad-hoc model that can also generalize different scenarios is essential. If we consider the transportation systems services like control of traffic signals [1,42,52,57], or of charging for electric vehicles [73], we can understand the importance and the utility of having a controlling agent that can learn all these situations directly and generalize for other cases [1].

However, RL has still some flaws, e.g., for traffic signal timing, the problem lies in the exponentially expanding complexity of the state set dimension [52,57]. For these reasons, more complex approaches have to be applied: Deep Reinforcement Learning (DRL). This leads also to a complication in the interpretability and explainability of the actions done by the agents [110]. The importance of choosing the correct decision is fundamental since errors can be extremely costly in these environments. Therefore, having an explanation of how the agent comes up with that choice can make a difference in increasing transparency and trust. This brings us to the explosion in the usage of Explainable Artificial Intelligence (XAI) and, in particular for our aim, Explainable Reinforcement Learning (XRL) [21].

In the following, a structured overview of the most important advances in XRL of the last years is presented to identify different research streams.

5.1.1 Selection of Literature

The widespread use of AI in different fields and the necessity to provide explanations for the results obtained by these algorithms have increased exponentially the survey papers that try to include the newest researches on this subject [2,12,19,22,29,30,68, 74,79,84,94]. As can be noticed, XAI is a rapidly expanding field. However, we still have limited knowledge of what is happening in the majority of the applications. In particular, this great inflation on the theme of explainability has involved also XRL in

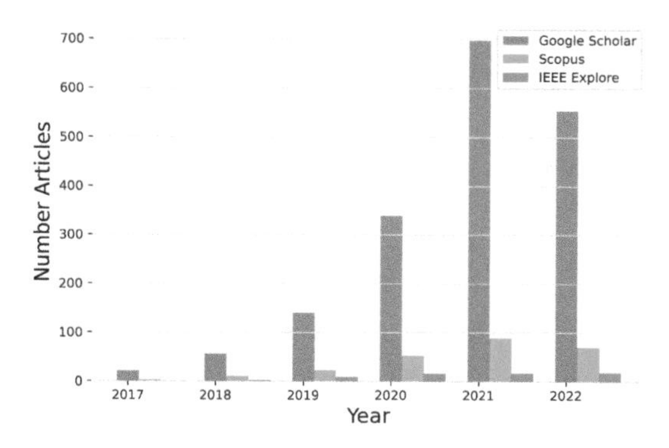

Figure 5.1 Bar Plot of the Articles found for each year using different search engines (until September 2022).

recent years, as shown in Figure 5.1. Also, in this case, a high number of review papers are written to summarize the improvements. The main problem is that there is still not a well-defined taxonomy that can efficiently classify all the methods presented, and for this reason, the number of survey papers trying to find a unique classification is exploding together with the research in this field [3, 21, 27, 38, 48, 71, 82, 83, 107]. For this reason, it is fundamental to select only a subset of all the research papers. In particular, we will focus on analysing the most cited articles in the last six years, i.e., from 2017 to 2022. The collection of methods is gathered using different keywords combinations of "Reinforcement Learning" and other specific terms ("Explainable," "Explanation," "Visualization," and "Information Visualization") in the following search engines: Google Scholar, Scopus, and IEEE Xplore. From these resources, backward research was performed, to not forget important methods that could not be found using the preceding keywords.

5.1.2 Contribution and Structure of the Chapter

The contribution of this chapter is related to the analysis and description of the most relevant advancements in the topic of XRL in the last six years. In particular, new classification and terminology are proposed together with an evaluation of the quality of each method. Therefore, the advantages and flaws are individually summarized.

The chapter is structured as follows: after a brief introduction on what RL is and on the used terminology, we will focus on the major characteristics that can help us to find a general classification for the XRL algorithms, defining then the proposed arrangement. Lastly, all the papers relevant to this review are analysed in detail, proposing, in the end, some possible future works and open questions.

5.2 REINFORCEMENT LEARNING

In RL, there is an agent, which is the learner and decision-maker: it has to understand how to use the information of the environment to choose the best action. Everything that interacts with the agent is called the environment. More in detail, at each discrete time step $t = 0, 1, 2 \ldots$ the agent receives the actual state $s_t \in \mathbb{S} \subseteq \mathbb{R}^n$ from the environment and chooses an action $a_t \in \mathbb{A} \subseteq \mathbb{R}^m$, where $\mathbb{S}$ and $\mathbb{A}$ are the state and action spaces respectively [90]. The agent will receive a numerical reward $r_t \in \mathbb{R}$ as a consequence of the action, and the environment update to a new state s_{t+1}. The agent has to maximize, by deciding the correct actions, the sum of rewards $R_t = \sum_{i=0}^{\infty} \gamma^i r_{t+i+1}$ where $\gamma \in [0, 1)$ is a discounting factor. An additional property is that the probability of each value for s_t and r_t depends only on the state s_{t-1} and action a_{t-1}, and not on earlier states and actions. This is the Markov property and characterizes the Markov Decision Process (MDP) framework. A MDP is a tuple $(\mathbb{S}, \mathbb{A}, \mathbb{T}, R)$, where $\mathbb{S}$ and $\mathbb{A}$ are previously defined, $\mathbb{T}$ is the transition probability of the environment, and $R : \mathbb{S} \times \mathbb{A} \rightarrow \mathbb{R}$ is the reward function such that $R(s_t, a_t) = r_{t+1}$. The MDP is the basis of RL. A particular class of RL methods is the *model-free* which do not need any knowledge of $\mathbb{T}$, as the *model-based* do, but they consider only the Q-value function $Q_\pi(s, a) = \mathbb{E}_\pi\left[\sum_{k=0}^{\infty} \gamma^k r_{t+k+1} | s_t = s, a_t = a\right]$ under the policy π; simply defined as: $\pi(a|s) = P(a_t = a|s_t = s)$. In other cases, the state-value function $V_\pi(s) = \mathbb{E}_\pi\left[\sum_{k=0}^{\infty} \gamma^k r_{t+k+1} | s_t = s\right]$ is used. After this brief background description (for further information we refer to [90]), we can now focus on the useful terms used for characterizing the explanations in the rest of the chapter.

5.3 EXPLAINABILITY AND INTERPRETABILITY

It is fundamental to focus on the terminology used for this paper since in previous works a uniform vocabulary is missing. This inconsistency can create misunderstanding if comparing different studies in the literature [71]. In particular, in many papers "Explainability" and "Interpretability" are used as synonyms [64, 67, 71, 76, 94] while for many others there are major differences that distinguish one to the other [9, 79, 109]. In general, the main focus is on the consideration of *interpretable* and *explainable* models: a model is *interpretable* if intrinsically it is possible to understand it without any further technique, so it is a passive quality; while a model is *explainable* if it is possible to apply an external algorithm that can generate a reasonable explanation, in this case, an active characteristic. For the purposes of this paper, we will consider the quality of interpretability and explainability as equivalent since the final task is always to increase trust and confidence in the action chosen by the agent. Moreover, we will define in the next section two different categories that relate to the previously described differences. Furthermore, as can be noticed, the notions of interpretability and explainability could be subjective and, therefore, dependant on the observer [27].

However, this is not the only ambiguity in the terminology for XRL. Together with the previous two definitions, also the case of "Transparency" is quite controversial. Usually, this definition can be seen as a synonym of interpretability [54] and in other cases, it is considered completely different [106]. There are also comprehensibility, simulatability and decomposability that are used with different acceptions [27, 54]. Therefore, to not increase the difficulty in reading this survey, we define as follows the specific terms: a model is *reliable* if the explanation generated can be understood by a human user, while the *performances* and *efficiency* will be related to the accuracy of the method.

5.4 PROPOSED CLASSIFICATION

In this section, we analyze the usual characteristics that are considered in the classification of XRL methodologies. In the first place, we consider in this analysis two basic factors: scope and time. For the scope component, there are two possible types: *global* (G) and *local* (L). The global models have to explain all the logic behind the general behaviour and policy [2, 71]. On the other hand, local explanations have to target a specific decision [2,48]. In the case of the time aspect, we outline the moment when the explanation is generated considering two instants: we call *intrinsic* (I) the models that are inherently interpretable or self-explanatory, e.g. decision trees; and we define *post-hoc* (P) the methods that need a simplified surrogate model to provide explanations for the original method [48, 71].

These are the classical factors used for the characterization of each model, but they are not sufficient for a clear definition of the class. For this reason, different authors proposed additional information. In our case, we focus also on the typology of the RL method applied, not remaining only on the surface considering the classification of *model-agnostic* and *model-specific* [3, 71], but going deep into the details of which particular method is considered.

A further fundamental aspect to bear in mind is in which way the explanation is presented to the user, or extracted from the model. It is one of the main reasons to classify a specific method in a particular class instead of in another one. Lastly, also the techniques applied for the generation of the models are relevant for recognising if a particular approach is a member of a class or not. For these reasons, our groups are method-output category based: centred on the approach used for the creation of the explanation and the type of outcome.

Therefore, we decided to consider these features for the classification of the surveyed papers counting a total of eight different classes. A selection of the papers that were analysed in detail is presented in Table 5.1. Furthermore, for the explanation of each group, we considered the frequently cited papers, for the category, found in the last six years, i.e., the pillars of the respective methodologies, hence the ones where we should focus more of our attention. In addition, for the newer articles, we considered the ones contributing to the main papers of their category that can open interesting new roads for the research. Moreover, we decided to represent two particular qualities that are, from our point of view, fundamental for future works: we checked if a methodology tested also their explanation through a human study and if the proposed approach is completely automatic. These qualities are indispensable for the creation of reliable models since they should be used by human users and without any intervention from the exterior that could compromise their real efficiency.

In the following sections, we describe the eight different classes, starting with the Relational and Symbolic class.

TABLE 5.1 Selection of the Principal Papers for the Reviewed Literature in the Field of XRL Following the Proposed Classification Rules

Reference	Year	Time	Scope	Agent	Explanation	Human	Automatic
[111] Zambaldi et al.	2018	I	G	AC	Graph/Importance	×	✓
[59] Lyuet al.	2019	I	G/L	DQN	Simbolic Policies	×	✓
[60] Ma et al.	2021	I	L	DQN	Importance	×	✓
[23] Erwing et al.	2018	I	L	SARSA	Values/Bar chart	×	∼
[46] Juozapaitis et al.	2019	I	L	Q-learn./DQN	Values/Bar chart	×	∼
[77] Rietz et al.	2022	I	G/L	DQN	Values/Bar chart	×	∼
[87] Shu et al.	2017	I	G/L	A2C	Hierarchical Graph	×	×
[11] Beyret et al.	2019	I	G	PG	Heatmaps	×	✓
[99] van der Waa et al.	2018	P	G/L	Value-based	Consequences	✓	✓
[20] Davoodi & Komeili	2021	P	G/L	Alg.-agnostic	Heatmaps	×	✓
[31] Guo et al.	2021	P	G	Q-learn./PG	Importance	×	✓
[62] Madumal et al.	2020	P	L	Model-free	Causal Expl.	✓	×
[61] Madumal et al.	2020	P	L	Model-free	Distal Expl.	✓	×
[37] Herlau & Larsen	2022	P	G	Policy	Feature Imp.	×	∼
[25] Gajcin & Dusparic	2022	P	L	DQN	Causal relation	×	×
[72] Pynadath et al.	2018	P	L	Model-based	Written Expl.	✓	×
[92] Tabrez et al.	2019	P	G	POMDP	Written Expl.	✓	✓
[17] Cruz et al.	2022	P	L	Not specified	Written Expl.	✓	✓
[28] Greydanus et al.	2018	P	G	A3C	Saliency maps	✓	✓
[43] Iyer et al.	2018	P	L	DQN	Saliency maps	✓	✓
[32] Gupta et al.	2019	P	L	Not Specified	Saliency maps	✓	✓
[41] Huber et al.	2019	P	L	Dueling DQN	Saliency maps	×	✓
[45] Joo & Kim	2019	P	L	A3C	Saliency maps	×	✓
[104] Wang et al.	2020	P	L	DQN	SHAP	×	✓
[78] Rizzo et al.	2019	P	L	PG	SHAP	×	✓
[103] Wang et al.	2018	P	L	DQN	Visual anal.	✓	✓
[18] Dao et al.	2018	P	G	Double DQN	Snapshots	×	✓
[65] Mishra et al.	2018	P	G	Double DQN	Snapshots	×	✓
[86] Sequeira & Gervasio	2020	P	G	Q-learn.	Short Video	✓	✓
[5] Amir & Amir	2018	P	G	Q-values/policy	Importance	×	✓
[40] Huang et al.	2018	P	G	SAC	Importance	✓	∼
[57] Liu et al.	2018	I	G/L	DQN	Tree/maps	×	✓
[88] Silva et al.	2020	I	G/L	Q-learn./PG	Tree	∼	✓
[16] Coppens et al.	2019	P	G/L	PPO	Heatmaps	×	×
[100] Vasic et al.	2022	I	G	DQN	Tree	×	✓
[63] McCalmon et al.	2022	P	G	DQN/PPO	Written Expl.	✓	∼
[35] Hein et al.	2017	I	G	Model-based	Fuzzy rules	×	∼
[36] Hein et al.	2018	P	G	Model-based	Boolean/Eq.	×	✓
[39] Huang et al.	2020	I	G	Model-based	If-Then rule	×	✓
[102] Verma et al.	2018	I	G	Policy	Policy	×	∼

The symbol $\sim$ indicates that it is not clear if that property holds.

5.4.1 Relational and Symbolic

In this section, we analyse the methods that apply relational and symbolic representation for states and actions in RL. Therefore, the focus is on logical encoding which can lead to the generation of better explanations.

One of the basic papers in this context is [111]. The authors used Relational Reinforcement Learning (RRL) in the DRL context to increase the efficiency, generalization capability and interpretability of these methods. RRL consists in translating states, actions and policies using first-order (or relational) language. A benefit of this approach is that we can include background information directly: indeed, if we have a particular known rule, we can encode it by a logical fact. An articulated architecture is proposed consisting of convolutional layers and a relational module composed of multi-head dot-product attention (MHDPA), or self-attention [101], blocks. Good computational results are obtained, but, for the generalization capability, has been found a problem when the architecture becomes bigger. This issue could be caused by overfitting in the training phase.

Following the previous idea, [59] decided to apply a symbolic representation to DRL to improve task-level interpretability. This new framework called Symbolic Deep Reinforcement Learning consists of three components: a planner, a controller, and a meta controller. The planner has the task to plan the actions of the agent for achieving further goals using some prior symbolic representations [51]. The controller is a usual DRL method: Deep Q-Network (DQN) [66], which is trained using the regular reward of the environment. In fact, in this case, there are two different classes of rewards: an intrinsic reward, which is the classical one, and an extrinsic reward, learned by the meta controller, which measures the performance of the controller to suggest different intrinsic goals. The advantages of using a methodology like this are the opportunity to easily learn sub-policies for each sub-task and also the sub-tasks can be identified with more accuracy. The main disadvantage is the computational time for the training of an efficiently interpretable plan. Despite [59] used simple environments (Taxi Domain and ATARI) for the experimentation, they needed more than a million episodes for the identification of all the sub-policies and sub-tasks and, in some cases, not all the sub-tasks were learned.

The direct improvement of the previous methodologies is presented in [60]. In this case, the authors proposed a Neural-Symbolic Reinforcement Learning approach to enhance the transparency and interpretability of DRL methods. This method consists of three main components, one of which has been already presented in the analysis of [111]. Firstly the symbolic state is represented as a matrix and then is given as input to the attention module MHDPA. At the end of this process, there are the attention weights on logical rules of different lengths. The initial matrix of the symbolic state and these weights are then processed in the second component: the reasoning module, which aims to provide reasoning on the actual information. Finally, these results are sent to a Multi-Layer Perceptron (MLP), called the policy module, to create the action predicate. In the end, the chosen action is based on the value of these resulting action predicate matrices. In this way, it is possible to increase interpretability by visualizing the relational path to different goals. However, this approach can also be improved through more expressive rules like trees and junction rules.

5.4.2 Reward Decomposition

In the following, we analyse the articles relative to Reward Decomposition, a technique, introduced in [81], used to divide the classical reward into different types to extract more interpretability from the policy learned by the agent; as already seen in [59].

Similarly to the previous class of methods, [23] defined multiple semantically relevant rewards and the concept of reward difference explanations. In this method, the Q-function that is considered is the sum of all the different components trained using the respective reward obtained after the decomposition. In detail, a SARSA agent [80] is trained for each reward type, leading to the definition of multiple Q-functions, that are finally considered as one by summing up all their values. The benefit of this approach is its simplicity and, at the same time, its efficiency. However, in the majority of the cases, the main component of this explanation is dependent on only one reward, which makes the evaluation of all the other factors not optimal.

The authors from [46] improved the idea exposed in [23], considering an off-policy algorithm that could learn the decomposed Q-functions in such a way that the sum of all these components could converge to the optimal value. If we consider the previous approach, learning the Q-function for each type of reward does not guarantee that the sum of these factors will converge to the correct general value function. Therefore, for the update of the decomposed Q-functions, they first evaluated the action that maximized the full value function, and, then, applied the update using that action in the usual formula for the Q-learning [105]. The main issue derives from the choice of reward shaping since it has the highest importance in finding optimal results. Usually, in many environments, there is a natural reward decomposition, but this fact can not be generalized.

In [77], the authors pointed out some flaws of the previously described approaches: the single-step explanation generated through reward decomposition can be misleading since they are not considering any future behaviour of the agent. They account only for the actual transition values. Consequently, they proposed to combine reward decomposition with a hierarchical method to correct this shortcoming. The framework introduced is based on two DQNs, a higher-level and a lower-level, trained using reward decomposition, following the idea of [49]. The higher-level DQN has to learn the sequences of goals that must be done to solve a task, while the lower-level DQN learns how to solve these goals. The main issue is the lack of human evaluation for these explanations. Indeed, the final goal is to increase transparency and trust in humans by generating explanations for the action done. Therefore, considering also how humans perceive this kind of justification is one of the major aspects to consider.

5.4.3 Hierarchical and Goal-Based

Now, we analyse the Hierarchical and Goal-based explanation approaches. As already introduced in the last method of the previous section, Hierarchical RL involves the application of two-layer policies: one layer for the definition of the sub-goals, and the second layer for finding the optimal policy to achieve the previously defined task [91].

The first approach, in chronological order, relative to this class is [87]. In this work, the authors presented a hierarchical approach for interpreting the policy obtained from training actor-critic agents. In their framework, they considered multitasking RL settings where each goal is identified using two word templates: one for the skill and the other for the item. The set of these goals is progressively increased as the agent learns how to complete them. The hierarchical technique appears in the choice of the policy to solve these tasks. The hierarchy is the following: a base policy for the already learnt tasks, an instruction policy that allow the actual policy and the base policy to communicate, an augmented policy that allows the actual policy to do the actions, and, lastly, a switch policy that decides whether to consider the base policy or the actual one. For the choice of the base policy, they considered the previous step trained policy. The main problem in this method is the need for a human that can decipher the environment to give to the agent the correct tasks to learn. In fact, for each episode, the goals are based on human instruction. On the other hand, the efficiency tests present positive results, and humans can also look at the typical hierarchical plan to understand better why the agent is doing that action.

A much more complex approach is presented by [11], which introduced a new hierarchical/goal-based method called *Dot-to-Dot* that can reach simultaneously good results comparable to the usual RL algorithms and interpretable representations of the decision-making process of the agent. This methodology relies on many different algorithms: the first is a Deep Deterministic Policy Gradients algorithm [53] trained using the Hindsight Experience Replay [7] to make every trace important in the training phase regardless of the achievement of the goals. Lastly, as in the previous cases, the hierarchical approach consists in dividing the architecture into a high level for the identification of the sub-goals and a lower level for the solution of them. The new idea introduced in [11] is to reach sub-goals that are nearby the actual state in sparse reward environments. Therefore, the sub-goals chosen are considered as a small perturbation of the previously achieved goal. The interpretability of the policies is expressed using heat maps of the Q-value function. This can be useful to understand the importance of doing the correct action in that precise state, but, generally, it gives no information or transparency on the quality of the action chosen.

5.4.4 Consequence-Based

In this section, we are going to define methods that can create explanations from the consequences of the actions chosen by the agents.

The authors in [99] improved the method firstly developed by [34], where the states are transformed into a more interpretable set of variables applying an appropriate function, and the rewards are explained using the action outcomes as input of a different novel function. The main benefit of this approach is that it can be generalized for all the RL algorithms. In particular, [99] studies how to use this framework to find an explanation for contrastive questions, since they are the most frequently used by humans [64]. Specifically, to this end, [99] used a policy foil that is obtained from the combination of the trained Q-value function and a function, learned from simulations, that represents the user preference in choosing specific actions. The explanation is then generated by the comparison of the different trajectories generated following the usual RL policy and the foil policy obtained by the foil Q-value function. Despite the independence of the RL algorithm applied in this approach, there is a central question regarding the reliability of the action chosen by the user and their subjectivity: not all human users have the same ideas when they have to decide how to solve a particular task. Therefore, learning the preference function could be biased.

Instead of considering a preference function for the actions, [20] proposed to evaluate the risk of the state considering the next n states visited by a trained agent. In particular, for the definition of the risk function, the feature importance is studied and from this analysis, they conveyed that not all the characteristics have the same usefulness for the interpretability of the model. Therefore, only a limited subset of the features are used for the creation of the risk function. The main problem of this approach is the difficulty of finding the right risk function given the environment. It is a completely post-hoc agent-agnostic approach that, however, depends principally on how the risk is defined. An advantage of this method is the possibility to generalize it for continuous problems without any loss.

In all the previous cases, we have always considered the Q-value function as available; however, this condition is not constantly satisfied. Therefore, without any knowledge of the Q-value function, [31] developed a complex framework that can output the final reward of an episode, and evaluate the most important states in the studied sequence. The structure of the methodology consists of the application of a Recurrent Neural Network (RNN) and an MLP for the embedding of the time steps and the episode, and a Gaussian Process for the generation of a latent representation of the complete episode which is used for predicting the final reward. The explanations derived consist of the longest continuous sequence of the K most important states in the episode leading to the final reward. This method is evaluated considering a fidelity interpretation: comparing the most critical time steps derived in different interpretation methods. This evaluation process is not as rigorous as considering specific metrics for the estimation of the quality of the methodology proposed.

5.4.5 Causal Models

Following the same ideas of the previous section, it is possible to generate explanations from the predictions of what will happen thanks to Causal models that can correlate features with each other.

A significant example is proposed in [62]. The authors decided to create a post-hoc approach for the generation of an explanation for model-free agents considering an action influence graph: a particular causal graph that represents features as nodes and actions as edges. Through this structure, together with an ML model, useful for the prediction of the outcomes of a particular action, it is possible to create minimal sufficient explanations [64]. The reliability of the explanations generated was tested through a human study obtaining satisfying results for the quality of completeness and sufficiency. This approach was later improved in [61] introducing the concept of distal action which is an action that has to be enabled by a series of previous movements. For the forecast of the distal action, they adopted RNNs with input the trace of the episodes. Although the accuracy of the prediction is high, the main issue in this method is the need for a human expert that can create a action influence graph. Therefore, this limitation nullifies the automation of the entire process. While there are methods that can help to learn causal graphs automatically, like in [69], to the best of our knowledge, there are no algorithms for the action influence graphs discovery.

In [37], the authors considered the problem of finding the causal representation in RL context applying mediation analysis [4, 70]. Adopting this methodology makes it possible to compare two policies, one trained to maximize the return, and the other that tries to make a variable true. In this way, by evaluating a particular value called Natural Indirect Effect [26] it is possible to understand the causal importance of that variable in achieving the final goal. Therefore, they consider a parsimonious causal graph that can give major insights into how important some features can be in solving the problem. However, it is difficult to understand how parsimony can be effective for more complex MDPs where the relation between variables and goals can be hard to identify directly.

One main issue relative to the creation of explanations through the causal graph is Causal Confusion: a phenomenon where the agent generalizes the relation between features even though it does not hold for every state. For this reason, [25] introduced an algorithm to detect causal confusion before the deployment of an agent. This approach consists of creating alternative environments from critical states of a specific policy in order to test its performance compared to a policy trained using only a subset of features. In this way, it is possible to discriminate the environments where causal confusion is detected. Therefore, by studying these particular situations, the authors could find the features subset useful for letting the starting policy act differently. Considering a derived policy that examines only subsets of features can help human users to better understand in what way these characteristics are connected. However, the analysis of the critical states and the identification of the subset of features useful for the decision of the correct action are processes that have to be automatized to increase the performance of the approach. To this end, we introduce the Human–Robot collaboration class in the next section, where the insights of humans can help the agents to perform better.

5.4.6 Human–Robot Collaboration

A different approach is to consider XRL as a tool for improving the collaboration between humans and robots. This relationship has been dramatically increased together with the potential errors caused by a misunderstanding between man and machine [97]. In the articles that we consider here, the theoretical focus is on the Partially Observable Markov Decision Processes (POMDPs), which represents an environment where it is not possible to obtain the full information of the states visited [90].

One of the first research that tried to fill the gap between humans and robots is [72]. They proposed an algorithm that follows the principles of Situation Awareness-based Agent Transparency (SAT) [14] for the generation of natural-language explanations from different analyses: beliefs, observations, and outcome likelihood. From this information, four different classes of explanations are presented and tested with a human study. The first and simplest one is the none explanation, where only the action is presented to the user. Then, explanations with an increasing number of information and SAT requirements are generated. From the human study, the last two explanations were the best in terms of recognition of possible fault decisions and then finding the best action for the given state. One interesting remark about this analysis is the importance of using a well-designed explanation since in the case of the second type of explanation, the results were comparable to the none explanation.

The authors from [92] proposed an algorithm to find and correct erroneous actions characterized by incomplete or incorrect information about the reward function analysed. The general idea is to generate different reward functions to compare the results achieved by the agent that tries to optimize its policy following these objective functions. The methodology consists of three components: first, the human user's knowledge of the reward function of the environment is tested, then a policy for choosing when to intervene is generated, and, lastly, the explanations for the corrections of the reward are created. One major problem that comes up in the second step is the fact that the space of the possible models that can fit the observed trajectories is exponentially exploding with the number of reward components. However, if only the trivial reward functions are considered, it is possible to prove the helpful, trustworthy, and more positive overall experience obtained from the explanations generated.

In [17], they tested what kind of explanations are the best with respect to how much the users are experts in a particular environment. In particular, they considered two explanations for local actions: the first and technical one, is based on the Q-value associated with the actual state and the considered action; while the second and more human-like explanation, is composed of the probability of success to complete the goal. In general, they proved a higher efficiency of a counterfactual answer than the standalone. Furthermore, the best explanations according to their human study are the probabilistic ones, for both groups. Only a small amount of experts evaluated the technical standalone explanations better than the others. However, this study was based only on two simple explanations in a modest scenario. Therefore, it is fundamental to test broader explanations for a complex environment where there are multiple goals to reach.

5.4.7 Visual-Based

One of the most popular approaches to XRL is using visual representation to mediate between the agent's actions and the human users. Visualization is one of the main modes for creating trust and interpretability for black-box ML methods [75].

The most prolific class of Visual-based explanation is saliency maps methods. Therefore, a selection of techniques using saliency maps is described in the followings. One of the first, and main papers conceiving saliency maps in XRL context is [28]. In their approach, the authors considered an actor-critic agent. Therefore, they created saliency maps for evaluating both the policy and the value function obtained from the training. As a saliency metric, they adopted, respectively, the squared magnitude of the difference between the policy or the value function, given a sequence of images and their perturbed versions. From this idea, it is also possible to slightly change the saliency metric by adopting soft or binary masks for the perturbation [108], or considering, instead of the policies and the value function, the Q-values as done by [43]. In these cases [28, 43], the approaches were tested with a human study, achieving good results in terms of accuracy and understandability for both experts and non-expert users. From the formulation applied by [43], the authors in [32] derived a more complex formulation that considers the softmax of the Q-value function and its respective Kullback-Leibler divergence [50]. In this way, the properties of specificity, focus on the effect of the perturbation for the given action that has to be explained, and relevance, the alteration in the expected reward should be lower for actions different from the considered, are fulfilled.

However, a usual situation is where a well-known concept in XAI is applied or extended to the XRL scenario. In fact, [41] based the saliency maps generated for the RL agent explanations on the layer-wise relevance propagation (LRP) method [10]. The authors in [41] proposed to consider an *argmax* version to obtain sparse saliency maps. They proved that this approach is compatible with any possible DQN implementation at that time. Also in the case of [45], they considered a well-known approach in XAI: Grad-CAM [85]. In this circumstance, the saliency maps are generated by first calculating the weighted gradient for each node in the CNN layer, global average pooling them, multiplying the results in the previous step to the feature maps and inputting it to an activation function to create the outcome. However, the interpretation of the saliency maps is difficult to understand in the case of non-experts in machine learning, since they are usually accentuating multiple causes in the input.

Further options for the generation of different saliency maps are through the integration of the concept of the SHAP value [58], as done in [104]. However, the importance, in this case, is not evaluated for the actual state and its perturbation, but, instead, on the contribution of the input's features to the background data. This kind of approach was tested in a different context and problem in [78]: traffic signal controls. The results obtained are satisfying in terms of explainability and consistency although some biases related to the traffic in the training phase occur. As noticed, saliency maps perform very well in augmenting the trust of the users through simple but effective visual representation. However, there are still flaws that can not be solved using these methods, e.g., the multiple influences highlighted in the input and the technicalities. Therefore, we now focus on solutions that try to counterbalance the problems that appear in these cases.

A methodology called *DQNViz*, that can give a comprehensive visual analysis of DQN agents, is introduced in [103]. This visual approach provides several graphic designs for the description of various levels of details, starting with an overall training view, epoch and episode level, and lastly a segment level. The flaws of this approach are related to the difficulty to generalize in a complex environment where the action space is large (e.g., Montezuma's Revenge Atari game). Another remark can be pointed out in the segment analysis module, where a dimensionality reduction is applied using Principal Component Analysis (PCA) and then testing also a non-linear method called t-SNE [98]. These algorithms capture more diverse features which results in less clear visualization maps.

This pitfall was also considered in other papers. In particular, the following approach [18, 65] was created to compensate for the preceding research that was principally focused on studying the salient pixels in the input images, as done by [110] applying the previously cited t-SNE method. [18,65] proposed to record the most important moments in the training of the agent to create a store of snapshots useful for understanding the choice of action in a particular context. In detail, for picking the correct image to save and use as an explanation, a Sparse Bayesian Learning algorithm (known also as Relevance Vector Machines) is applied [93]. Despite the sparse set of snapshots needing less memory and being easy to interpret for a human user, it is not often possible to rely on small storage of information for real-world applications. Therefore, this model and its sparsity have to be evaluated in a complex environment as well, e.g., a continuous state-action space.

A similar approach was developed by [86], where the authors extracted the most interesting elements in form of video clips that could summarize the key moments in a trajectory performed by the agent. The main problem of this technique depends on the impossibility to rely only on a small interesting element for providing a deep understanding of the actions of the agent. However, the methodology applied is straightforward, and considers, in order, frequency, execution certainty, transition value, and sequences for the detection of the most important moments. The theoretical background of this method is directly inspired by a different approach introduced by [5] relative to the last class of Policy simplification that we analyze in the following section.

5.4.8 Policy Simplification

The class of Policy simplification methods is composed of approaches that try to encapsulate the most important information in the trajectory chosen by the agents. An example is a previously cited work [5]. In this paper, the authors considered the difference, given the state, between the maximum and minimum value of the Q-function as a formulation of importance [95]. This methodology was tested in basic human-study tasks like agent selection and summary preferences, achieving good results. However, this approach relies principally on the Q-value and, for this reason, many complications arise. First, different agents could have various Q-value distributions over the state space, contributing to the creation of their own importance judgement. Furthermore, it may also happen that the worst-case scenario is never explored, thereby creating a wrong perception of that state. Other challenges were later described by the same author in [6] highlighting the importance of a rigorous and autonomous state encoding and presenting different evaluation methodologies.

Following the same concept, [40] used a different metric for recognizing critical states: the authors decided to use the difference between the maximum and the average of the Q-values in a particular state. Therefore, the set of critical states was defined by this idea. Subsequently, a human study is performed to evaluate this approach. Moreover, the technique of summarizing a policy through its critical states is promising since it can directly reveal its truthfulness.

A different way of summarizing policies is to use trees with different useful peculiarities for the approximation of the Q-function. In fact, given enough leaves, a regression tree can approximate a continuous function arbitrarily closely [13]. Using this property, [56] introduced Linear Model U-trees (LMUTs) extending the concept of the Continuous U-Trees (CUTs) [96]. In the proposed approach, it is possible to use a linear model inside each leaf. In this way, the tree has higher accuracy and interpretability than a normal function approximator. Despite the good result regarding the performance when compared to other tree methods and neural networks, the dimension of the U-trees is fundamental for interpretability: the smaller the structure, the more understandable the model is. Therefore, this methodology works well for small environments or problems but could lead to difficulties for the users in interpreting the results for high-dimensions situations.

Always in the same thread, [88] decided to consider a differentiable version of the classic decision tree introduced in [89], where the Boolean decisions are replaced by the use of a sigmoid function. The interpretability of the model was then tested in a small human study having only 15 participants. The results of this limited study showed better usability of decision trees and lists over full MLPs.

A different way to distil the information from a bigger model to a tree is by using the Soft Decision Trees (SDTs) [24]: a mixture of a usual classification model with a binary split and a neural network. Therefore, [16] decided to use this method for summarizing how deep policies take action, rather than approximating the Q-value function. The performance obtained by this approach, using a high depth (7), is comparable to the accuracy achieved by the DNNs. However, the explanations are generated considering heatmaps created from the chosen features in the tree, which could be difficult to understand in a non-spatial environment. Another drawback is the choice of depth since it must be set beforehand. More recent results using tree-based summarizing techniques apply Mixture of Experts [44] for the creation of a Mixture of Expert Trees [100], and a combination of a clustering technique that is called CLTree [55] together with a critical state analysis that was explained before [40] for the generation of differently sized explanations [63].

It is also possible to consider fuzzy rule sets for the distillation of policies as done in the following cases. The first paper in literature is [35] which considered the Particle Swarm Optimization [47]: an algorithm for the generation of interpretable policies. In the beginning, only a minimal number of rules are used. Subsequently, by increasing the number of rules and comparing the performances achieved it was possible to choose the best policy: the one that has greater accuracy and, at the same time, a lower number of rules. This approach reported high-quality results in easy system dynamics problems, while, on the other hand, it has mediocre performance in real-world dynamics. This is also connected to the necessity of a higher number of rules. Following this idea, the authors also implemented the genetic programming version [36]. The outcome of this approach is an explanation composed of Boolean terms and compact algebraic equations. Using this methodology, better results than NNs are achieved in the real dynamics, testifying that the equations obtained outperform the non-interpretable methods. However, the performances in a simple environment are slightly worse than the NNs. In the same thread, [39] applied a more complex model to implement interpretable fuzzy reinforcement learning consisting of clustering the states, identifying the favourable ones in order to memorize them, and, finally, using the fuzzy system AnYa [8] for the approximation of the state-action values.

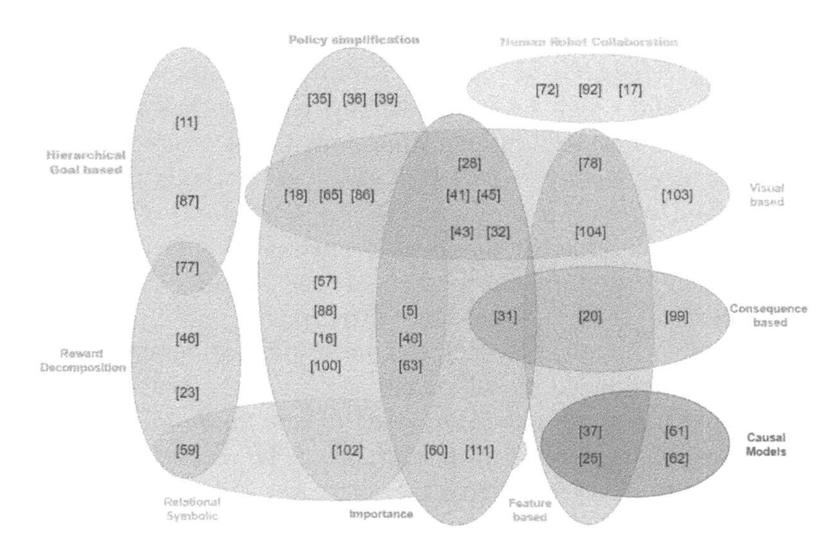

Figure 5.2 Classes of different methodologies found in XRL literature.

A last remark for the Policy simplification class is the work of [102], which introduced the Neurally Directed Program Search (NDPS) for finding the optimal policy that can be related to a predefined syntactic one. Therefore, NDPS is a guided direct policy search algorithm that wants to find a policy easy to understand. Unfortunately, in this way, it is possible to generate a high-level human-readable explanation but with a performance that can not be compared with the one achieved by DRL.

5.5 DISCUSSION

As we have already seen in different classes, the application of a particular methodology can not directly characterize the approach used, since multiple different-category methods can be applied together. For this reason, it is impossible to create a unique taxonomy by looking at the methodology employed. This evidence is also testified by Figure 5.2, where the different classes are represented. In this scenario, there are plenty of intersections indicating how intricate an XRL mechanism can be. Therefore, we decided to consider a classification, which can include this kind of junction point, instead of a taxonomy, that should not have any commonalities.

Another aspect that stands out from the analysis and Table 5.1, is how the human study is not much considered in these researches (only 35% of papers analysed), while the automation feature is mostly present (85%). However, we should focus on developing methods that are understandable to humans. Therefore, in order to check the reliability of an explanation, a user study should be performed.

The last remark concerns the scalability and generalizability of these approaches. In substance, if we have a method that can work in a particular simplified environment, we would like to use it also in a real-world problem without making major changes or retraining the agent. This is a general problem in RL, but for the particular case of XRL, it becomes essential to find a solution, otherwise, these approaches would be inconsistent and all the main objectives will fail. Therefore, future research avenues should try to solve the previously exposed problems, giving at the same time an evaluation for the explanation generated and considering also an automated approach increasing the accuracy of the methodologies and preventing possible human errors.

5.6 CONCLUSION AND FUTURE WORKS

In this chapter, we revised the last six years' literature, gathering the most relevant methods and advances achieved in XRL. In particular, we tried to provide greater clarity in the terminology used in this field since there are plenty of contradictory cases. One of which is the attempt to find a taxonomy, instead of simply a classification, using the methodology applied for the creation of the explanations. Therefore, we proposed a classification based on the approaches considered for the generation of the explanation and on the kind of explanation. Moreover, we collected for each category the papers that were central in terms of impact (older) and improvements (newer papers). In this way, it is possible to obtain a structured overview of the most important advances in XRL of the last years and the different research streams.

Possible future works can be in this direction: generating human-validated explanations and applying these approaches to real-world situations. In fact, a further aspect to bear in mind is the final aim of this approach: being applied in complex scenarios, e.g., in ITS, medicines and economics. On the point of the creation of an unambiguous taxonomy, the focus should not be anymore on the kind of method used but on consideration of different qualities as proposed here: human acceptance, automatic explanation generation and type of explanation.

Bibliography

[1] Baher Abdulhai, Rob Pringle, and Grigoris J Karakoulas. Reinforcement learning for true adaptive traffic signal control. *Journal of Transportation Engineering*, 129(3):278–285, 2003.

[2] Amina Adadi and Mohammed Berrada. Peeking inside the black-box: A survey on explainable artificial intelligence (xai). *IEEE Access*, 6:52138–52160, 2018.

[3] Alnour Alharin, Thanh Nam Doan, and Mina Sartipi. Reinforcement learning interpretation methods: A survey. *IEEE Access*, 8:171058–171077, 2020.

[4] Duane F Alwin and Robert M Hauser. The decomposition of effects in path analysis. *American Sociological Review*, pages 37–47, 1975.

[5] Dan Amir and Ofra Amir. Highlights: Summarizing agent behavior to people. In *Proceedings of the 17th International Conference on Autonomous Agents and MultiAgent Systems*, pages 1168–1176. International Foundation for Autonomous Agents and Multiagent Systems, 2018.

[6] Ofra Amir, Finale Doshi-Velez, and David Sarne. Summarizing agent strategies. *Autonomous Agents and Multi-Agent Systems*, 33(5):628–644, 2019.

[7] Marcin Andrychowicz, Filip Wolski, Alex Ray, Jonas Schneider, Rachel Fong, Peter Welinder, Bob McGrew, Josh Tobin, OpenAI Pieter Abbeel, and Wojciech Zaremba. Hindsight experience replay. *Advances in Neural Information Processing Systems*, 30, 2017.

[8] Plamen Angelov and Ronald Yager. Simplified fuzzy rule-based systems using non-parametric antecedents and relative data density. In *2011 IEEE Workshop on Evolving and Adaptive Intelligent Systems (EAIS)*, pages 62–69. IEEE, 2011.

[9] Alejandro Barredo Arrieta, Natalia Diaz-Rodriguez, Javier Del Ser, Adrien Bennetot, Siham Tabik, Alberto Barbado, Salvador Garcia, Sergio Gil-Lopez, Daniel Molina, Richard Benjamins, et al. Explainable artificial intelligence (xai): Concepts, taxonomies, opportunities and challenges toward responsible ai. *Information fusion*, 58:82–115, 2020.

[10] Sebastian Bach, Alexander Binder, Gregoire Montavon, Frederick Klauschen, Klaus-Robert Muller, and Wojciech Samek. On pixel-wise explanations for non-linear classifier decisions by layer-wise relevance propagation. *PloS One*, 10(7):e0130140, 2015.

[11] Benjamin Beyret, Ali Shafti, and A Aldo Faisal. Dot-to-dot: Explainable hierarchical reinforcement learning for robotic manipulation. In *2019 IEEE/RSJ International Conference on Intelligent Robots and Systems (IROS)*, pages 5014–5019. IEEE, 2019.

[12] Nadia Burkart and Marco F Huber. A survey on the explainability of supervised machine learning, 2021.

[13] Probal Chaudhuri, Min-Ching Huang, Wei-Yin Loh, and Ruji Yao. Piecewise-polynomial regression trees. *Statistica Sinica*, pages 143–167, 1994.

[14] Jessie YC Chen, Shan G Lakhmani, Kimberly Stowers, Anthony R Selkowitz, Julia L Wright, and Michael Barnes. Situation awareness-based agent transparency and human-autonomy teaming effectiveness. *Theoretical issues in ergonomics science*, 19(3):259–282, 2018.

[15] Jianyu Chen, Shengbo Eben Li, and Masayoshi Tomizuka. Interpretable end-to-end urban autonomous driving with latent deep reinforcement learning. *IEEE Transactions on Intelligent Transportation Systems*, 2021.

[16] Youri Coppens, Kyriakos Efthymiadis, Tom Lenaerts, Ann Nowe, Tim Miller, Rosina Weber, and Daniele Magazzeni. Distilling deep reinforcement learning policies in soft decision trees. In T. Miller, R. Weber, & D. Magazzeni (Eds.), *Proceedings of the IJCAI 2019 Workshop on Explainable Artificial Intelligence*, (pp. 1–6). ULB - UNIVERSITÉ LIBRE DE BRUXELLES, 2019.

[17] Francisco Cruz, Charlotte Young, Richard Dazeley, and Peter Vamplew. Evaluating human-like explanations for robot actions in reinforcement learning scenarios. *arXiv preprint arXiv:2207.03214*, 2022.

[18] Giang Dao, Indrajeet Mishra, and Minwoo Lee. Deep reinforcement learning monitor for snapshot recording. In *2018 17th IEEE International Conference on Machine Learning and Applications (ICMLA)*, pages 591–598. IEEE, 2018.

[19] Arun Das and Paul Rad. Opportunities and challenges in explainable artificial intelligence (xai): A survey. *arXiv preprint arXiv:2006.11371*, 2020.

[20] Omid Davoodi and Majid Komeili. Feature-based interpretable reinforcement learning based on state-transition models. In *2021 IEEE International Conference on Systems, Man, and Cybernetics (SMC)*, pages 301–308. IEEE, 2021.

[21] Richard Dazeley, Peter Vamplew, and Francisco Cruz. Explainable reinforcement learning for broad-xai: A conceptual framework and survey. *arXiv preprint arXiv:2108.09003*, 2021.

[22] Filip Karlo Dosilovic, Mario Brcic, and Nikica Hlupic. Explainable artificial intelligence: A survey. In *2018 41st International Convention on Information and Communication Technology, Electronics and Microelectronics (MIPRO)*, pages 0210–0215. IEEE, 2018.

[23] Martin Erwig, Alan Fern, Magesh Murali, and Anurag Koul. Explaining deep adaptive programs via reward decomposition. In *IJCAI/ECAI Workshop on Explainable Artificial Intelligence*, 2018.

[24] Nicholas Frosst and Geoffrey Hinton. Distilling a neural network into a soft decision tree. *arXiv preprint arXiv:1711.09784*, 2017.

[25] Jasmina Gajcin and Ivana Dusparic. Reccover: Detecting causal confusion for explainable reinforcement learning. *arXiv preprint arXiv:2203.11211*, 2022.

[26] Hector Geffner, Rina Dechter, and Joseph Y Halpern. Probabilistic and causal inference: The works of judea pearl. ACM, 2022.

[27] Claire Glanois, Paul Weng, Matthieu Zimmer, Dong Li, Tianpei Yang, Jianye Hao, and Wulong Liu. A survey on interpretable reinforcement learning. *arXiv preprint arXiv:2112.13112*, 2021.

[28] Samuel Greydanus, Anurag Koul, Jonathan Dodge, and Alan Fern. Visualizing and understanding atari agents. In *International Conference on Machine Learning*, pages 1792–1801. PMLR, 2018.

[29] Riccardo Guidotti, Anna Monreale, Salvatore Ruggieri, Franco Turini, Fosca Giannotti, and Dino Pedreschi. A survey of methods for explaining black box models. *ACM Computing Surveys*, 51, 8 2018.

[30] David Gunning. Explainable artificial intelligence (xai). *Defense Advanced Research Projects Agency (DARPA), nd Web*, 2(2):1, 2017.

[31] Wenbo Guo, Xian Wu, Usmann Khan, and Xinyu Xing. Edge: Explaining deep reinforcement learning policies. *Advances in Neural Information Processing Systems*, 34:12222–12236, 2021.

[32] Piyush Gupta, Nikaash Puri, Sukriti Verma, Sameer Singh, Dhruv Kayastha, Shripad Deshmukh, and Balaji Krishnamurthy. Explain your move: Understanding agent actions using focused feature saliency. *arXiv preprint arXiv:1912.12191*, 2019.

[33] Ammar Haydari and Yasin Yilmaz. Deep reinforcement learning for intelligent transportation systems: A survey. *IEEE Transactions on Intelligent Transportation Systems*, 23:11–32, 1 2022.

[34] Bradley Hayes and Julie A Shah. Improving robot controller transparency through autonomous policy explanation. In *2017 12th ACM/IEEE International Conference on Human-Robot Interaction, HRI*, pages 303–312. IEEE, 2017.

[35] Daniel Hein, Alexander Hentschel, Thomas Runkler, and Steffen Udluft. Particle swarm optimization for generating interpretable fuzzy reinforcement learning policies. *Engineering Applications of Artificial Intelligence*, 65:87–98, 2017.

[36] Daniel Hein, Steffen Udluft, and Thomas A Runkler. Interpretable policies for reinforcement learning by genetic programming. *Engineering Applications of Artificial Intelligence*, 76:158–169, 2018.

[37] Tue Herlau and Rasmus Larsen. Reinforcement learning of causal variables using mediation analysis. 2022.

[38] Alexandre Heuillet, Fabien Couthouis, and Natalia Díaz-Rodríguez. Explainability in deep reinforcement learning, 2021.

[39] Jianfeng Huang, Plamen P Angelov, and Chengliang Yin. Interpretable policies for reinforcement learning by empirical fuzzy sets. *Engineering Applications of Artificial Intelligence*, 91:103559, 2020.

[40] Sandy H Huang, Kush Bhatia, Pieter Abbeel, and Anca D Dragan. Establishing appropriate trust via critical states. In *2018 IEEE/RSJ International Conference on Intelligent Robots and Systems (IROS)*, pages 3929–3936. IEEE, 2018.

[41] Tobias Huber, Dominik Schiller, and Elisabeth Andre. Enhancing explainability of deep reinforcement learning through selective layer-wise relevance propagation. In *Joint German/Austrian Conference on Artificial Intelligence (Kunstliche Intelligenz)*, pages 188–202. Springer, 2019.

[42] Julián Hurtado-Gómez, Juan David Romo, Ricardo Salazar-Cabrera, Álvaro Pachón de la Cruz, and Juan Manuel Madrid Molina. Traffic signal control system based on intelligent transportation system and reinforcement learning. *Electronics (Switzerland)*, 10, 10 2021.

[43] Rahul Iyer, Yuezhang Li, Huao Li, Michael Lewis, Ramitha Sundar, and Katia Sycara. Transparency and explanation in deep reinforcement learning neural networks. In *Proceedings of the 2018 AAAI/ACM Conference on AI, Ethics, and Society*, pages 144–150, 2018.

[44] Robert A Jacobs, Michael I Jordan, Steven J Nowlan, and Geoffrey E Hinton. Adaptive mixtures of local experts. *Neural Computation*, 3(1):79–87, 1991.

[45] Ho-Taek Joo and Kyung-Joong Kim. Visualization of deep reinforcement learning using grad-cam: how ai plays atari games? In *2019 IEEE Conference on Games (CoG)*, pages 1–2. IEEE, 2019.

[46] Zoe Juozapaitis, Anurag Koul, Alan Fern, Martin Erwig, and Finale Doshi-Velez. Explainable reinforcement learning via reward decomposition. In *IJCAI/ECAI Workshop on Explainable Artificial Intelligence*, 2019.

[47] James Kennedy and Russell Eberhart. Particle swarm optimization. In *Proceedings of ICNN'95-International Conference on Neural Networks*, volume 4, pages 1942–1948. IEEE, 1995.

[48] Agneza Krajna, Mario Brcic, Tomislav Lipic, and Juraj Doncevic. Explainability in reinforcement learning: perspective and position. *arXiv preprint arXiv:2203.11547*, 2022.

[49] Tejas D Kulkarni, Karthik Narasimhan, Ardavan Saeedi, and Josh Tenenbaum. Hierarchical deep reinforcement learning: Integrating temporal abstraction and intrinsic motivation. *Advances in Neural Information Processing Systems*, 29, 2016.

[50] Solomon Kullback and Richard A Leibler. On information and sufficiency. *The Annals of Mathematical Statistics*, 22(1):79–86, 1951.

[51] Joohyung Lee, Vladimir Lifschitz, and Fangkai Yang. Action language bc: Preliminary report. In *IJCAI*, pages 983–989. Citeseer, 2013.

[52] Li Li, Yisheng Lv, and Fei Yue Wang. Traffic signal timing via deep reinforcement learning. *IEEE/CAA Journal of Automatica Sinica*, 3:247–254, 7 2016.

[53] Timothy P Lillicrap, Jonathan J Hunt, Alexander Pritzel, Nicolas Heess, Tom Erez, Yuval Tassa, David Silver, and Daan Wierstra. Continuous control with deep reinforcement learning. *arXiv preprint arXiv:1509.02971*, 2015.

[54] Zachary C Lipton. The mythos of model interpretability: In machine learning, the concept of interpretability is both important and slippery. *Queue*, 16(3):31–57, 2018.

[55] Bing Liu, Yiyuan Xia, and Philip S Yu. Clustering through decision tree construction. In *Proceedings of the Ninth International Conference on Information and knowledge management*, pages 20–29. Association for Computing Machinery, 2000.

[56] Guiliang Liu, Oliver Schulte, Wang Zhu, and Qingcan Li. Toward interpretable deep reinforcement learning with linear model u-trees. In *Joint European Conference on Machine Learning and Knowledge Discovery in Databases*, pages 414–429. Springer, 2018.

[57] Xiao-Yang Liu, Zihan Ding, Sem Borst, and Anwar Walid. Deep reinforcement learning for intelligent transportation systems. *arXiv preprint arXiv:1812.00979*, 2018.

[58] Scott M Lundberg and Su-In Lee. A unified approach to interpreting model predictions. *Advances in neural information processing systems*, 30, 2017.

[59] Daoming Lyu, Fangkai Yang, Bo Liu, and Steven Gustafson. Sdrl: interpretable and data-efficient deep reinforcement learning leveraging symbolic planning. In *Proceedings of the AAAI Conference on Artificial Intelligence*, volume 33, pages 2970–2977. AAAI Press, 2019.

[60] Zhihao Ma, Yuzheng Zhuang, Paul Weng, Hankz Hankui Zhuo, Dong Li, Wulong Liu, and Jianye Hao. Learning symbolic rules for interpretable deep reinforcement learning. *arXiv preprint arXiv:2103.08228*, 2021.

[61] Prashan Madumal, Tim Miller, Liz Sonenberg, and Frank Vetere. Distal explanations for explainable reinforcement learning agents. *arXiv preprint arXiv:2001.10284*, 2020.

[62] Prashan Madumal, Tim Miller, Liz Sonenberg, and Frank Vetere. Explainable reinforcement learning through a causal lens. In *Proceedings of the AAAI conference on artificial intelligence*, volume 34, pages 2493–2500, 2020.

[63] Joe McCalmon, Thai Le, Sarra Alqahtani, and Dongwon Lee. Caps: Comprehensible abstract policy summaries for explaining reinforcement learning agents. In *Proceedings of the 21st International Conference on Autonomous Agents and Multiagent Systems*, pages 889–897, 2022.

[64] Tim Miller. Explanation in artificial intelligence: Insights from the social sciences. *Artificial intelligence*, 267:1–38, 2019.

[65] Indrajeet Mishra, Giang Dao, and Minwoo Lee. Visual sparse Bayesian reinforcement learning: a framework for interpreting what an agent has learned. In *2018 IEEE Symposium Series on Computational Intelligence (SSCI)*, pages 1427–1434. IEEE, 2018.

[66] Volodymyr Mnih, Koray Kavukcuoglu, David Silver, Andrei A Rusu, Joel Veness, Marc G Bellemare, Alex Graves, Martin Riedmiller, Andreas K Fidjeland, Georg Ostrovski, et al. Human-level control through deep reinforcement learning. *Nature*, 518(7540):529–533, 2015.

[67] C Molnar. Interpretable machine learning: A guide for making black box models explainable. Published online, 2019.

[68] Gregoire Montavon, Wojciech Samek, and Klaus-Robert Muller. Methods for interpreting and understanding deep neural networks. *Digital Signal Processing*, 73:1–15, 2018.

[69] Nick Pawlowski, Daniel Coelho de Castro, and Ben Glocker. Deep structural causal models for tractable counterfactual inference. *Advances in Neural Information Processing Systems*, 33:857–869, 2020.

[70] Judea Pearl. The causal mediation formula—a guide to the assessment of pathways and mechanisms. *Prevention Science*, 13(4):426–436, 2012.

[71] Erika Puiutta and Eric Veith. Explainable reinforcement learning: A survey. pages 77–95, 2020.

[72] David V Pynadath, Michael J Barnes, Ning Wang, and Jessie YC Chen. Transparency communication for machine learning in human-automation interaction. In *Human and machine learning*, pages 75–90. Springer, 2018.

[73] Tao Qian, Chengcheng Shao, Xiuli Wang, and Mohammad Shahidehpour. Deep reinforcement learning for ev charging navigation by coordinating smart grid and intelligent transportation system. *IEEE Transactions on Smart Grid*, 11:1714–1723, 2020.

[74] Gabriëlle Ras, Ning Xie, and Derek Doran. Explainable deep learning: A field guide for the uninitiated, 2022.

[75] Marco Tulio Ribeiro, Sameer Singh, and Carlos Guestrin. “why should i trust you?” explaining the predictions of any classifier. In *Proceedings of the 22nd ACM SIGKDD International Conference on Knowledge Discovery and Data Mining*, pages 1135–1144. Association for Computing Machinery. 2016.

[76] Marco Tulio Ribeiro, Sameer Singh, and Carlos Guestrin. Model-agnostic interpretability of machine learning. *arXiv preprint arXiv:1606.05386*, 2016.

[77] Finn Rietz, Sven Magg, Fredrik Heintz, Todor Stoyanov, Stefan Wermter, and Johannes A Stork. Hierarchical goals contextualize local reward decomposition explanations. *Neural Computing and Applications*, pages 1–12. Springer, 2022.

[78] Stefano Giovanni Rizzo, Giovanna Vantini, and Sanjay Chawla. Reinforcement learning with explainability for traffic signal control. In *2019 IEEE Intelligent Transportation Systems Conference (ITSC)*, pages 3567–3572. IEEE, 2019.

[79] Cynthia Rudin. Stop explaining black box machine learning models for high stakes decisions and use interpretable models instead. *Nature Machine Intelligence*, 1:206–215, 2019.

[80] Gavin A Rummery and Mahesan Niranjan. *On-line Q-learning using connectionist systems*, volume 37. Citeseer, 1994.

[81] Stuart J Russell and Andrew Zimdars. Q-decomposition for reinforcement learning agents. In *Proceedings of the 20th International Conference on Machine Learning (ICML-03)*, pages 656–663. AAAI Press, 2003.

[82] Fatai Sado, C Kiong Loo, Matthias Kerzel, and Stefan Wermter. Explainable goal-driven agents and robots-a comprehensive review and new framework. *arXiv preprint arXiv:2004.09705*, 180, 2020.

[83] Tatsuya Sakai and Takayuki Nagai. Explainable autonomous robots: a survey and perspective. *Advanced Robotics*, 36:219–238, 2022.

[84] Wojciech Samek and Klaus Robert Müller. Towards explainable artificial intelligence, 2019.

[85] Ramprasaath R Selvaraju, Michael Cogswell, Abhishek Das, Ramakrishna Vedantam, Devi Parikh, and Dhruv Batra. Grad-cam: Visual explanations from deep networks via gradient-based localization. In *Proceedings of the IEEE international conference on computer vision*, pages 618–626. IEEE, 2017.

[86] Pedro Sequeira and Melinda Gervasio. Interestingness elements for explainable reinforcement learning: Understanding agents' capabilities and limitations. *Artificial Intelligence*, 288:103367, 2020.

[87] Tianmin Shu, Caiming Xiong, and Richard Socher. Hierarchical and interpretable skill acquisition in multi-task reinforcement learning. *arXiv preprint arXiv:1712.07294*, 2017.

[88] Andrew Silva, Matthew Gombolay, Taylor Killian, Ivan Jimenez, and Sung-Hyun Son. Optimization methods for interpretable differentiable decision trees applied to reinforcement learning. In *International conference on artificial intelligence and statistics*, pages 1855–1865. PMLR, 2020.

[89] Alberto Suárez and James F Lutsko. Globally optimal fuzzy decision trees for classification and regression. *IEEE Transactions on Pattern Analysis and Machine Intelligence*, 21(12):1297–1311, 1999.

[90] Richard S Sutton and Andrew G Barto. *Reinforcement learning: An introduction.* MIT Press, 2018.

[91] Richard S Sutton, Doina Precup, and Satinder Singh. Between mdps and semi-mdps: A framework for temporal abstraction in reinforcement learning. *Artificial intelligence*, 112(1–2):181–211, 1999.

[92] Aaquib Tabrez, Shivendra Agrawal, and Bradley Hayes. Explanation-based reward coaching to improve human performance via reinforcement learning. In *2019 14th ACM/IEEE International Conference on Human-Robot Interaction (HRI)*, pages 249–257. IEEE, 2019.

[93] Michael E Tipping. Sparse Bayesian learning and the relevance vector machine. *Journal of Machine Learning Research*, 1:211–244. International Foundation for Autonomous Agents and Multiagent Systems, 2001.

[94] Erico Tjoa and Cuntai Guan. A survey on explainable artificial intelligence (xai): Toward medical xai. *IEEE Transactions on Neural Networks and Learning Systems*, 32:4793–4813, 2021.

[95] Lisa Torrey and Matthew Taylor. Teaching on a budget: Agents advising agents in reinforcement learning. In *Proceedings of the 2013 International Conference on Autonomous Agents and Multi-agent Systems*, pages 1053–1060. International Foundation for Autonomous Agents and Multiagent Systems, 2013.

[96] William TB Uther and Manuela M Veloso. Tree based discretization for continuous state space reinforcement learning. *Aaai/iaai*, 98:769–774, 1998.

[97] Peter Vamplew, Richard Dazeley, Cameron Foale, Sally Firmin, and Jane Mummery. Human-aligned artificial intelligence is a multiobjective problem. *Ethics and Information Technology*, 20(1):27–40, 2018.

[98] Laurens Van der Maaten and Geoffrey Hinton. Visualizing data using t-sne. *Journal of Machine Learning Research*, 9(11):2579–2605, 2008.

[99] Jasper van der Waa, Jurriaan van Diggelen, Karel van den Bosch, and Mark Neerincx. Contrastive explanations for reinforcement learning in terms of expected consequences. *arXiv preprint arXiv:1807.08706*, 2018.

[100] Marko Vasic, Andrija Petrovic, Kaiyuan Wang, Mladen Nikolic, Rishabh Singh, and Sarfraz Khurshid. Moet: Mixture of expert trees and its application to verifiable reinforcement learning. *Neural Networks*, 151:34–47, 2022.

[101] Ashish Vaswani, Noam Shazeer, Niki Parmar, Jakob Uszkoreit, Llion Jones, Aidan N Gomez, Lukasz Kaiser, and Illia Polosukhin. Attention is all you need. *Advances in Neural Information Processing Systems*, 30, 2017.

[102] Abhinav Verma, Vijayaraghavan Murali, Rishabh Singh, Pushmeet Kohli, and Swarat Chaudhuri. Programmatically interpretable reinforcement learning. In *International Conference on Machine Learning*, pages 5045–5054. PMLR, 2018.

[103] Junpeng Wang, Liang Gou, Han-Wei Shen, and Hao Yang. Dqnviz: A visual analytics approach to understand deep q-networks. *IEEE Transactions on Visualization and Computer Graphics*, 25(1):288–298, 2018.

[104] Yuyao Wang, Masayoshi Mase, and Masashi Egi. Attribution-based salience method towards interpretable reinforcement learning. In *AAAI Spring Symposium: Combining Machine Learning with Knowledge Engineering (1)*, 2020.

[105] Christopher JCH Watkins and Peter Dayan. Q-learning. *Machine Learning*, 8(3):279–292, 1992.

[106] Adrian Weller. Transparency: motivations and challenges. In *Explainable AI: Interpreting, Explaining and Visualizing Deep Learning*, pages 23–40. Springer, 2019.

[107] Lindsay Wells and Tomasz Bednarz. Explainable ai and reinforcement learning—a systematic review of current approaches and trends. *Frontiers in Artificial Intelligence*, 4, 2021.

[108] Zhao Yang, Song Bai, Li Zhang, and Philip HS Torr. Learn to interpret atari agents. *arXiv preprint arXiv:1812.11276*, 2018.

[109] Hao Yuan, Haiyang Yu, Shurui Gui, and Shuiwang Ji. Explainability in graph neural networks: A taxonomic survey. *arXiv preprint arXiv:2012.15445*, 2020.

[110] Tom Zahavy, Nir Ben-Zrihem, and Shie Mannor. Graying the black box: Understanding dqns. In *International Conference on Machine Learning*, pages 1899–1908. PMLR, 2016.

[111] Vinicius Zambaldi, David Raposo, Adam Santoro, Victor Bapst, Yujia Li, Igor Babuschkin, Karl Tuyls, David Reichert, Timothy Lillicrap, Edward Lockhart, et al. Deep reinforcement learning with relational inductive biases. In *International Conference on Learning Representations*, 2018.

CHAPTER 6

Road-Traffic Data Collection: Handling Missing Data

Abdelilah Mbarek
LISAC Laboratory, Faculty of Sciences Dhar El Mahraz, Sidi Mohamed Ben Abdellah University, Fez, Morocco

Mouna Jiber
Ministry of Equipment and Water, Fez, Morocco

Ali Yahyaouy and Abdelouahed Sabri
LISAC Laboratory, Faculty of Sciences Dhar El Mahraz, Sidi Mohamed Ben Abdellah University, Fez, Morocco

CONTENTS

DOI: 10.1201/9781003324140-6

ROAD TRAFFIC IS a complex phenomenon and its collection follows a spatio temporal process. Its management has attracted the attention of academia and industry because it often suffers from data gaps caused by the temporary deployment of sensors, malfunctioning detectors, and poor communication systems. In Morocco, road traffic data management is managed by the "National Center for Road Studies," which is attached to the Ministry of Equipment and Water. The center collects a large amount of data on the traffic flow on the main roads in Morocco. To obtain the traffic volume AADT (Annual Average Daily Traffic), which is the common traffic indicator, it is necessary to obtain continuous data. However, it is not possible to collect data on all roads because of the cost involved. Therefore, the center uses periodic counters with a counting time of 8 days per semester (16 days per year). These data, together with those obtained from the permanent counter, are then used to calculate the estimated annual traffic characteristics and seasonal variations. In this context, and to fully exploit the history of road traffic information and its spatio-temporal correlation, this chapter presents an approach based on artificial intelligence methods to estimate the missing values related to road flow in Morocco. The simulation results show that the adopted approach allows for estimating the missing data with high accuracy. The present work will contribute to improving the quality of the road traffic collection because the missing data are likely to have an important weight and therefore cannot be ignored during the statistical analysis of road traffic.

6.1 INTRODUCTION

The growing size of cities and the increased mobility of the population have led to a rapid increase in the number of vehicles on the roads and eventually the traffic data. Road traffic management has become necessary to maintain the normal flow of people and goods. However, road traffic management is now facing challenges such as traffic congestion and accidents. To address these problems, practitioners have developed road network strategies for road planning and traffic control. The key element in the development of such a strategy is the analysis of the rapidly growing traffic data collected and the development of an accurate traffic prediction system. Traffic flow prediction consists in predicting the next state (volume, speed, density, or behavior) of the traffic flow based on historical and/or real-time data. It is the essence of Intelligent Transportation Systems (ITS) and is essential to the decision-making process.

In recent years, road traffic management has experienced a major evolution worldwide, thanks to the emergence of new technologies, such as the Internet of Things, Intelligent Transportation Systems, and Business Intelligence Systems. There are constant components introduced into global transportation systems and new techniques applied to operate and maintain these systems safely and economically. The primary purpose of these systems is to assist in the implementation of key transportation strategies. ITS involves the collection, processing, and analysis of data to provide an effective decision-making tool [1].

The development of these systems is highly dependent on the quality and quantity of road traffic data. Typically, traffic data, such as vehicle speed or traffic flow, is collected using fixed sensors strategically placed along the road network [2]. Data collection through cell phones and in-vehicle GPS is now another source of data collection that can provide accurate real-time information on a large road network while overcoming some of the problems associated with fixed detectors. Rapid advances in Artificial Intelligence (AI) have led to a major shift in traffic management. AI can now predict and control the flow of people, objects, vehicles, and goods at different points in the transportation network with high accuracy. In addition to providing better service to citizens, AI also helps reduce accidents by optimizing flows at intersections, as well as improving safety during periods when roads are closed due to construction or other events. Moreover, AI's ability to process and analyze large amounts of data has enabled efficient data analysis, such as prediction and possibly reconstruction of missing data [3]. Machine Learning (ML) is a sub-field of AI that aims to analyze and interpret patterns and data structures to enable training, reasoning, and decision-making without human interaction. ML is a computer programming technique that uses statistical probabilities to give computers the ability to learn on their own without explicit programming, it also uses programs that adapt each time they are exposed to different types of input data. ML teaches computers to learn and subsequently act and react as humans do, improving their learning style and knowledge autonomously over time.

Researchers have paid much attention to the field of road traffic analysis and road safety because of their importance. They have considered many appropriate methods, ranging from traffic models to statistical or Machine Learning approaches. The fundamental challenge of road safety is to predict the possible future states of road traffic with extreme accuracy. Traffic flow prediction helps to prevent the occurrence of incidents such as traffic jams or other road anomalies [2]. In addition, it can potentially help anticipate road infrastructure improvement. There are many methods for training and predicting road traffic, such as heuristics based on a statistical analysis of historical data. They are easy and convenient for prediction, such as the KNN algorithm, linear and nonlinear regression, and linear autoregressive models such as Autoregressive Integrated Moving Average (ARIMA) [4, 5]. In statistical techniques, ARIMA models are, in theory, the most general class of models for forecast ing time series data. They can potentially and effectively predict future data due to their convenience and accuracy, low input data requirements, and computational simplicity [6, 7]. For regression problems, the Support Vector Regression (SVR) model has proven to be a powerful tool for real data estimation. As a supervised learning algorithm in Machine Learning, the SVR model uses a symmetric loss function, which penalizes boundary decisions. In this study, hourly road traffic data collected from four adjacent stations in the city of Marrakech are exposed to the problem of missing data. This problem can be caused by failures in the acquisition systems, such as the failure of one of the sensors during a certain period of time, resulting in a loss of data that affects the reliability of the subsequent prediction and decision systems.

From this perspective, and in order to address these limitations, a model was proposed to reconstruct the missing traffic data of a section from the records of the sensors of the neighboring sections. To this end, the three most well-known machine learning methods in the literature were developed to predict the hourly traffic flow of the defective sensor, based on adjacent ones. The study is based on historical data collected from the years 2014 to 2017 of four national road sections at the entry of the city of Marrakech where the studied sensors are installed. The rest of the paper is organized as follows: Section 6.2 describes the used data and methods; Section 6.3 provides and discusses the obtained results and Section 6.4 concludes the paper.

6.2 METHODS

Road traffic is an important consideration in all stages of a Moroccan road project, including feasibility studies, technical design studies, operation, and pavement sizing. As a result, controlling this parameter, as well as its evolution and composition, is central to the government's concerns.

6.2.1 Data Description

The data used in this study were collected by the Road Directorate service of the Ministry of Transport and Water. The Road Directorate installs sensors that collect information on road traffic throughout the Moroccan territory. Since the 1970s, the Roads Directorate has been implementing a road counting system to meet this demand. For the census of road traffic, two types of sensors are used:

- A single-loop sensor that can only record vehicle flow in all combined circulation.
- Multi-loop sensors that can record vehicle throughput, speed, and length (for each direction of traffic and lane).

Two types of counts are used, permanent and temporary (periodic) ones, and are set up to monitor the traffic. Permanent counts collect road traffic on a permanent basis in fixed stations while temporary counts are carried out by mobile counters for a specific period of the year. Permanent counts are performed on a continuous basis within properly classified road links. Using multi-loop sensors installed at fixed points throughout the road or highway network, this type of count is carried out automatically 24 hours a day, seven days a week. It is important that some traffic volume data, such as AADT, be obtained on a regular basis. However, due to the costs involved, it is not possible to collect data on all roads continuously. Various periodic counts are performed in order to make reasonable estimates of annual traffic statistics for an area. They are performed using mobile single- or multiple-loop sensors installed in periodic stations and typically operating at an 8-hour interval. It should be noted that each periodic counting station is linked to a permanent station, allowing for adjustments as needed. It is therefore critical to emphasize that the location of permanent and/or periodic stations is determined by the amount of traffic passing through the station, regardless of the road network's current classification.

The preliminary dataset consists of a set of separate files, each sensor having its corresponding files. Among the information in the files are the start and end date of the recording, the road number, the number of lanes, the class of the vehicle, and the hourly traffic records, the columns show the classes of vehicles, and the rows show

TABLE 6.1 Sample Data from a Generated File for a Two-Lane Road with Vehicle Classes

Date	Hour	Lane 1	Class 1	Class 2	Class 3	Class 4	Lane 2	Class 1	Class 2	Class 3	Class 4
01/01/16	22:00	181	85	45	34	17	202	46	63	52	41
01/01/16	23:00	389	182	152	42	13	136	46	41	25	24
02/01/16	00:00	418	156	124	87	51	118	42	45	19	12
02/01/16	01:00	438	205	140	52	41	134	52	43	24	15
02/01/16	02:00	316	132	85	67	32	370	196	92	45	37

TABLE 6.2 Sample Data from the Final Table for Analysis

Date	Lane 1 Station 39	Lane 2 Station 39	Lane 1 Station 74	Lane 2 Station 74	Lane 1 Station 137	Lane 2 Station 137	Lane 1 Station 138	Lane 2 Station 138
2014-01-07 13:00	371	370	436	411	463	463	299	303
2014-01-07 14:00	370	364	471	471	405	405	274	360
2014-01-07 15:00	356	379	489	426	400	400	315	400
2014-01-07 16:00	476	378	563	433	412	412	302	494
2014-01-07 17:00	506	364	583	389	498	498	294	377

the number of vehicles counted. Table 6.1 gives a brief description of the structure of the recording files.

6.2.2 Data Preprocessing

The collected dataset studied contains the traffic flow recorded over four years from 2014 to 2017 related to the four road stations (138, 137, 74 and 39) and are composed of a set of separate files. The resulting files from sensors were then processed to make them functional. Initially, we combined the files from each sensor into a single file that contains road traffic records, the date, and the time of records. Next, we have considered the sum of the hourly traffic of each lane for all vehicle classes. After that, we merged the four files (one for each road station) into a single file. Table 6.2 describes the structure of the resulting table; the columns indicate the road traffic of each lane of the different stations.

The data are then preprocessed by first removing the extreme or outlier data (outliers). This is justified by the fact that outliers may reduce the quality of the results. The method used to measure the dispersion of the data is the "Interquartile range" (IQR). It allows the elimination of values that are strongly distant from other observations. Figure 6.1 shows the distribution of road traffic data for each station before and after the application of the IQR method.

As shown in the right figure in Figure 6.1, the extreme values (indicated by circles in figure a) have been removed. In addition, only station 74 and direction 1 of stations 137 and 39 have outliers. The figure also indicates that all stations have 50% of the traffic recorded between values 40 and 460 Veh/h.

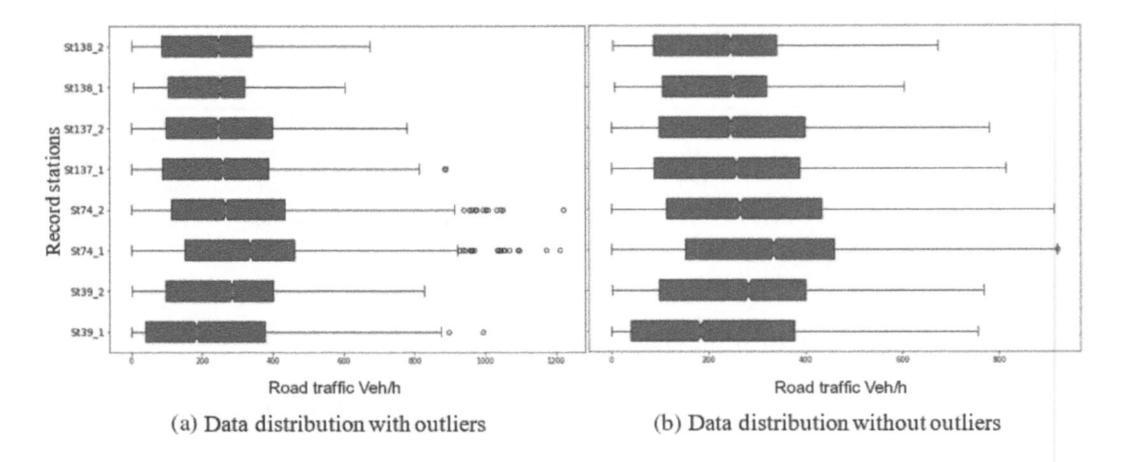

(a) Data distribution with outliers (b) Data distribution without outliers

Figure 6.1 Data distribution with and without outliers.

6.2.3 Forecasting Models

6.2.3.1 *Linear Regression – Ordinary Least Squares (OLS)*

In machine learning, a linear regression model is a regression model that seeks to establish a linear relationship between a variable, called the explained variable, and one or more variables called the explanatory variable. OLS (Ordinary Least Squares) regression is a technique for estimating the coefficients of linear regression that describe the relationships between one or more quantitative variables and a variable. Least squares refer to the minimum squared error. This form of analysis estimates the coefficients of the linear equation, involving one or more independent variables that best predict the value of the dependent variable. linear regression fits a straight line or surface that minimizes the differences between the predicted and actual output values. There are simple linear regression calculators that use the "least squares" method to discover the best-fitting line for a paired data set. For a model with p explanatory variables, the statistical model for OLS regression is written:

$$Y = \beta_0 + \sum_{j=1}^{p} \beta_i X_j + \epsilon \tag{6.1}$$

where Y denotes the dependent variable, β_0 is the constant of the model, X_j denotes the jth explanatory variable of the model ($j = 1$ to p), and ϵ is a random error of mean 0 and variance σ^2. Given n observations, the estimate of the value of the variable Y for observation i is given by the following equation:

$$Y_i = \beta_0 + \sum_{j=1}^{p} \beta_i X_{ij} \ (i = 1...n) \tag{6.2}$$

linear regression is one of the most commonly used interpretable models for modeling only regression problems. The linear nature of the model makes it easy to interpret. linear regression as its name suggests is a linear model, i.e., the association between the features and the output is linearly modeled. This linear relationship has a direct interpretation of the obtained weights. It is also a model with monotonicity constraints, i.e., it ensures that the relationship between a feature and the output always goes in the same direction over the entire range of the feature: an increase in the value of the feature either always leads to an increase or always to a decrease in the output. Monotonicity is useful for model interpretation because it makes it easier to understand a relationship.

There is a Confidence Interval (CI) for each weight in the linear equation. A CI is an interval for estimating the true weight with a certain degree of confidence. For example, a CI of 95% of an estimated weight means that if we repeated the estimation so in 95% of cases, the estimated weight will be within this interval.

6.2.3.2 SVR

Support Vector Regression (SVR) is a regression algorithm, which applies a technique similar to Support Vector Machines (SVM) for regression analysis. However, regression data contains continuous real numbers. To model such data, SVR adjusts the regression line so that it approximates the best values with a given margin called ϵ-tube (epsilon-tube, where ϵ identifies a tube width). SVR attempts to fit the hyperplane that has a maximum number of points while minimizing the error. The ϵ-tube takes into account the complexity of the model and the error rate, which is the difference between the actual and predicted outputs. The constraints of the SVR problem are described as follows:

$$|y_i - (w_i x_i + b)| \leq \epsilon \tag{6.3}$$

The hyperplane equation is:

$$y = wx + b \tag{6.4}$$

The objective function of the SVR is to minimize the coefficients:

$$min \frac{1}{2} \|w\|^2 \tag{6.5}$$

6.2.3.3 SARIMA

One of the most widely used time series forecast ing methods is the ARIMA method. It stands for Auto-Regressive Integrated Moving Average. It is a model that predicts the future values of a time series on some aspects of the statistical structure of the observed series. From its name, we can divide the model into smaller components as follows:

- *AR*: An autoregressive model that represents a type of random process. The output of the model depends linearly on its previous value, i.e., on some lagged data point or the number of past observations.
- *MA*: A moving average model whose output depends linearly on current and past observations of a stochastic term [8].

- I: Integrated here means the differentiation step to generate stationary time series data, i.e., the removal of seasonal and trend components.

The ARIMA model is usually denoted as ARIMA (p, d, q) and the parameters p, d, and q are defined as follows:

- p: The lag order or the number of time lags of the autoregressive model $AR(p)$.
- d: The degree of differentiation or the number of times the data has been subtracted from the past value.
- q: The order of the moving average model $MA(q)$.

SARIMA or Seasonal ARIMA similarly uses past values but also takes seasonality patterns into account. Because SARIMA incorporates seasonality as a parameter, it is much more powerful than ARIMA in predicting complex data.

6.3 EXPERIMENTS AND RESULTS

The simulations are performed using Python 3.6 on a 2.5 GHz Intel Core i5-7200U machine.

6.3.1 Experiment Setup

This study aims to train a model that can learn and predict the data of the sensor (#138), based on the data collected by three nearby sensors (sensors #29, 74, and 137). First, three methods well-known in the literature (OLS, SVR, and SARIMA) are applied and detailed in Section 6.2 of this manuscript to study the correlation between the sensor data. This would allow for the reconstruction of the missing data of a road section from the historical traffic collection of its nearby sensors. Indeed, the sensors often suffer from failures that generate a large amount of missing data. This would affect the quality and reliability of the traffic data analysis. Figure 6.2 shows the traffic flow of each station in chronological order.

The first observation indicates that the stations have close curves. Generally, they are high during the day and low at night. Figure 6.3 shows the correlation between the different stations through a heat map. A correlation heat map is assisted by a color bar making the data easily readable and understandable. It allows us to visualize the correlation values by using colors; in this, to represent more common values or higher activities, darker colors are mainly used, and to represent less common values or activities, brighter colors are used.

6.3.2 Evaluation Metrics

The quality of the models is measured using the following error metrics most commonly used in the literature:

- MSE: Mean Square Error:

$$RMSE = \frac{1}{N}\sum_{i=1}^{N}(x_i - \hat{x}_i)^2 \tag{6.6}$$

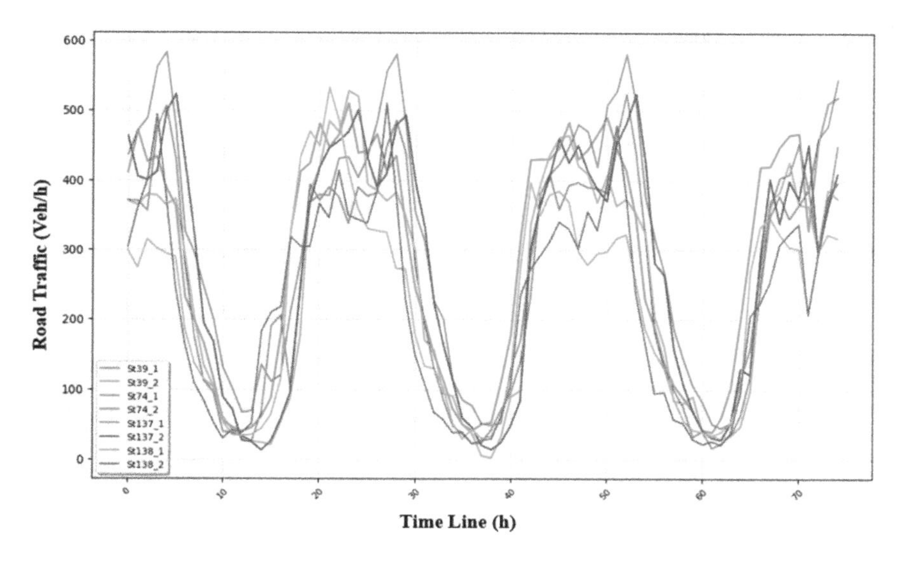

Figure 6.2 Traffic flow of each station in chronological order.

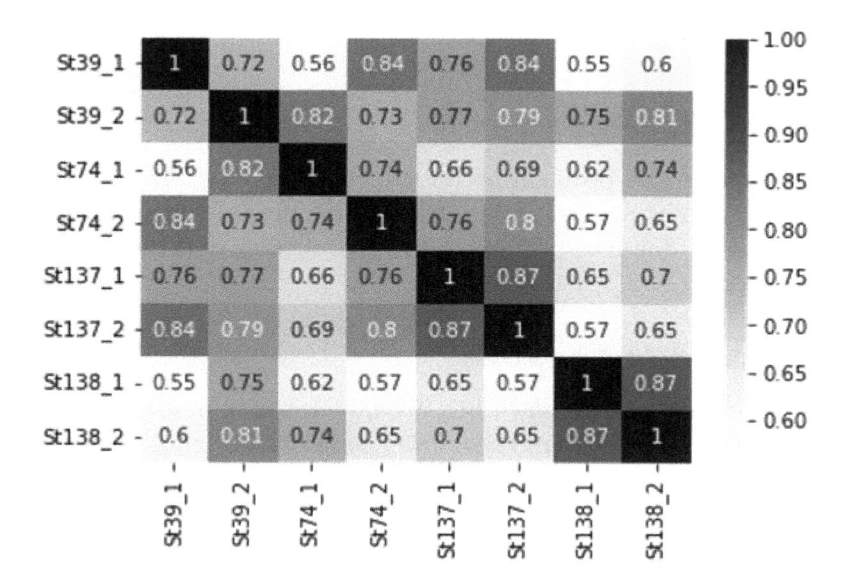

Figure 6.3 Correlation between the different stations.

- RMSE: Root Mean Square Error:

$$RMSE = \sqrt{\frac{1}{N}\sum_{i=1}^{N}(x_i - \hat{x}_i)^2} \tag{6.7}$$

- MAE: Mean Absolute Error:

$$MAE = \frac{1}{N}\sum_{i=1}^{N}|x_i - \hat{x}_i| \tag{6.8}$$

- MAPE: Mean Absolute Percentage:

$$MAE = \frac{1}{N}\sum_{i=1}^{N}\frac{|x_i - \hat{x}_i|}{x_i} \tag{6.9}$$

For N observations, x_i and $\hat{x}_i$ represent, respectively, the ith real value and its predicted value by the model.

- R-Squared:

$$R^2 = 1 - \frac{SSE}{SST} \tag{6.10}$$

where SSE is the squared sum of the errors, and SST is the squared sum of the variance.

- Adjusted R-squared:

$$\bar{R}^2 = 1 - (1 - R^2)\frac{n-1}{n-p-1} \tag{6.11}$$

where n and p are the numbers of instances and features, respectively.

6.4 DISCUSSION

This study addresses the problem of missing data in road traffic and proposes a model to predict accurately. To validate the effectiveness of the proposed model, this study first focuses on the interpretability of three models and then in a second part compares the models.

6.4.1 The Interpretability of the Models

Many problems are solved with black-box machine learning models. They become the source of information, rather than the data. The interpretability allows us to extract this additional insight and to understand the reason for the model's decisions.

6.4.1.1 Linear Regression Model Interpretation

linear regression is one of the most commonly used interpretable models. The results of the linear regression are shown in Figure 6.4. We will then discuss the interpretability of this model. There is a positive correlation between all stations lane1_39, lane1_74, lane2_74, lane1_137, lane1_138, and the target station lane2_138. This indicates that an increase in traffic at these stations leads to an increase in traffic at the target station. For example, taking the lane1_39 station, an increase in traffic in lane 1 of road 39 by 100 vehicles, implies an increase of 19 vehicles in lane 2 of road 138. Conversely, there is a negative correlation between all lane2_137 and lane2_39 stations and the target station lane2_138. This indicates that an increase in traffic at these stations leads to a decrease in traffic at the target station. For example, taking the lane2_39 station, an increase in traffic in lane 2 of road 39 by 100 vehicles,

F-statistic:	1.763e+04
Prob (F-statistic):	0.00
Log-Likelihood:	-28781.
AIC:	5.758e+04
BIC:	5.762e+04

	coef	std err	t	P>|t|	[0.025	0.975]
lane1_39	0.1910	0.022	8.865	0.000	0.149	0.233
lane2_39	-0.0748	0.021	-3.522	0.000	-0.116	-0.033
lane1_74	0.0766	0.013	5.734	0.000	0.050	0.103
lane2_74	0.0998	0.012	8.002	0.000	0.075	0.124
lane1_137	0.0449	0.011	4.077	0.000	0.023	0.066
lane2_137	-0.0006	0.013	-0.048	0.961	-0.025	0.024
lane1_138	0.5747	0.009	64.138	0.000	0.557	0.592

Omnibus:	666.615	Durbin-Watson:	0.626
Prob(Omnibus):	0.000	Jarque-Bera (JB):	1140.641
Skew:	0.845	Prob(JB):	2.06e-248
Kurtosis:	4.513	Cond. No.	35.3

Figure 6.4 OLS Regression results.

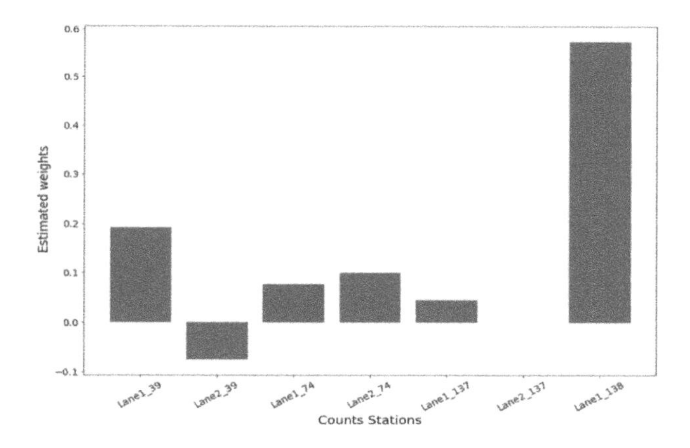

Figure 6.5 Histogram of estimated weights by linear regression.

implies a decrease of approximately 7 vehicles in lane 2 of road 138. The R-squared measurement (R-squared) tells us that 95.9% of the total variance of the target outcome is explained by the OLS model. The Histogram plot of the estimated weights (Figure 6.5) shows that lane1_138 has a strong positive effect on the predicted number of vehicles in lane2_138, followed by lane1_38, lane2_74, lane1_74, and lane1_137 in descendent order. There is also a weak negative correlation between lane2_39 and lane2_138. The weight of the lane2_137 feature is very close to zero, which means that the effect of lane2_137 is not statistically significant.

6.4.1.2 SARIMA Model Interpretation

Figure 6.6 shows the summary of the SARIMA model. The coefficient column where the values under "Coef" are the weights of the terms in the model equation. We

No. Observations:	7603
Log Likelihood	-40693.355
AIC	81406.711
BIC	81476.104
HQIC	81430.520

	coef	std err	z	P>|z|	[0.025	0.975]
ar.L1	1.8194	0.007	276.063	0.000	1.807	1.832
ar.L2	-0.9009	0.007	-133.232	0.000	-0.914	-0.888
ma.L1	-0.7944	0.012	-68.642	0.000	-0.817	-0.772
ma.L2	-0.2270	0.014	-16.184	0.000	-0.254	-0.199
ma.L3	0.2406	0.014	17.083	0.000	0.213	0.268
ma.L4	0.2495	0.013	18.664	0.000	0.223	0.276
ma.L5	-0.0598	0.014	-4.346	0.000	-0.087	-0.033
ma.L6	-0.3884	0.011	-34.999	0.000	-0.410	-0.367
ar.S.L8	-0.4153	0.012	-34.222	0.000	-0.439	-0.391
sigma2	2648.91	38.849	68.185	0.000	2572.77	2725.06

Ljung-Box (L1) (Q):	0.55	Jarque-Bera (JB):	249.24
Prob(Q):	0.46	Prob(JB):	0.00
Heteroskedasticity (H):	1.14	Skew:	0.24
Prob(H) (two-sided):	0.00	Kurtosis:	3.75

Figure 6.6 SARIMA results.

remark that all the coefficients of the AR and MA terms have the value p less than 0.05 in the column $p > |z|$, which means that they are all highly significant. To analyze the results of the SARIMA model, it is necessary to analyze the assumptions of the model using the Ljung-Box, chi-square, and residual autocorrelation statistics; assess whether each term is significant using the p-values; and identify the accuracy of the model using the mean square error. To make sure that each term in the model is statistically significant, each term must have a p-value of less than 0.05, to reject the null hypothesis with statistically significant values. The null hypothesis in this section is that each coefficient is not statistically significant. Results show that all terms are statistically significant. Next, there is the need to ensure that our model meets the assumption that the residuals are independent, known as white noise by using Ljung-Box. The Ljung-Box (L1) (Q) test statistic at lag 1 shows the Prob(Q) is 0.55, and the p-value is 0.46. Since the probability is greater than 0.05, the null hypothesis that the residuals are independent cannot be rejected.

The Heteroskedasticity tests that the error residuals have the same variance. The summary statistics show a test statistic of 1.14 and a p-value of 0.00, which means that we reject the null hypothesis and that our residuals have variance. The Jarque-Bera test verifies the normality of the errors. It tests the null hypothesis that the data are normally distributed. The test statistic is 249.24 with a probability of 0. This means that we reject the null hypothesis and that the data are not normally distributed. Furthermore, as shown by the Jarque-Bera test, the distribution has a slight positive skewness and kurtosis.

The log-likelihood, AIC, BIC, and HQIC are used to compare the models to each other so that better models have lower values. The log-likelihood function identifies a distribution that best fits the sampled data. The Akaike Information Criterion (AIC) is used to determine the strength of the linear regression model. The Bayesian

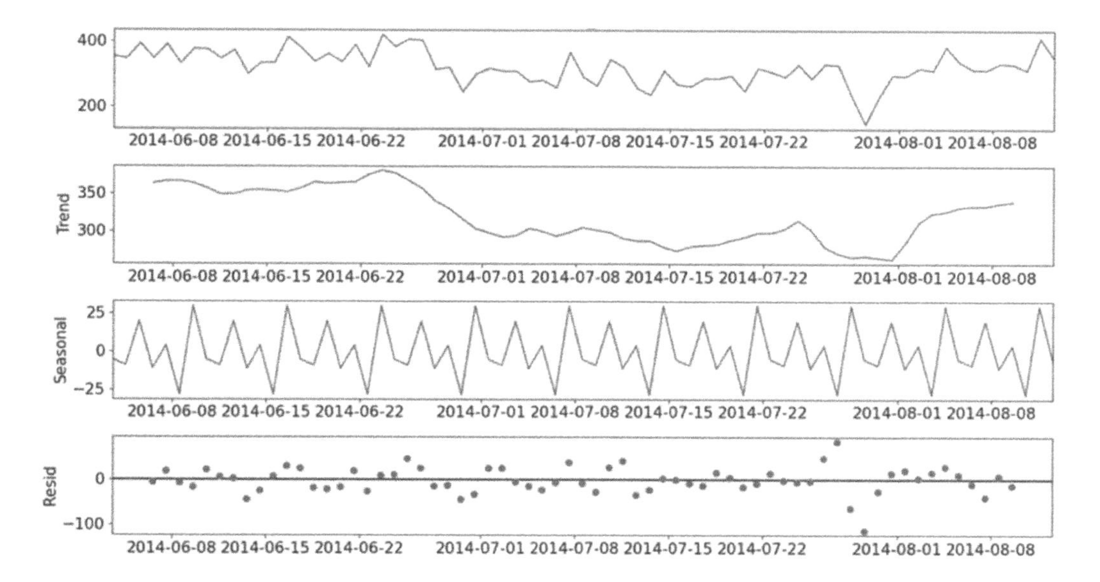

Figure 6.7 Results of Seasonal and Trend decomposition using Loess.

information criterion (BIC) is also used to judge the goodness of data fit. The Hannan-Quinn Information Criterion (HQIC) is another model selection criterion, used as an alternative to the AIC and BIC.

The decomposition of the studied data into the trend, seasonal component, and residual is shown in Figure 6.7. This decomposition allows us to see if data present seasonality. Points to ponder from Figure 6.7. by the Seasonal and Trend decomposition using Loess (STL) decomposition is that the model exhibits weekly seasonality, which means that the traffic flow presents a weekly cycle. The studied data present a downward and upward trend. The traffic flow remains stable in the range of 200 and 400 Veh/h. From the residual plot, we can see that the residuals are also relevant, presenting periods of high variability in the summer periods.

6.4.1.3 SVR Model Interpretation

The most important features of a model can be visualized by plotting the SHAP values of each feature for each sample. The plot in Figure 6.8. sorts the features by the sum of the magnitudes of the SHAP values across all samples, and uses the SHAP [8] values to show the distribution of impacts of each feature on the model output. Each point is a traffic record, the y-axis ranks the stations in increasing order of effect (most significant at the top of the axis) and the colors indicate the value of the traffic (low traffic has a blue color and high traffic has a red color). The x-axis represents the SHAP values, which are the impacts on the model output. 0 indicates no effect, a positive value indicates that the correlation is positive, and a negative value indicates that there is a negative correlation.

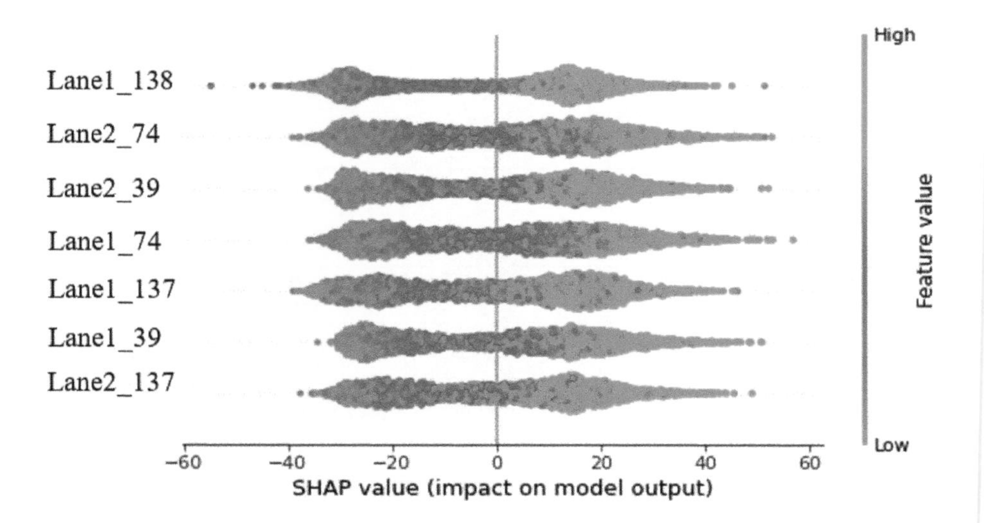

Figure 6.8 SHAP output for SVR.

According to Figure 6.8 (SHAP), all the stations have a significant influence on the SVR model. The most significant is lane1_138 and the least significant is lane2_137. We can see that all the features have the same range of SHAP values on the x-axis, which means that SVR gives similar importance to all the features for the final prediction. The SVR model considers that all features have a positive correlation with the output. An increase in the input stations leads to an increase in the output station.

6.4.2 Comparison between Models

To compare one model to another we used the metrics MSE, RMSE, MAE, MAPE, and R-squared. Table 6.3 shows that the SARIMA model performs better than the linear regression and the SVR model, with much smaller errors. However, the OLS and SVR models explain 62.96 and 26.6 of the variances of the output variable, which shows that OLS is more accurate than SVR. As we have seen before, SVR gives the same importance to all stations in the prediction, but OLS gives different importance, which indicates that each station has its impact on the model. On the other hand, from the prediction results of all models over 48 hours (Figure 6.9), it can be seen that the SARIMA model was able to predict the test data with higher accuracy than the linear regression (OLS) and SVR models.

The SARIMA model is clearly a good fit for the data. The SARIMA model is expected to perform better given the obvious seasonality of the data. Also, when the trend changes, the SARIMA model quickly finds the right trend. Compared to SVR and OLS, the fit is quite good, which implies that the SARIMA model can learn seasonality better. Therefore, the SARIMA model is the most appropriate to use in seasonal data.

TABLE 6.3 Evaluation of the OLS, SARIMA, and SVR Models

Metrics \ Model	OLS	SARIMA	SVR
MSE	7973.5129	**1639.1981**	8052.0512
RMSE	89.2945	**40.4870**	89.7332
MAE	64.9580	**29.5850**	65.4944
MAPE	0.4072	**0.2515**	0.4152
Adj. R-squared	62.96	**92.72**	26.60

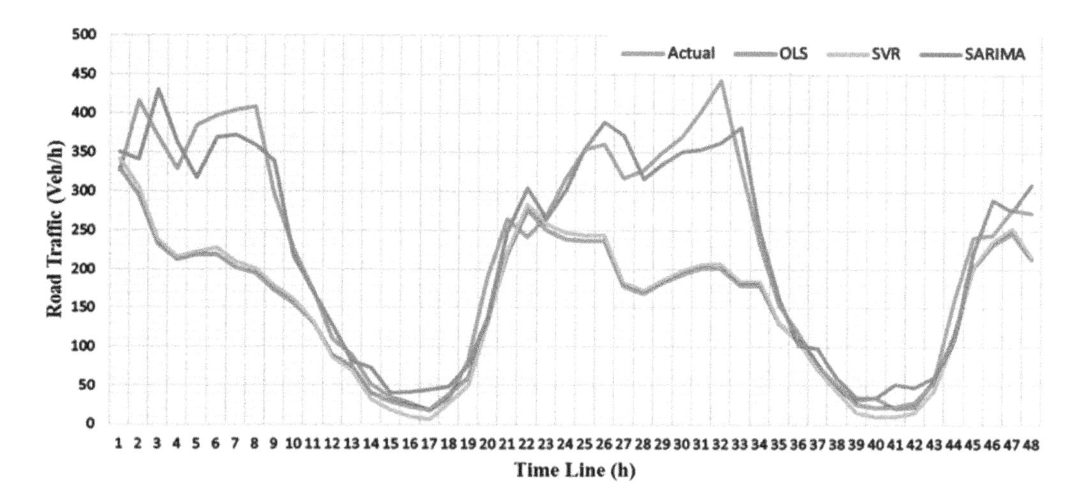

Figure 6.9 Results of traffic prediction.

6.5 CONCLUSION

In this paper, we discussed and used three machine learning algorithms to design a prediction model of hourly road traffic data. The study uses historical data from the years 2014 to 2017 from four road sections in Marrakech, Morocco. The main contribution of our work is to fully exploit the historical road traffic information and its spatio-temporal correlation, to predict the lost data of a sensor based on its surrounding stations. The models used are evaluated by well-known error measures in literature and the simulation results showed that SARIMA performs better in terms of accuracy and stability. The proposed traffic prediction model will be used to reconstruct lost or destroyed data with predicted values. This fills the gap in the traffic matrix used to calculate the annual average daily traffic (AADT), which is the common traffic indicator.

It would be interesting to work on data collected from real-time sensors. These have now been implemented by the Ministry of Equipment and Water, but are not yet deployed throughout the country. Intelligent analysis of this data would allow the development of short-term traffic flow prediction models. This will allow road users to have advanced traffic information to plan their trips and manage their time efficiently by avoiding the most congested hours.

Bibliography

[1] M. Jiber, I. Lamouik, Y. Ali, and M. A. Sabri, "Traffic flow prediction using neural network," 2018 Int. Conf. Intell. Syst. Comput. Vision, ISCV 2018, vol. 2018-May, pp. 1–4, 2018, doi: 10.1109/ISACV.2018.8354066.

[2] M. Jiber, A. Mbarek, A. Yahyaouy, M. A. Sabri, and J. Boumhidi, "Road traffic prediction model using Extreme Learning Machine: the case study of Tangier, Morocco," Inf. 2020, Vol. 11, Page 542, vol. 11, no. 12, p. 542, Nov. 2020, doi: 10.3390/INFO11120542.

[3] A. Mbarek, M. Jiber, A. Yahyaouy, and A. Sabri, "Road traffic mortality in Morocco: Analysis of statistical data," 2020 Int. Conf. Intell. Syst. Comput. Vis., pp. 1–7, Jun. 2020, doi: 10.1109/ISCV49265.2020.9204325.

[4] Y. Zhang and Y. Liu, "Comparison of parametric and nonparametric techniques for non-peak traffic forecast ing," World Acad. Sci. Eng. Technol., vol. 39, no. 3, pp. 242–248, 2009.

[5] B. M. Williams, M. Asce, L. A. Hoel, and F. Asce, "Modeling and Forecasting Vehicular Traffic Flow as a Seasonal ARIMA Process: Theoretical Basis and Empirical Results", `https://doi.org/10.1061/(ASCE)0733-947X(2003)129:6(664)`.

[6] H. Moeeni and H. Bonakdari, "Impact of Normalization and Input on ARMAX-ANN Model Performance in Suspended Sediment Load Prediction," Water Resour. Manag., vol. 32, no. 3, pp. 845–863, Feb. 2018, doi: 10.1007/s11269-017-1842-z.

[7] H. Moeeni and H. Bonakdari, "Forecasting monthly inflow with extreme seasonal variation using the hybrid SARIMA-ANN model," Stoch. Environ. Res. Risk Assess., vol. 31, no. 8, pp. 1997–2010, Oct. 2017, doi: 10.1007/s00477-016-1273-z.

[8] S. M. Lundberg, P. G. Allen, and S.-I. Lee, "A Unified Approach to Interpreting Model Predictions," Adv. Neural Inf. Process. Syst., vol. 30, 2017.

CHAPTER 7

Explainability of Surrogate Models for Traffic Signal Control

Pawel Gora

Faculty of Mathematics, Informatics, and Mechanics of the University of Warsaw, Poland
TensorCell, Poland

Dominik Bogucki and M. Latif Bolum

TensorCell, Poland

CONTENTS

TRAFFIC SIGNAL CONTROL is one of the most important means of managing and optimizing road transport in cities. Developing a new generation of efficient traffic signal control algorithms is challenging and it is expected that methods based on AI may bring advantages. However, explainability and interpretability of these techniques are often limited and sometimes it might be difficult even for traffic engineers to understand the decisions taken by such systems. In this chapter we focus on explainability of one of the AI-based techniques used for traffic signal-control which employs evolutionary algorithms to find heuristically optimal signal settings for given traffic conditions, and surrogate models based on graph neural networks to evaluate the quality of signal settings much faster than by traffic simulations. The

DOI: 10.1201/9781003324140-7

considered surrogate model is a graph neural network in which the topology of connections between neurons is built based on the topology of the road network graph (neurons correspond to intersections with traffic signals, or road segments between the intersections). Therefore, it can be suspected that analyzing behaviour of these neural networks may also bring better understanding of the urban road traffic and the spatio-temporal impact of traffic conditions on various intersections. In order to investigate the effects of input features on the output of the graph neural networks, we calculated the Shapley values and applied the ZORRO method. Thanks to applying these techniques, it was possible to assess importance of intersections and their impact on the times of waiting at red signals in the considered areas for points from randomly generated datasets and on datasets with 500 settings found using genetic algorithm (considered as close to local optima). It turned out that both methods produce quite consistent results. Thanks to them, it was possible to identify the most critical intersections in the road network topology which might be important from the traffic engineering perspective in tasks related to traffic signal control.

7.1 INTRODUCTION

Large road traffic congestion in cities is a serious commercial and civilizational problem causing wasted time, energy and environmental pollution. Therefore, it is important to manage it properly, so it is natural that many traffic management strategies and systems are developed. Among them are adaptive traffic signal control systems which aim to adjust signal settings based on traffic conditions. These tools often use advanced algorithms based on artificial intelligence and techniques like evolutionary algorithms [7, 14, 15] or reinforcement learning [4, 16]. The explainability and interpretability of these techniques are often limited and sometimes it might be difficult even for traffic engineers to understand the decisions taken by such systems (however, it is also important to know that there exist many non-AI rule-based traffic signal control systems in which the decisions taken by controllers are clear and hand-crafted by traffic engineers, e.g., [8, 12]).

In this chapter, we focus on explainability of one of the AI-based techniques used for traffic signal-control, presented in [13], which employs evolutionary algorithms to find heuristically optimal signal settings for given traffic conditions, and surrogate models based on machine learning to evaluate the quality of signal settings much faster than by traffic simulations. The considered surrogate model is a graph neural network in which the topology of connections between neurons is built based on the topology of the road network graph (neurons correspond to intersections with traffic signals, or road segments between the intersections). Therefore, it can be suspected that analyzing behaviour of these neural networks may also bring better understanding of the urban road traffic and the spatio-temporal impact of traffic conditions on various intersections.

In order to investigate the effects of input features on the output of the graph neural networks, we calculated the Shapley values [11] and applied the ZORRO method [6].

Section 7.2 introduces the surrogate models based on graph neural networks that were studied in the presented research. Section 7.3 explains two techniques that were used to investigate importance of input features to the studied surrogate models: calculating the Shapley values and applying the ZORRO method. Section 7.4 presents the conducted experiments. Finally, Section 7.5 concludes the chapter.

7.2 SURROGATE MODELS

Surrogate modelling is a technique aiming to build simpler (meta)models of other, more advanced and complex models [10, 17]. It can be applied in situations in which evaluation of advanced models is computationally too expensive, so it might be better (in terms of time, costs etc.) to build a simpler model able to approximate the advanced model much easier/faster. Therefore, the outcomes of a surrogate model might be only an approximation of the advanced model, but if the error of approximation is acceptable, such a model can be very useful.

In our case, we used metamodels approximating outcomes of time-consuming traffic simulations run on realistic maps of Warsaw taken from the OpenStreetMap service [2]. These models were graph neural networks described in [13], in which inputs correspond to the traffic signal settings, outputs to the outcomes of the traffic simulation (in our case: the total time of waiting at red signals in a given area), while neurons correspond to road network intersections or road segments between them, and connections between neurons in consecutive layers exist only if there are direct connections between corresponding intersections and road segments in the road network graph (for details, see [13]).

The models we used were trained on datasets generated by running computer simulations using the Traffic Simulation Framework – the goal of simulations was to compute the total times (in seconds) of waiting on red signals for different traffic signal settings in a given area during simulations of 10 minutes of traffic in Warsaw. More precisely, the inputs were vectors representing offsets of traffic signal settings in a given area, each offset (time in seconds from the beginning of the simulation to the first switch from the red signal state to the green signal state) was an integer number from the set $\{0, 1, 2, \ldots, 119\}$ (it was assumed that durations of red and green signal phases are constant and equal to 62 seconds and 58 seconds, respectively). For each signal setting / input vector, the traffic conditions (i.e., number of cars, start and end points, routes, times of departure, and other traffic model settings) were the same, so that signal settings were the only factor having impact on the simulation's output.

The length of each input vector was the number of intersections in a given area - in our experiments, we considered three districts in Warsaw: Centrum (with 11 intersections), Ochota (with 21 intersections), and Mokotów (with 42 intersections). These districts and the considered intersections are visualized in Figures 7.1–7.3.

After generating the datasets composed of about 10^5 pairs (signal setting, times of waiting at red signals) for each district, it was possible to train graph neural network models described in [13]. For each of the three considered districts in Warsaw, we used five graph neural network models that performed best (in terms of mean absolute percentage error on the test set) in experiments described in [13]. For Centrum, we

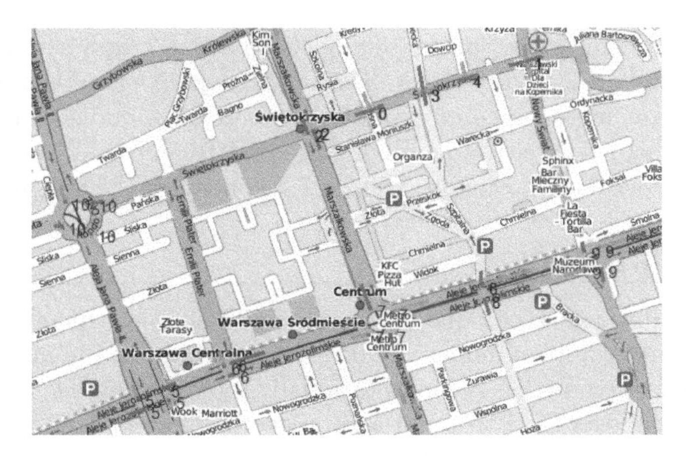

Figure 7.1 Visualization of intersections for the Centrum district. Numbers indicate indices of intersections.

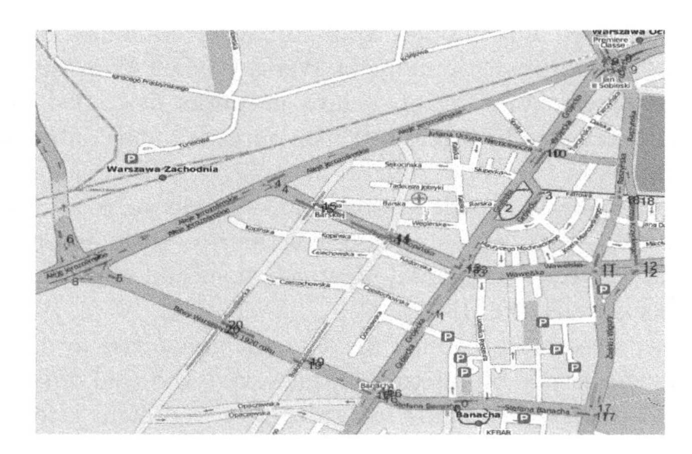

Figure 7.2 Visualization of intersections for the Ochota district. Numbers indicate indices of intersections.

selected the following models: cent_2_3, cent_3_4, cent_3_5, cent_4_4, and cent_4_5. For Ochota, we used: ocho_2_5, ocho_3_3, ocho_3_4, ocho_4_4, and ocho_4_5. For Mokotów, we used: moko_2_4, moko_2_3, moko_2_5, moko_3_4, and moko_3_3. The first part of each model's name corresponds to the name of the district, the next two parts (the numbers) correspond to the number of hidden layers (not counting input and output layers) in the graph neural network, and to the number of channels per layer, respectively.

Having the 15 models trained and performing well on the test set, it was possible to investigate the importance of their features (offsets of signals for all intersections) by calculating the Shapley values and using the ZORRO method for a dataset composed of randomly generated signal settings. It was also possible to apply these trained metamodels as evaluators of qualities of traffic signal settings in optimization algorithms (like genetic algorithms) aiming to find signal settings giving relatively small times of waiting on red signals. Finally, we were able to investigate the importance

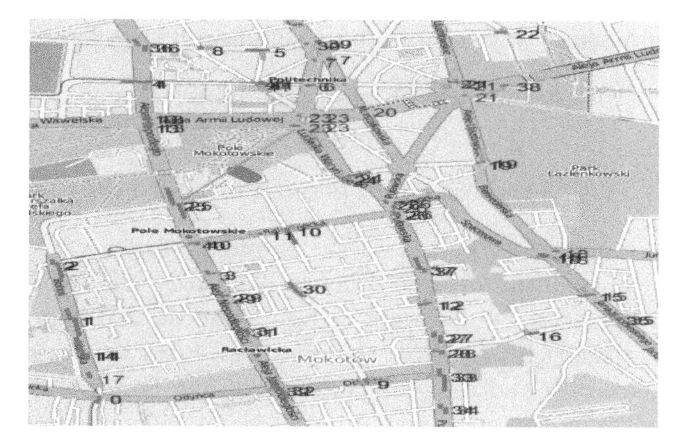

Figure 7.3 Visualization of intersections for the Mokotów district. Numbers indicate indices of intersections.

of features by calculating the Shapley values and using the ZORRO method for a dataset composed of signal settings considered as good by the optimization algorithm (see 7.4 for details).

7.3 EXPLAINABILITY METHODS

7.3.1 Shapley Value

The Shapley value [11] was developed and named after Lloyd S. Shapley, an American mathematician and Nobel Prize winner in Economics. The value is commonly used in the cooperative games theory. The method attributes a value of a given player (in our case: feature) for all possible other-players-collaborations up to the great all-players-coalition. The formula below presents the Shapley value for the i-th player:

$$\varphi_i(v) = \frac{1}{n!} \sum_{S \subseteq N \setminus \{i\}} |S|!\,(n - |S| - 1)!\,(v(S \cup \{i\}) - v(S)) \tag{7.1}$$

where:

- N is the set of n players {1,2,...,n};
- v is a function (the so-called characteristic function) that maps subsets of players to the real numbers, $v : 2^N \longrightarrow \mathbb{R}$, with $v(\emptyset) = 0$, where $\emptyset$ is the empty set – it has the following meaning: if S is a coalition of players, then $v(S)$, called the worth of coalition S, describes the total expected sum of payoffs the members of S can obtain by cooperation;
- $\frac{1}{n!}$ is the reciprocal of the number of all possible coalitions – it averages the score among the players;
- $S \subseteq N \setminus \{i\}$ is a possible coalition without the i-th player;

- $|S|!$ - the number of ways of forming a coalition S that player i will join;
- $(n - |S| - 1)!$ is the number of ways remaining players can be added to the coalition;
- $v(S \cup \{i\}) - v(S)$ is the value that player i brings to the coalition S.

Shapley value can be interpreted as the average marginal value that the given player brings to the coalition of other players. The player is assigned zero Shapley value, when on average he/she does not bring any additional gain to the coalitions. The stand alone values-rich players or the ones who bring synergies to the coalition are assigned bigger scores and they are considered more important players.

In machine learning, this method can be used to assess the importance of input features in deriving the final prediction of the model. Single feature is treated as a player joining the coalition of the features and the Shapley value can be assigned to it.

7.3.2 ZORRO Method

ZORRO [6] is an instance-level model exploration algorithm that uses a perturbation-based approach. Given an input, ZORRO uses a hard mask to identify important features. The assumption ZORRO makes is that if a feature has low importance, changing its value with a random noise from input space should change the prediction less compared to a feature with high importance.

To achieve an explanation, ZORRO starts with an empty list of features and uses a greedy algorithm to include features. In each step, the feature with the highest fidelity score is selected by the greedy algorithm. Adding a new feature is repeated until a user-specified fidelity score has been achieved. Selected features are replaced by random noise, hence structure of the model is untouched and no training is necessary to generate explanations.

The fidelity score is a measure of how much the altered feature changed the output compared to the output from the unaltered feature. In our case, we used absolute percentage change from the original prediction. We also discarded the greedy algorithm and focused on the importance per individual feature (intersection).

Also, one of the ZORRO advantages is being able to check importance near local minima. In contrary, methods like SA [5] and Guided BP [5] are gradient-based methods that use gradients (but with respect to input instead of model parameters) to calculate importance, so they cannot give proper importance close to local minima where gradients are close to 0.

7.4 EXPERIMENTS

7.4.1 Setup of Experiments

Experiments were carried out using 15 different graph neural network models trained to predict outcomes of traffic simulations (calculating the total time of waiting at red

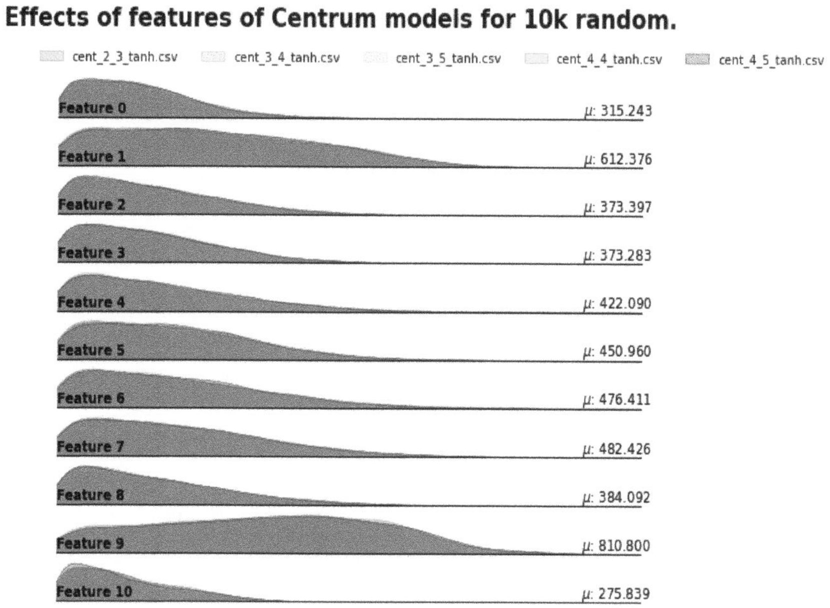

Figure 7.4 A plot of Shapley value estimates for 5 models for Centrum district.

signals for all vehicles for a given traffic light setting) in a specific district in Warsaw. For the road network graph, we considered 3 districts in Warsaw: Centrum, Ochota, and Mokotów. Every district had 5 models and there were slight differences in terms of their structure, i.e., in terms of the number of layers and number of channels per layer. Models for Centrum had 11 input features, for Ochota – 21, and for Mokotów – 42. Every feature in a model corresponds to an intersection in the considered district (for details, see Section 7.2 or [13]).

For these experiments, we used 2 datasets: 1) A dataset with 10^4 randomly generated data points (traffic signal setting) for every district, 2) A dataset with 500 data points (for each district) found using a genetic algorithm aiming to find good signal settings (these data points are considered as relatively good).

7.4.2 Results of Experiments

7.4.2.1 Shapley Value

Shapley values were calculated using the Python SHAP library [9] for neural networks (model agnostic approach). The number of all possible permutations for all the features and 10^4 simulations is huge, so in order to simplify computations, the SHAP library uses random sampling of permutations to calculate the Shapley values estimates. In our experiments, the size of the random sample was set to 10 * number of features. The permutation sampling was repeated for each of 10^4 model outputs.

Figures 7.4–7.6 present the kernel density estimate plots for calculated Shapley values for Centrum, Ochota and Mokotów, respectively, for the dataset with 10^4

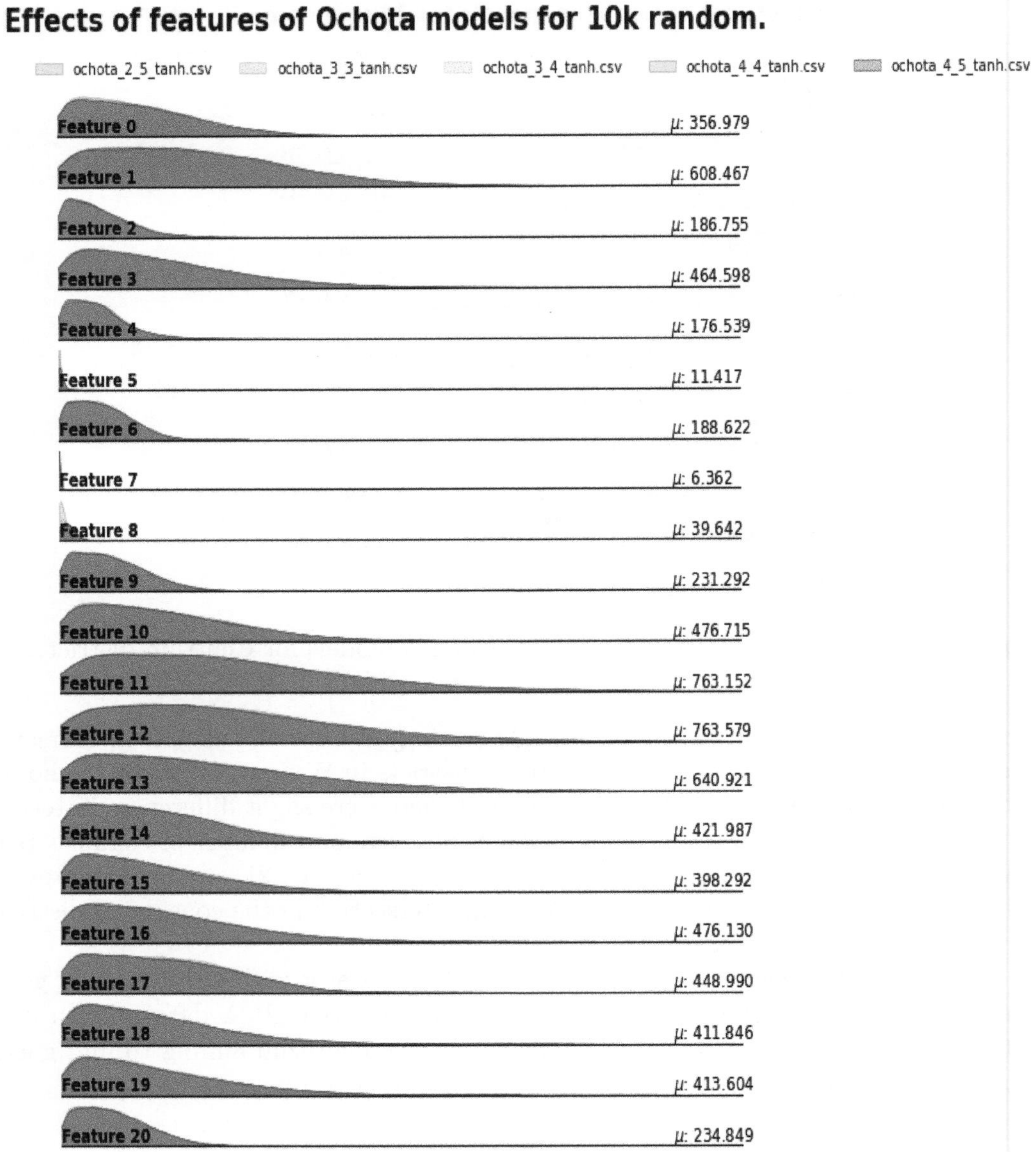

Effects of features of Ochota models for 10k random.

ochota_2_5_tanh.csv ochota_3_3_tanh.csv ochota_3_4_tanh.csv ochota_4_4_tanh.csv ochota_4_5_tanh.csv

Feature 0 μ: 356.979
Feature 1 μ: 608.467
Feature 2 μ: 186.755
Feature 3 μ: 464.598
Feature 4 μ: 176.539
Feature 5 μ: 11.417
Feature 6 μ: 188.622
Feature 7 μ: 6.362
Feature 8 μ: 39.642
Feature 9 μ: 231.292
Feature 10 μ: 476.715
Feature 11 μ: 763.152
Feature 12 μ: 763.579
Feature 13 μ: 640.921
Feature 14 μ: 421.987
Feature 15 μ: 398.292
Feature 16 μ: 476.130
Feature 17 μ: 448.990
Feature 18 μ: 411.846
Feature 19 μ: 413.604
Feature 20 μ: 234.849

Figure 7.5 A plot of Shapley value estimates for Ochota district.

points (traffic signal settings) for each district. Kernel density estimate is a way to generate a distribution for a given dataset [1]. Given a finite dataset, kernel density estimate places a distribution over each data point and sums it up to create the underlying distribution. In our case, for each feature we have 10^4 values (one value for each point from the dataset).

To generate these plots we used one of the Python's data visualization libraries, Seaborn, and its seaborn.kdeplot function [3] with default parameters. For each

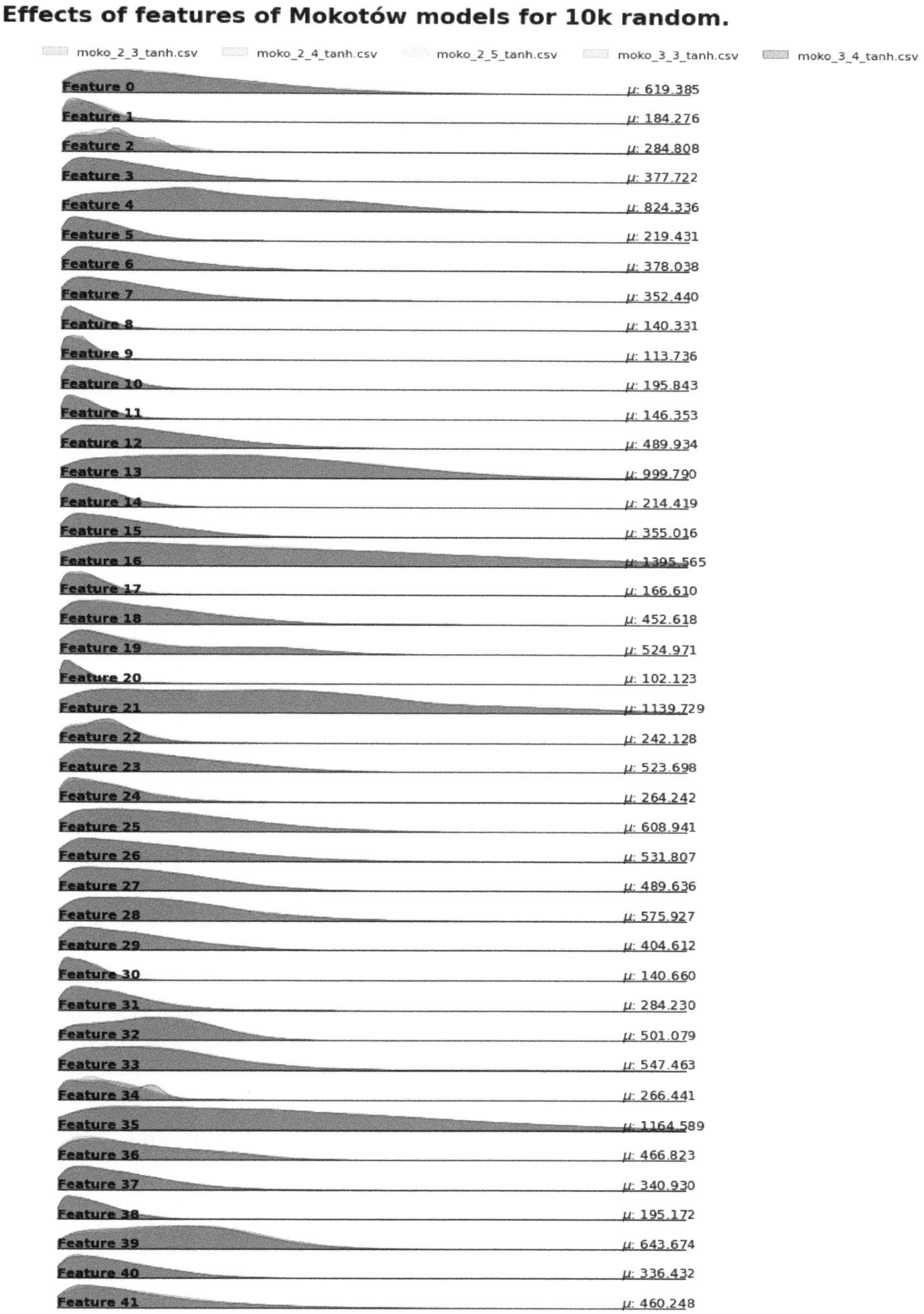

Effects of features of Mokotów models for 10k random.

moko_2_3_tanh.csv moko_2_4_tanh.csv moko_2_5_tanh.csv moko_3_3_tanh.csv moko_3_4_tanh.csv

Feature 0 μ: 619.385
Feature 1 μ: 184.276
Feature 2 μ: 284.808
Feature 3 μ: 377.722
Feature 4 μ: 824.336
Feature 5 μ: 219.431
Feature 6 μ: 378.038
Feature 7 μ: 352.440
Feature 8 μ: 140.331
Feature 9 μ: 113.736
Feature 10 μ: 195.843
Feature 11 μ: 146.353
Feature 12 μ: 489.934
Feature 13 μ: 999.790
Feature 14 μ: 214.419
Feature 15 μ: 355.016
Feature 16 μ: 1395.565
Feature 17 μ: 166.610
Feature 18 μ: 452.618
Feature 19 μ: 524.971
Feature 20 μ: 102.123
Feature 21 μ: 1139.729
Feature 22 μ: 242.128
Feature 23 μ: 523.698
Feature 24 μ: 264.242
Feature 25 μ: 608.941
Feature 26 μ: 531.807
Feature 27 μ: 489.636
Feature 28 μ: 575.927
Feature 29 μ: 404.612
Feature 30 μ: 140.660
Feature 31 μ: 284.230
Feature 32 μ: 501.079
Feature 33 μ: 547.463
Feature 34 μ: 266.441
Feature 35 μ: 1164.589
Feature 36 μ: 466.823
Feature 37 μ: 340.930
Feature 38 μ: 195.172
Feature 39 μ: 643.674
Feature 40 μ: 336.432
Feature 41 μ: 460.248

Figure 7.6 A plot of Shapley value estimates for Mokotów district.

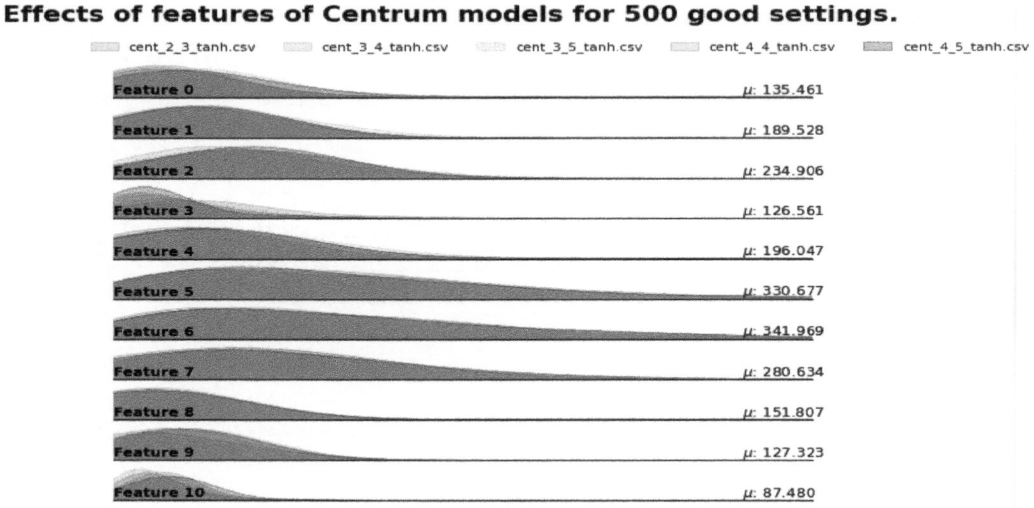

Effects of features of Centrum models for 500 good settings.

cent_2_3_tanh.csv cent_3_4_tanh.csv cent_3_5_tanh.csv cent_4_4_tanh.csv cent_4_5_tanh.csv

Feature 0 μ: 135.461
Feature 1 μ: 189.528
Feature 2 μ: 234.906
Feature 3 μ: 126.561
Feature 4 μ: 196.047
Feature 5 μ: 330.677
Feature 6 μ: 341.969
Feature 7 μ: 280.634
Feature 8 μ: 151.807
Feature 9 μ: 127.323
Feature 10 μ: 87.480

Figure 7.7 A plot of features Shapley value estimates for Centrum district near local minima.

feature, we have a separate plot (feature numerations correspond to indices of features) and same features from different models are plotted over each other with different colors. Legends on the plots show which color is for which model (the plots for all models for a given district are very similar, so these plots largely overlap).

For each plot, the X axis corresponds to the absolute value of the Shapley value, while Y axis corresponds to the likelihood. Beside the plot, we also calculated the average of absolute Shapley values for the whole dataset (denoted as μ). Intuitively, the greater the absolute Shapley value is, the more important is the feature.

We can observe that in the case of Centrum district, the most important features (intersections) seem to be 1 and 9, while the least important are 0, 2, 3, 8, and 10. In the case of Ochota district, the most important features are 1, 11, 12, and 13, while least important are 5, 7, and 8. In the case of Mokotów district, the most important features are 13, 16, 21, and 35, while the least important are features 8, 9, and 20.

Later, we did the same for the second dataset consisting of points (traffic signal settings) close to local minima, that were found using a genetic algorithm (as described in [13]). Plots with kernel density estimate plots and the average values of the absolute Shapley value are presented in Figures 7.7–7.9. It has turned out that in this case, the most important features for Centrum are 2, 5, 6, and 7, while the least important are 0, 3, and 10. For Ochota – the most important are 1, 3, 11, and 13, while the least important are 5, 7, and 8. For Mokotów – the most important are 12 and 26, while the least important are 2, 5, 9, 17, 22, and 25.

7.4.2.2 ZORRO Method

To visualize results of the ZORRO method we also used kernel density estimate plots for the dataset with 10^4 randomly selected points (Figures 7.10–7.12). On the X axis,

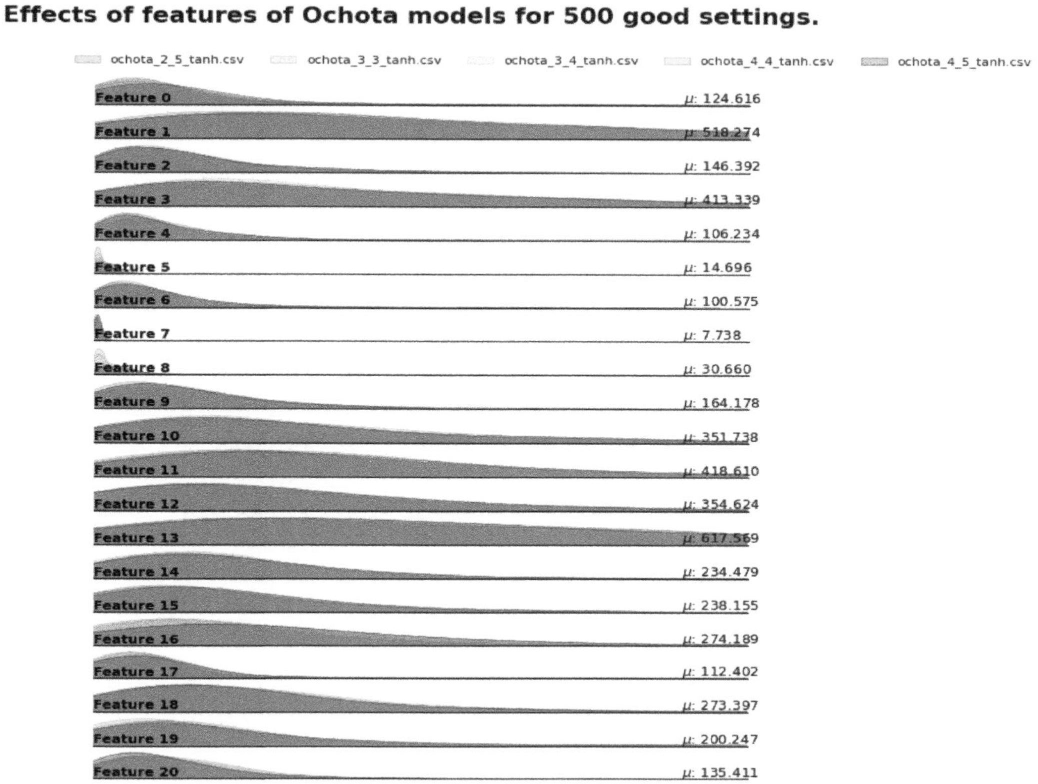

Effects of features of Ochota models for 500 good settings.

ochota_2_5_tanh.csv ochota_3_3_tanh.csv ochota_3_4_tanh.csv ochota_4_4_tanh.csv ochota_4_5_tanh.csv

Feature 0 μ: 124.616
Feature 1 μ: 518.274
Feature 2 μ: 146.392
Feature 3 μ: 413.339
Feature 4 μ: 106.234
Feature 5 μ: 14.696
Feature 6 μ: 100.575
Feature 7 μ: 7.738
Feature 8 μ: 30.660
Feature 9 μ: 164.178
Feature 10 μ: 351.738
Feature 11 μ: 418.610
Feature 12 μ: 354.624
Feature 13 μ: 617.569
Feature 14 μ: 234.479
Feature 15 μ: 238.155
Feature 16 μ: 274.189
Feature 17 μ: 112.402
Feature 18 μ: 273.397
Feature 19 μ: 200.247
Feature 20 μ: 135.411

Figure 7.8 A plot of features Shapley value estimates for Ochota district near local minima.

we have fidelity scores for different points (traffic signal settings) from the dataset, and the Y axis is the likelihood. For each point, the fidelity score is calculated as the absolute percentage change of the output value of the considered graph neural network model, considering the original value and the value obtained by random noise introduced to the considered feature – the value (offset) of the considered feature (intersection) was randomly changed to another value from the set $\{0, 1, \ldots, 119\}$.

Every feature in the plot has a μ value that shows the mean importance of each feature. The bigger this number, the larger importance that feature has. This value is calculated as the mean fidelity score for the considered feature, where the mean is calculated over fidelity scores for different traffic signal settings from the dataset.

We can observe that in the case of Centrum district, the most important features (intersections) seem to be 5, 6, and 9, while the least important are 0, 2, 3, 8, and 10. In the case of Ochota district, the most important features are 1, 11, 12, and 13, while least important are 5, 7, and 8. In the case of Mokotów district, the most important features are 16 and 35, while the least important are features 1, 2, 8, 9, 10, 11, 17, 20, 22, 30, 34, and 38.

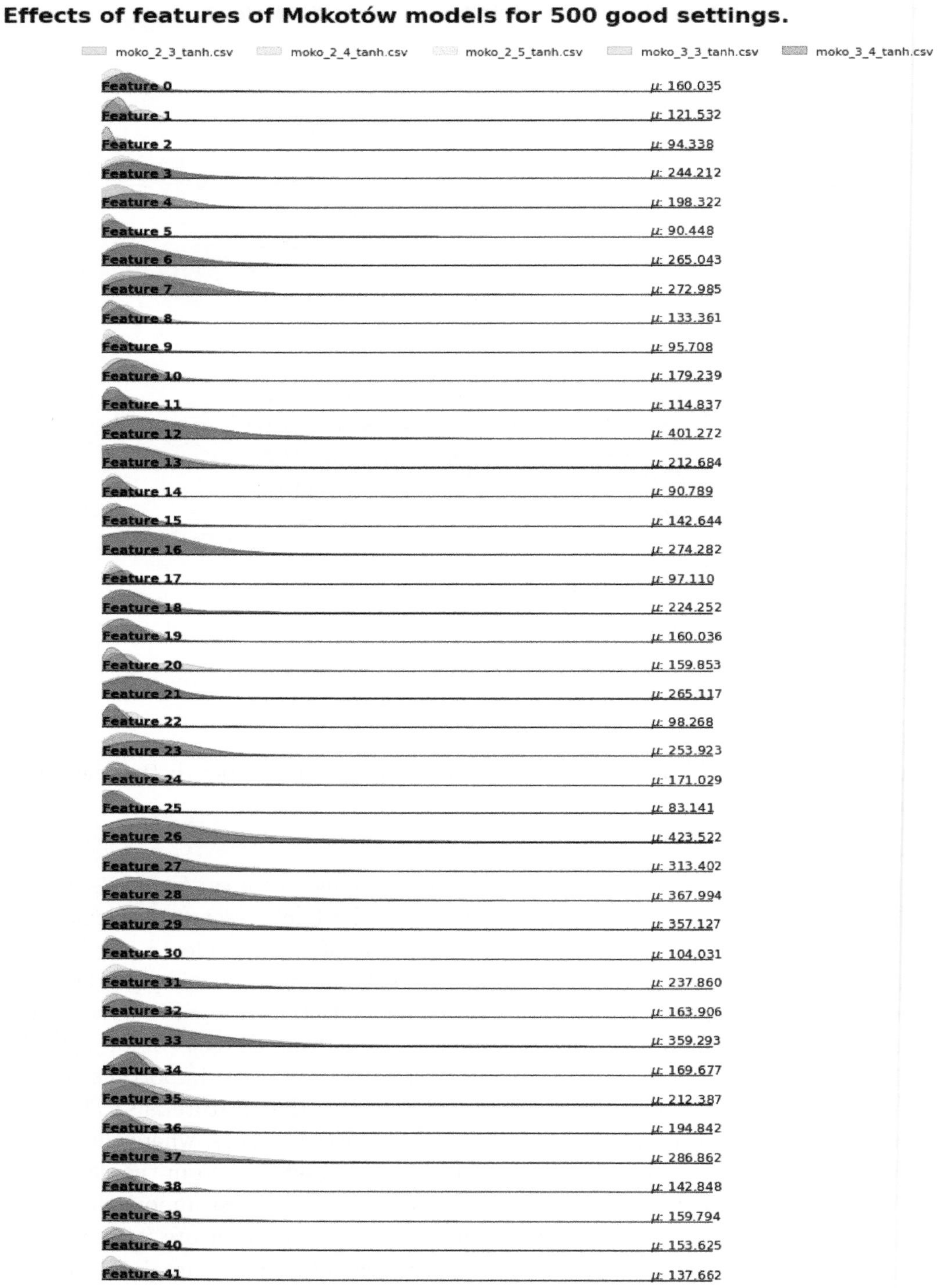

Effects of features of Mokotów models for 500 good settings.

moko_2_3_tanh.csv moko_2_4_tanh.csv moko_2_5_tanh.csv moko_3_3_tanh.csv moko_3_4_tanh.csv

Feature 0 μ: 160.035
Feature 1 μ: 121.532
Feature 2 μ: 94.338
Feature 3 μ: 244.212
Feature 4 μ: 198.322
Feature 5 μ: 90.448
Feature 6 μ: 265.043
Feature 7 μ: 272.985
Feature 8 μ: 133.361
Feature 9 μ: 95.708
Feature 10 μ: 179.239
Feature 11 μ: 114.837
Feature 12 μ: 401.272
Feature 13 μ: 212.684
Feature 14 μ: 90.789
Feature 15 μ: 142.644
Feature 16 μ: 274.282
Feature 17 μ: 97.110
Feature 18 μ: 224.252
Feature 19 μ: 160.036
Feature 20 μ: 159.853
Feature 21 μ: 265.117
Feature 22 μ: 98.268
Feature 23 μ: 253.923
Feature 24 μ: 171.029
Feature 25 μ: 83.141
Feature 26 μ: 423.522
Feature 27 μ: 313.402
Feature 28 μ: 367.994
Feature 29 μ: 357.127
Feature 30 μ: 104.031
Feature 31 μ: 237.860
Feature 32 μ: 163.906
Feature 33 μ: 359.293
Feature 34 μ: 169.677
Feature 35 μ: 212.387
Feature 36 μ: 194.842
Feature 37 μ: 286.862
Feature 38 μ: 142.848
Feature 39 μ: 159.794
Feature 40 μ: 153.625
Feature 41 μ: 137.662

Figure 7.9 A plot of features Shapley value estimates for Mokotów district near local minima.

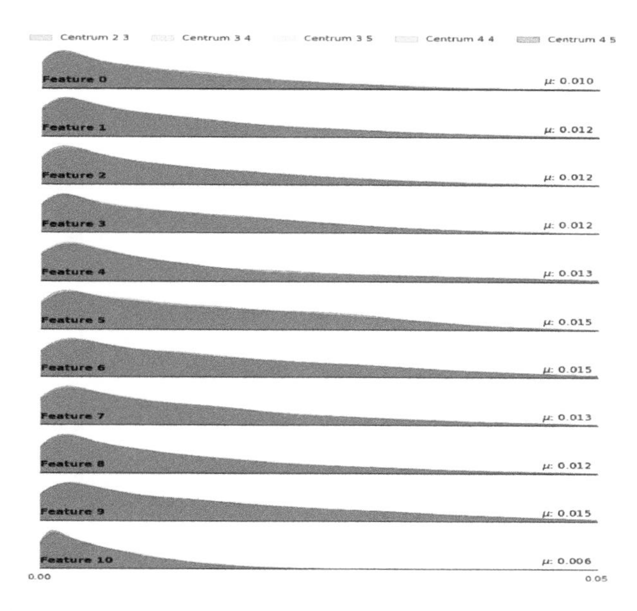

Figure 7.10 A plot of feature effects for five models for Centrum district based on the ZORRO method.

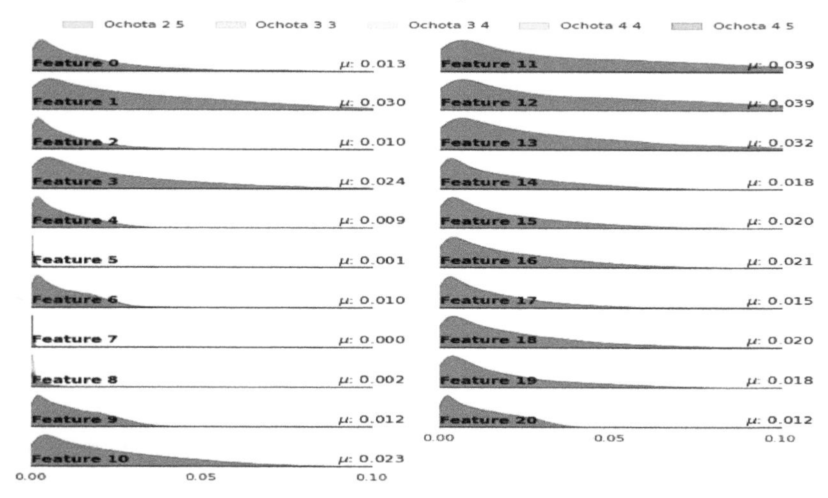

Figure 7.11 A plot of feature effects for five models for Ochota district based on the ZORRO method.

Later, we did the same for the second dataset consisting of traffic signal settings close to local minima found using a genetic algorithm (as described in [13]). In this case, kernel density estimate plots are presented in Figures 7.13–7.15. It has turned out that in this case, the most important features for Centrum are 5, 6, and 9, while the least important – 0 and 10. For Ochota – the most important are 1, 11, 12, and

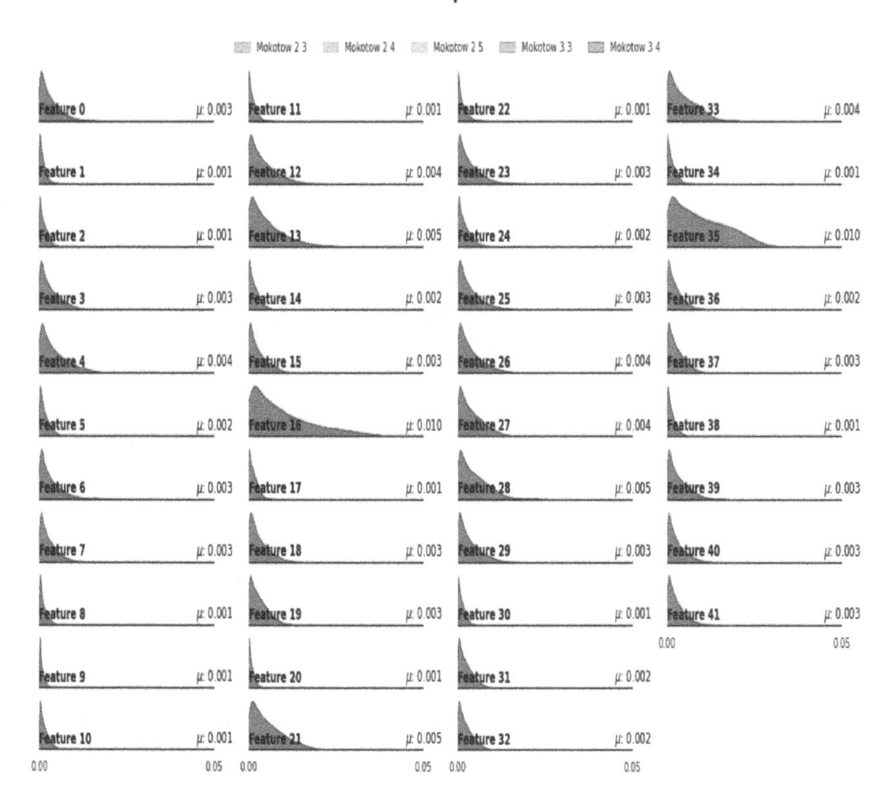

Figure 7.12 A plot of feature effects for five models for Mokotów district based on the ZORRO method.

13, while the least important are 5, 7, and 8. For Mokotów – the most important are 13, 16, 21, 26, and 35, while the least important are 1, 2, 5, 8, 9, 11, 17, 20, 22, 30, and 38.

7.4.2.3 Comparison of Results

Even thought the investigated methods, calculating Shapley value and applying ZORRO method, are different measures of feature importance, their conclusions are similar. In the case of Centrum district, both methods identified the Feature (intersection) 9 as the most important and intersections 0, 2, 3, 8, and 10 as the least important for the random dataset. For the set with local minima, features 5 and 6 are considered as the most important, while 0 and 10 as the least important.

In the case of Ochota, for the randomly generated dataset both methods identified features 1, 11, 12, and 13 as the most important, and features 5, 7, and 8 as the least important (both approaches are ideally consistent). For the set with local minima, features 1, 11, and 13 are considered as the most important according to both methods, while features 5, 7, and 8 as the least important.

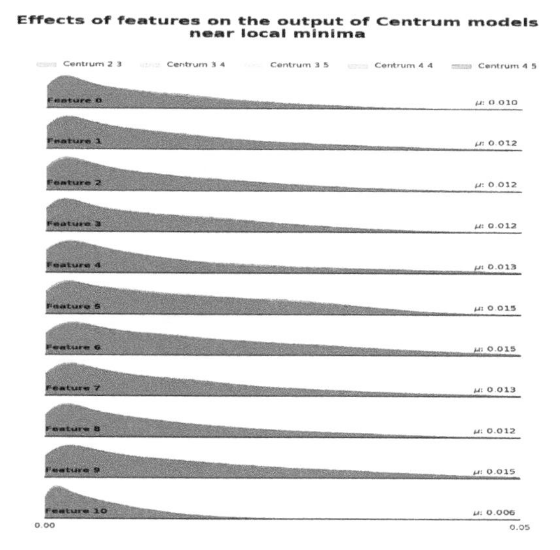

Figure 7.13 A plot of feature effects for five models for Centrum district near local minima based on the ZORRO method.

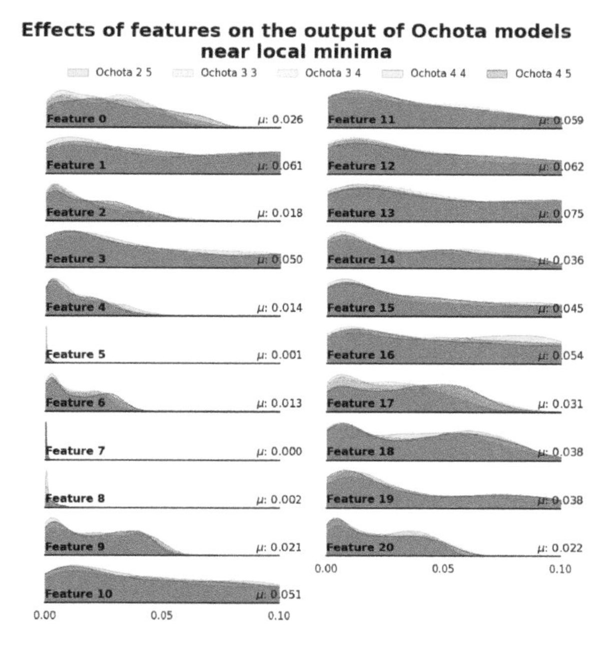

Figure 7.14 A plot of feature effects for five models for Ochota district near local minima based on the ZORRO method.

In the case of Mokotów, for the randomly generated dataset both methods identified features 16 and 35 as the most important and the features 8, 9, and 20 as the least important. For the set with local minima, the feature 16 is considered as the most important, while features 2, 5, 9, 17, and 22 as the least important.

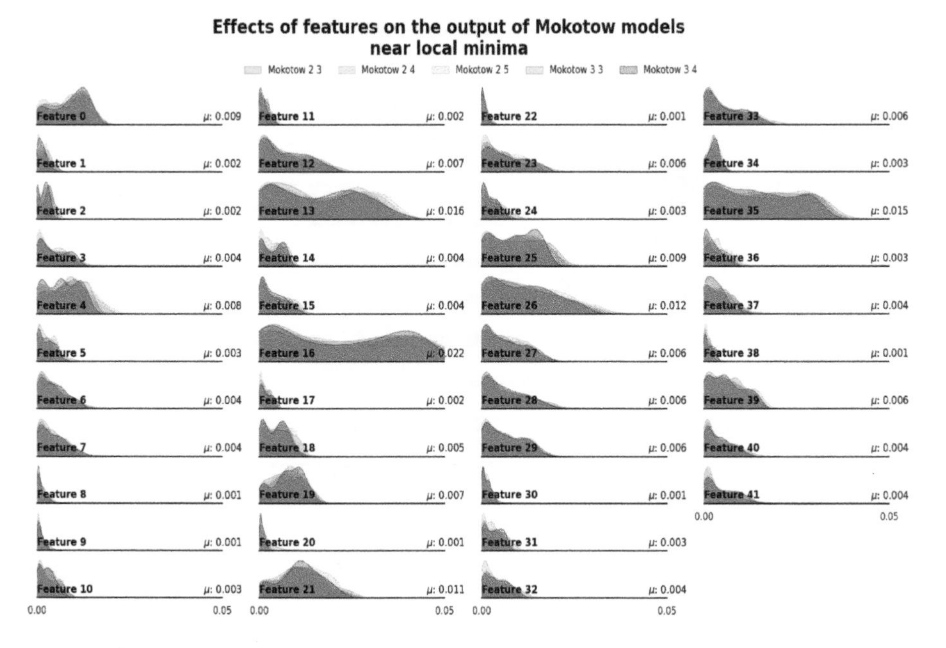

Figure 7.15 A plot of feature effects for five models for Mokotów district near local minima based on the ZORRO method.

There are differences in perceiving importance of some features (for example, in the case of Mokotów, the ZORRO method perceives quite many features as less important), but there are no features considered as important by one method and not important by another method. Therefore, both methods seem to be consistent.

7.5 CONCLUSIONS AND FUTURE RESEARCH

In this chapter, we presented two techniques for measuring importance of features in machine learning models, namely: calculating the Shapley value and using the ZORRO method. We applied them to assess importance of 15 models based on graph neural network architecture in which neurons correspond to intersections with traffic signals or road segments, while connections between neurons in consecutive layers exist if and only if there are direct connections between corresponding intersections and road segments in the road network graph. The models were earlier trained to approximate the total times of waiting at red signals in three districts in Warsaw (Centrum, Ochota and Mokotów) during 10 minutes of a realistic, microscopic traffic simulation, depending on the traffic signal setting input representing offsets of signals at considered intersections. Thanks to applying the Shapley value and ZORRO methods, it was possible to assess importance of intersections and their impact on the times of waiting at red signals in the considered areas on points from randomly generated datasets (with 10^3 points) and on datasets with 500 settings found using genetic algorithm (considered as close to local optima).

It turned out that both methods produce quite consistent results. Thanks to them, it was possible to identify the most critical intersections in the road network topology which might be important from the traffic engineering perspective in tasks related to traffic signal control.

As for the future research, we would like to investigate not only importance of a single feature / intersection, but also importance of subsets of features considered together, which may also give interesting results from the traffic engineering perspective, i.e., it may happen that the emergence of complex patterns, like traffic jams formation, is caused not by a single, poorly set signalization, but by not optimal coordination of traffic signals at several intersections.

Also, we are going to work on traffic signal control algorithms able to optimize traffic taking into account the importance of various features.

Bibliography

[1] Explanation of the kernel density distribution. https://mathisonian.github.io/kde. Online, accessed: 30.09.2022.

[2] Openstreetmap service. http://openstreetmap.org. Online, accessed: 30.09.2022.

[3] Seaborn.kdeplot. https://seaborn.pydata.org/generated/seaborn.kdeplot.html. Online, accessed: 30.09.2022.

[4] James Ault and Guni Sharon. Reinforcement learning benchmarks for traffic signal control. In *Proceedings of the 35th Neural Information Processing Systems (NeurIPS 2021) Track on Datasets and Benchmarks*, December 2021.

[5] Federico Baldassarre and Hossein Azizpour. Explainability techniques for graph convolutional networks. *CoRR*, abs/1905.13686, 2019.

[6] Thorben Funke, Megha Khosla, and Avishek Anand. Hard masking for explaining graph neural networks, 2021.

[7] Pawel Gora. A genetic algorithm approach to optimization of vehicular traffic in cities by means of configuring traffic lights. In *Emergent Intelligent Technologies in the Industry. Studies in Computational Intelligence*, Vol. 369, pp. 1–10. Springer, 2011.

[8] Jacques Lardoux, Raymundo Martinez, Chris White, N. Gross, Nick Patel, and Robert Meyer. Adaptive traffic signal control for tarrytown road in white plains, New York. 2014.

[9] Scott M Lundberg and Su-In Lee. A unified approach to interpreting model predictions. In I. Guyon, U. V. Luxburg, S. Bengio, H. Wallach, R. Fergus, S. Vishwanathan, and R. Garnett, editors, *Advances in Neural Information Processing Systems 30*, pages 4765–4774. Curran Associates, Inc., 2017.

[10] C. Osorio and M. Bierlaire. A surrogate model for traffic optimization of congested networks: an analytic queueing network approach. In *EPFL-REPORT-152480*, 2009.

[11] Lloyd S. Shapley. Notes on the n-person game – ii: The value of an n-person game, August 1951.

[12] Nishant Shokeen, Vardhan Goyal, Vardaan Bhave, and Narad Muni Prasad. Design of traffic signal using webster method. *International Journal of Scientific Engineering and Applied Science (IJSEAS)*, 7, 2021.

[13] Lukasz Skowronek, Pawel Gora, Marcin Mozejko, and Arkadiusz Klemenko. Graph-based sparse neural networks for traffic signal optimization. In *Proceedings of the 29th International Workshop on Concurrency, Specification and Programming (CS&P 2021)*, pages 145–155, 2021.

[14] Elena Sofronova. Evolutionary computations for traffic signals optimization. *Procedia Computer Science*, 186:802–811, 2021.

[15] Ayad Turky, Mohd Ahmad, and Mohd Yusoff. The use of genetic algorithm for traffic light and pedestrian crossing control. *International Journal of Computer Science and Network Security*, 9(2):88–96, 2009.

[16] Hua Wei, Nan Xu, Huichu Zhang, Guanjie Zheng, Xinshi Zang, Chacha Chen, Weinan Zhang, Yanmin Zhu, Kai Xu, and Zhenhui Li. Colight: Learning network-level cooperation for traffic signal control. In *Proceedings of the 28th ACM International Conference on Information and Knowledge Management*, CIKM'19, pages 1913–1922, New York, NY, USA, 2019. Association for Computing Machinery.

[17] Jie Zhang, Souma Chowdhury, Junqiang Zhang, Achille Messac, and Luciano Castillo. Adaptive hybrid surrogate modeling for complex systems. *AIAA J*, 51(3):643–656, 2013.

CHAPTER 8

Intelligent Techniques and Explainable Artificial Intelligence for Vessel Traffic Service: A Survey

Meng Joo Er, Huibin Gong, Chuang Ma, and Wenxiao Gao
Dalian Maritime University, China

CONTENTS

WITH THE RAPID development of the economy, water transportation has been increasingly busy, and there has been a strong demand for vessel traffic services (VTS) using artificial intelligence (AI) techniques. In order to make AI models more understandable and reliable, VTS based on explainable AI (XAI) techniques is one of the most important development directions in the recent times. In this context, this chapter summarizes intelligent techniques of VTS and applications of XAI techniques in VTS. Firstly, basic background of VTS is introduced. Next, on this

DOI: 10.1201/9781003324140-8

basis, this chapter reviews AI techniques in VTS in intelligent prediction and intelligent perception. Finally, XAI techniques are introduced, and their applications and necessity in intelligent water transportation are discussed.

8.1 INTRODUCTION

Water transportation has advantages of large capacity, little investment and low energy consumption, and is the main mode of transport for world trade. With rapid development of the shipping industry, the number of vessels is also increasing [79]. At the same time, the development of vessels towards high speed and large size demands for space in navigation and berthing, result in very crowded water traffic. As a result, difficulty of vessel maneuvering has increased tremendously and there has been a great concern on safety guarantee [14, 18].

In order to meet the demand of development of water transportation and realize modernization of water traffic management, port management departments of many countries have established vessel traffic services (VTS) on the shore. VTS can monitor vessel movement in the water area and provide information, advice or instructions to the vessels. VTS can effectively control traffic flow and reduce risks of vessel traffic accidents and environmental pollution while obtaining maximum port operation benefits. Therefore, establishment of VTS can bring about significant social and economic benefits [31, 45].

With the continuous progress of science and technology, and the demand of shipping, VTS has been constantly upgraded. Before 1990, the main aim of VTS was to improve efficiency and safety of water transportation. The coverage of water was limited to ports and canals, narrow channels of rivers and estuaries. And radar and high-frequency radio were the main technologies. Since 1990, VTS has been continuously upgraded. In terms of functions, water environment protection has been added, and in terms of area coverage, it has been extended to near port areas, coastal routes and straits with high density of vessels, and other accident-prone areas. Various information collection and processing technologies leveraging computers have been adopted. At present, VTS is usually composed of a radar system, closed circuit television (CCTV), very high frequency (VHF), automatic identification system (AIS), radar data processing system, hydro-meteorological system, traffic display system, multi-sensor integrated processing system and data management information system, etc. The main functions of VTS include collection and evaluation of vessel traffic data, providing information services and navigation assistance services to vessels participating in the VTS, implementing traffic organization for vessels in the VTS area, supervising vessels to comply with navigation rules, and providing joint operation support to relevant activities in the VTS area [23, 40, 53]. Figure 8.1 shows the schematic diagram of the composition of VTS.

With the rapid development of technologies such as computer, AI and microwave communication, and in the context of busy water traffic management business needs and continuous growth of commercial demand, users have created more demand for VTS. Therefore, construction of informationized and intelligent VTS have become an urgent need for the construction of modern ports [68]. Under these factors, more and

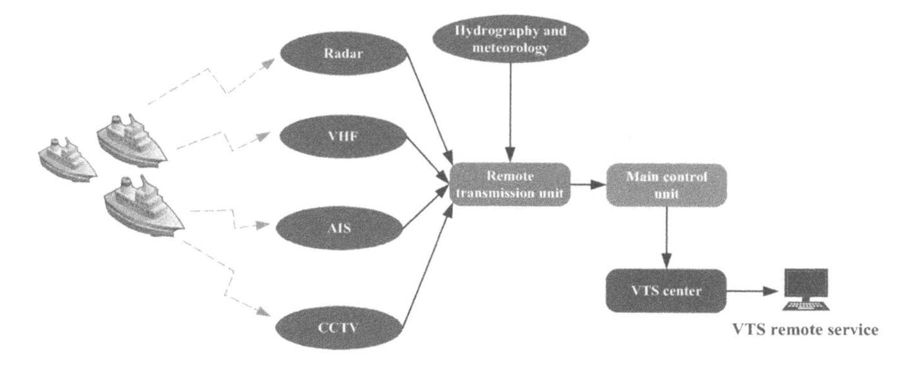

Figure 8.1 Schematic diagram of the composition of VTS.

more intelligent techniques for VTS have been proposed by researchers, such as water prediction of traffic flow, prediction of water traffic situation, vessel identification, capture of environmental information, decision-making in navigation safety, etc. [36, 37, 65].

Applications of these AI techniques can generate some useful information, which can help VTS operators, captains, pilots and other staff members to make decisions. However, there are unexplainable black-box problems in current end-to-end AI techniques, which makes it difficult for users to trust and understand such AI systems, especially when autonomous decision-making may have uncontrollable risks. At the same time, AI systems may also have inherent biases, generate false alarms from time to time, and have uncertain security risks. These factors make it difficult for users to act on decisions that are not explained. In order to enable AI techniques better assist humans and be more widely used and popularized, it is necessary to transform the black-box problem in AI techniques into a transparent process, so as to establish trust and understanding between humans and machines. Therefore, explainable AI (XAI) came into being [35, 48, 58]. Similarly, XAI is needed in future VTS to make VTS more intelligent and more reliable, so as to realize intelligent management of water transportation.

For the survey of VTS, the development history, composition and structure of VTS, and key technologies of VTS are introduced in [19]. However, there is no survey of AI techniques in VTS. Moreover, vessel motion prediction techniques in VTS are reviewed in [66]. Although some XAI techniques in Air Traffic systems and road traffic systems have been reviewed [24, 70], there is no survey of XAI techniques in VTS so far.

Motivated by the aforementioned background, this chapter surveys intelligent techniques in VTS as well as XAI techniques, among other things. The main contributions of this chapter are as follows:

1. Intelligent techniques in VTS are summarized for the first time, which can provide reference for researchers engaged in VTS.

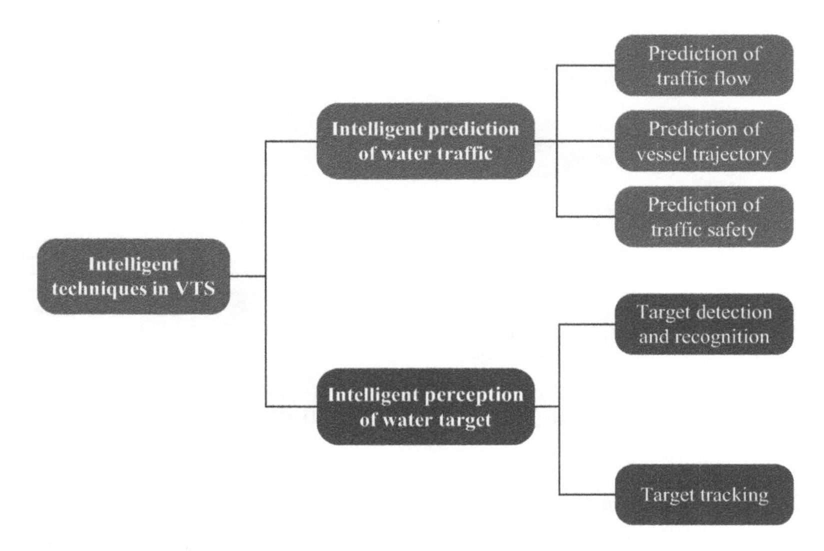

Figure 8.2 Classification diagram of current research works.

2. Some VTS schemes based on XAI, which play a pivotal role in promoting the development of VTS based on XAI, are reviewed and summarized for the first time.

The remainder of this chapter is organized as follow: In Section 8.2, state-of-the-art research on intelligent techniques in VTS are presented. Section 8.3 introduces basic concepts of XAI and summarizes some implementation schemes of XAI in VTS. In Section 8.4, conclusions are drawn.

8.2 INTELLIGENT TECHNIQUES IN VTS

In this section, AI techniques in VTS are reviewed. The review includes intelligent prediction of water traffic and intelligent perception of water target. The classification diagram of current research works is shown in Figure 8.2.

8.2.1 Intelligent Prediction of Water Traffic

Intelligent prediction techniques of water traffic include prediction of traffic flow, prediction of vessel trajectory and prediction of traffic safety. These intelligent prediction techniques can show vessel traffic situations on water and provide support for decision-making.

8.2.1.1 Prediction of Traffic Flow

Research on prediction of vessel flow can provide some references for the management team to ensure safety of water traffic and formulate pertinent policies. Therefore, research on prediction of traffic flow is of great significance.

Existing methods for prediction of vessel traffic flow mainly include regression analysis, Kalman filter and neural networks. He et al. [21] used the improved Kalman model combined with regression analysis to make short-term prediction of vessel traffic flow, and their experimental results proved that the method had good performance. Wang et al. [60] designed a hybrid model based on discrete wavelet decomposition and the PROPHET framework for prediction of vessel traffic flow, where wavelet decomposition was mainly used to complete predictions through training and reconstruction of prediction results. Back propagation neural networks (BPNN) is the most widely used prediction method. Tian et al. [56] established a vessel traffic flow prediction model based on BPNN, in which traffic flow data of the deepwater channel of the Yangtze Estuary were used as training samples.

In addition, in order to improve the prediction accuracy, combinations of an intelligent optimization algorithm and the prediction model are also widely used. Fan et al. [16] used particle swarm optimization algorithm to optimize the BPNN-Markov ship traffic flow prediction model, which overcame the problem of insufficient whitening coefficient selection in Markov model. Zhang et al. [77] established an improved BPNN prediction model based on adaptive particle swarm optimization algorithm, which realized the prediction of total ship traffic volume in the designated port area. Li et al. [32] proposed a chaotic cloud simulated annealing method to optimize parameters of the support vector regression model and applied this method to prediction of vessel traffic flow. Some researchers have adopted deep learning techniques to improve spatio-temporal correlation of prediction and accuracy of prediction results. Xie et al. [71] proposed a deep learning model based on long short-term memory networks for predicting inland river vessel traffic flow, where the time window method was used to increase time correlation. Liang et al. [34] proposed a vessel traffic flow prediction scheme based on a spatiotemporal multigraph convolution network, which mainly included a spatial multigraph convolution layer and two time gated convolution layers. Accuracy and robustness of the scheme have been proved by real AIS data experiments.

8.2.1.2 Prediction of Vessel Trajectory

Trajectory prediction is the focus of many researchers. As an important part of data mining, track predictions can provide accurate location services and play an important role in analyzing vessel motion behavior, judging vessel traffic risk, and planning intelligent vessel collision avoidance path.

Vessel trajectory prediction techniques generally include the Kalman filter algorithm, particle filter algorithm, and Bayesian algorithm and neural network. Perera et al. [44] proposed a vessel state estimation method based on an extended Kalman filter for the prediction of vessel motion trajectory. Mazzarella et al. [41] proposed a Bayesian trajectory prediction algorithm based on a particle filter which used historical vessel trajectory data. Xiao et al. [67] adopted machine learning, physical motion modeling and particle filtering algorithms to design a new prediction method for vessel track and navigation state based on AIS data.

Most of the above studies need to build complex kinematics model, which requires large amount of data. The neural network algorithm can greatly solve this problem. Capobianco et al. [9] designed a vessel trajectory prediction model based on the codec recurrent neural network, which combined the short-term memory recurrent neural network. Suo et al. [55] proposed a deep learning framework and Gate Recurrent Unit model for vessel trajectory, in which the DBSCAN algorithm, correction algorithm, and recurrent neural network were used.

Real-time prediction of multiple vessel tracks is very challenging. Gan et al. [17] adopted K-Means algorithm to classify historical track data, and then built a neural network model to realize prediction of multiple vessel tracks. Xiao et al. [69] proposed a vessel cluster trajectory prediction scheme based on AIS data, which relied on an open-source code framework. At the same time, in order to reduce the computation time, a new task-based load balancing strategy is proposed. Wang et al. [62] proposed a generative countermeasure network with an attention module and interaction module to predict the track of multiple vessels. Some indicators of the scheme were verified by using real historical data of AIS of Zhoushan Port, which proved the efficiency of the scheme.

8.2.1.3 Prediction of Traffic Safety

As early as the 1930s, the United States began to do relevant works on prediction of vessel safety traffic. With the continuous development of science and technology, people's awareness of safety engineering is also enhanced, which drives people to gradually increase research on this issue.

Water traffic accidents are mainly caused by people, vessels, environment and other factors. At first, experts analyzed single factors, especially human factors [5]. The analysis and research of single factor can reveal the main causes of the accident. However, the impact of other factors on the accident was ignored. Hence, researchers began to pay attention to the analysis of multiple factors. For accident analysis and prediction, it mainly includes fuzzy logic, neural networks, time-series analysis and support vector machine. Kao et al. [27] proposed a scheme based on fuzzy logic, which can accurately predict the time and location of vessel collision by using AIS data. Su et al. [54] proposed a fuzzy decision-making method based on the knowledge base of international rules for collision avoidance at sea, which improved operator's decision-making ability for collision avoidance by providing a fuzzy monitoring system in VTS. Based on processing of the original data, Rekha et al. [49] used a support vector machine to predict grounding accidents, and this method achieved a good prediction effect. Rahman et al. [47] used a Bayesian network to analyze the relation vessel between influencing factors of water traffic accidents and accident probability, which effectively predicted the number of different types of accidents. Debnath et al. [12] quantitatively predicted collision accidents in the channel of Singapore Port based on the time series prediction method. Kim et al. [29] proposed system architecture for adaptive visualization of monitoring services, in which decision trees method and data transmitted by vessel sensors at sea were used to assess collision risks between vessels.

8.2.2 Intelligent Perception of Water Target

In recent years, with the application of radar, AIS, GPS and other auxiliary technologies, the ability of VTS in vessel classification, vessel identification, vessel tracking, and other aspects has been greatly enhanced, providing more information for VTS in decision-making. However, these technologies also have their own limitations. It is possible to achieve intelligent perception of water target through video images and data fusion of various systems [6].

8.2.2.1 Target Detection and Recognition

Due to the limitations of both AIS and radar, target detection and recognition based on computer vision have become the main research trend. Wawrzyniak et al. [63] designed a vessel identification scheme, which mainly identified the type of vessel according to the visible inscription (name, registration number) on the hull. They also developed a vessel identification system using a neural network and rough set theory. This system can be used for systems with service-oriented architecture by identifying vessels from video data streams [64]. Zhang et al. [75] proposed an automatic recognition method for moving targets on water, where the motion displacement between images was obtained by calculating the relation vessel and correlation between background images and factor points to achieve automatic recognition of moving targets. Shi et al. [51] used the morphological filtering method of multiple structural elements to quickly detect vessel targets in order to effectively suppress background noise such as sleeping wave clutter.

With the complexity of the target, traditional algorithms have been unable to meet the needs. The emergence of deep learning methods has greatly improved this phenomenon. You et al. [74] proposed the deep convolutional neural network framework of Saliency-Fater R-CNN, which has solved the problem of multi-scale vessel detection. Zhang et al. [78] proposed a multi-spectral image vessel detection method based on cascade convolutional neural network to improve the detection accuracy. Qi et al. [46] proposed an improved Faster R-CNN algorithm for vessel target detection, which reduced the search scale of target detection and improved the detection accuracy while shortening the detection time. Liu et al. [7] proposed a vessel identification method based on Darknet network model and YOLOv3 algorithm, which can identify a variety of vessel types and important parts in monitoring of water.

8.2.2.2 Tracking of Vessel Target

For tracking of vessel target, classical methods include particle filter, mean shift, and Kalman filter. Kim et al. [30] designed an image-based vessel tracking system using a Bayesian classifier and Kalman filtering algorithm, which can effectively overcome marine environmental noise, bad weather and waves. Chen et al. [11] proposed an integrated vessel tracking scheme based on a multi-view learning algorithm and wavelet filtering model. In this scheme, target tracking is mainly achieved by extracting vessel contour and correcting abnormal oscillation.

However, classical tracking algorithms cannot solve more complex situation well. Hence, they have been combined with deep learning techniques. Yang et al. [73] proposed a single target tracking network model based on depth learning and a tracking and counting method of virtual detection line for inland waterway vessel flow statistics and conducted experiments in actual application scenarios.

The intelligent perception system has many sensors and complex sensing data, so data fusion techniques have been used to improve perception efficiency. Generally speaking, AIS is fused with radar. Zhang et al. [76] designed a reasonable network structure based on multi-sensor information fusion theory and the BPNN principle and proposed a BPNN-based radar and AIS target association algorithm. He et al. [20] proposed a radar track fusion method based on AIS to solve the problem of difficult data association and low accuracy when tracking dense targets on the sea. Huang et al. [22] proposed a multi-vessel-target tracking technique based on an improved single shot multi-box detector and the DeepSORT algorithm, where a data fusion algorithm was used to integrate AIS information into video display.

8.3 XAI AND ITS APPLICATIONS IN VTS

Considering current development of AI techniques and their popularity in VTS, explainability of AI techniques in intelligent water transportation is now discussed. This section will review the following: (1)What is XAI ? (2)Why is XAI needed? (3)Classification of XAI methods. (4)Applications of XAI in VTS.

8.3.1 What Is XAI and Why Is XAI Needed?

Thanks to its powerful computing power, AI techniques have exhibited superior performance in massive data processing, analysis and decision-making. AI techniques have been widely used in various fields, such as aerospace [2, 70], industrial automation systems [25, 80], medical [28, 57] and so on. With the continuous development of science and technology, the ability of AI techniques to learn and adapt has also been continuously strengthened [4]. In the process of increasing popularity at all levels of people's production and life, complexity of the algorithm has also increased exponentially.

As the mainstream of modern AI techniques, deep neural network techniques often have tens of millions of parameters. Unfortunately, they belong to the black-box method [10]. As far as sensitive areas such as data privacy and autonomous driving are concerned, the black-box method has some accountability, ethical and judicial issues. On the other hand, when the audience is different, the XAI needs to focus on goals and priorities are also different. For visual analysis, the focus is on providing explanations through visual analysis of the model or displaying significance of different features. The foundations of human understanding and structure of explanations are also of great importance to specialists in psychological research [43].

In summary, to better popularize and apply AI techniques, it is crucial to understand how these algorithms work and how decisions are made. This results in XAI, a field dedicated to developing a set of theories and techniques to parse models and generating intuitive explanations to help people who need them to understand and predict AI's behavioral decisions.

8.3.2 Classification of XAI Methods

At present, there is no universal classification and evaluation standard that can convince everyone in the XAI field. However, in most studies, explainability, interpretability, and model transparency are frequently mentioned [25]. Here, we review and discuss these concepts.

- *Explainability:* Explainability is an active feature that is used to describe the working mechanism inside a model.
- *Interpretability:* It is a passive characteristic that describes the ability of the model to be understood by people.
- *Transparency:* A model can be considered transparent if it can be understood by people without any manipulation. In the field of XAI, it is generally believed that the more transparent the model or algorithm, it is better.

According to different principles, XAI can be classified into different categories. Depending on the scope of the work, explanations can be global explanation, local explanation, or a combination of both. Local explanations focus on generating explanations for a single sample, and their essence is to explore how a specific sample affects a single output, such as LIME [50] and SHAP [38]. Global explanations act on groups of data to provide explanations for the operation of the entire model. This approach strives to simplify complex deep models into interpretable linear models through various methods such as class model visualization [52] and global attribution mapping [26]. Maksymiuk et al. [39] divided XAI into model interpretable group, model specific group and model agnostic group. Model interpretable groups generally represent traditional machine learning that are easy to explain, such as decision trees and generalized additive models. Model specific group refers to using a specific method to explain a specific model, which has poor generalization and is not applicable to other models. The model agnostic group is also called post-model interpretable method, which usually mines model-independent explanations. For example, LIME is dedicated to mining importance of super pixels in the source image to the class output, ensuring that a single data can play an outstanding role in prediction. Furthermore, LIME is extended to a version that can be used for global interpretation [15]. This type of method can be carried out when the model is completely unknown, i.e., this type of method is suitable for most deep neural network models and is currently the most active XAI technique, commonly used in land traffic management [1] and Air Traffic management [70].

8.3.3 Applications of XAI in VTS

Similar to land-based intelligent transportation and air-based intelligent transportation, the field of water-based intelligent transportation will involve a large amount

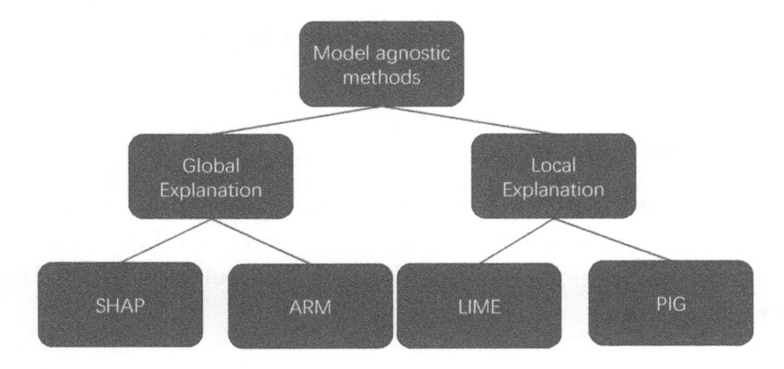

Figure 8.3 Most popular model agnostic explainable methods [3].

of data and traffic management operation. It is necessary to ensure that the information represented by massive data can be understood by the operators, and let them make reasonable decision based on the obtained information while trusting the model. Therefore, it is imperative to promote applications and popularization of XAI in the maritime field. At present, XAI has already had some applications in intelligent water transportation.

Considering that features such as tracks are often used to classify vessel types, to understand the meaning behind the decision, Burkart et al. [8] introduced existing methods for interpretation of vessel classification task and conducted experiments with an explanatory case for the first time with the participation of human experts. Anneken et al. [3] highlighted a data fusion technique that can help improve the efficiency of anomaly detection and discussed how to help people in the marine field understand and predict behavior and decisions of autonomous systems by introducing model agnostic XAI methods as shown in Figure 8.3. Veerappa et al. [3] confirmed that the XAI method applicable to images, text and tables can also be used for multivariate time series. Furthermore, the effect of different types of explanations on sequence classifiers is verified, which makes the popularization of XAI methods in sensitive fields such as water transportation strongly supported [58]. In the field of traffic safety, XAI also has some related applications. Veitch et al. [59] proposed a human-centric concept through a combination of analogy, visualization and mental simulation, which achieved good results in solving the problem of interactions between unmanned surface vehicles and different users. Furthermore, to ensure that vessels comply with important maritime policies, Yan et al. [72] developed an unsupervised anomaly detection model to help ports predict vessel detention. In this paper, SHAP was used to analyze and visualize the influence of different input characteristics on decision-making, and simulation experiments were carried out with real data of actual ports.

In addition, XAI also has some achievements in the field of marine target intelligent perception, which is mainly reflected in the enhancement of marine target detection and recognition technique by the attention mechanism [42]. To solve the problem of marine infrared small target detection without prior knowledge, Dong et al. [13] improved the visual attention model combined with an anti-vibration pipeline filtering algorithm and achieved good results. Li et al. [33] added a channel attention mechanism module to YOLOv5 to improve its recognition ability of effective features. Furthermore, Wang et al. [61] made multiple adjustments to the architecture of YOLOv5, where the channel attention mechanism module and the iterative refinement mechanism were combined to improve the network efficiency. However, these attention mechanisms are mainly used to help the model identify and acquire important features, and do not greatly enhance the interpretability of the model itself.

Finally, information acquisition techniques used by AIS occupy a vital part in VTS and is also widely used in vessels [20, 22, 76]. Some studies related to interpretability in the marine field [8, 58] are based on the AIS dataset. Although there is little work at present, it may be a very promising direction to conduct interpretable and relevant research on the data (such as AIS) obtained by intelligent sensing methods.

8.4 CONCLUSIONS

Based on the review of preliminaries of VTS and state of the arts of intelligent techniques, this chapter discusses applications of XAI in VTS. Currently, VTS with AI techniques is far more efficient and intelligent than before, but their huge amount of information and opaque decisions also create problems for real world operators. XAI techniques can provide interpretations in the form of visualisations, texts or rules to help users understand and process the information in the vast amount of data obtained more rationally as well as improve system's decision-making. This greatly increases user's trust and acceptance of the VTS and enables intelligent systems to provide better decision-making services to users and increase efficiency. On the other hand, maritime is a sensitive area, and given the risk of failure, the parsing of models and the transparency of decisions can be very helpful in clarifying the assumption of responsibility, as well as the development of relevant legislation.

Although XAI techniques are currently receiving little attention in the study of VTS, they have made their marks in both land and air intelligent transportation. Considering that intelligent traffic systems in different fields have relatively similar problems, such as the processing of large amounts of information and human-computer interactions between vehicles and operators, we believe that in the future VTS will develop towards more interpretable, more trustworthy, safer, more efficient and more ethical intelligent traffic management. We sincerely hope that this chapter will serve as a good reference for researchers and developers working on VTS.

Bibliography

[1] Ashraf Abdul, Jo Vermeulen, Danding Wang, Brian Y Lim, and Mohan Kankanhalli. Trends and trajectories for explainable, accountable and intelligible systems: An hci research agenda. In *Proceedings of the 2018 CHI Conference on Human Factors in Computing Systems*, pages 1–18. Association for Computing Machinery, 2018.

[2] Natalia Alexandrov. Explainable AI decisions for human-autonomy interactions. In *Proceedings of 17th AIAA Aviation Technology, Integration, and Operations Conference*, page 3991. American Institute of Aeronautics and Astronautics, Inc., 2017.

[3] Mathias Anneken, Manjunatha Veerappa, and Nadia Burkart. Anomaly detection and XAI concepts in swarm intelligence. 2021. https://www.sto.nato.int/publications/STO%20Meeting%20Proceedings/STO-MP-SCI-341/MP-SCI-341-05.pdf.

[4] Alejandro Barredo Arrieta, Natalia Diaz-Rodriguez, Javier Del Ser, Adrien Bennetot, Siham Tabik, Alberto Barbado, Salvador Garcia, Sergio Gil-Lopez, Daniel Molina, Richard Benjamins, Raja Chatila, and Francisco Herrera. Explainable artificial intelligence (XAI): Concepts, taxonomies, opportunities and challenges toward responsible AI. *Information Fusion*, 58:82–115, 2020.

[5] HP Berg. Human factors and safety culture in maritime safety. *Marine Navigation and Safety of Sea Transportation: STCW, Maritime Education and Training (MET), Human Resources and Crew Manning, Maritime Policy, Logistics and Economic Matters*, 107:107–115, 2013.

[6] Domenico D Bloisi, Fabio Previtali, Andrea Pennisi, Daniele Nardi, and Michele Fiorini. Enhancing automatic maritime surveillance systems with visual information. *IEEE Transactions on Intelligent Transportation Systems*, 18(4):824–833, 2016.

[7] Liu Bo, Shengzheng Wang, Zhao Jiansen, and Mingfeng Li. Ship tracking and recognition based on darknet network and YOLOv3 algorithm. *Journal of Computer Applications*, 39(6):1663–1668, 2019.

[8] Nadia Burkart, Marco F Huber, and Mathias Anneken. Supported decision-making by explainable predictions of ship trajectories. In *Proceedings of International Workshop on Soft Computing Models in Industrial and Environmental Applications*, pages 44–54. Springer, 2020.

[9] Samuele Capobianco, Leonardo M Millefiori, Nicola Forti, Paolo Braca, and Peter Willett. Deep learning methods for vessel trajectory prediction based on recurrent neural networks. *IEEE Transactions on Aerospace and Electronic Systems*, 57(6):4329–4346, 2021.

[10] Davide Castelvecchi. Can we open the black box of AI? *Nature News*, 538(7623):20, 2016.

[11] Xinqiang Chen, Huixing Chen, Huafeng Wu, Yanguo Huang, Yongsheng Yang, Wenhui Zhang, and Pengwen Xiong. Robust visual ship tracking with an ensemble framework via multi-view learning and wavelet filter. *Sensors*, 20(3):932, 2020.

[12] Ashim Kumar Debnath and Hoong Chor Chin. Navigational traffic conflict technique: a proactive approach to quantitative measurement of collision risks in port waters. *The Journal of Navigation*, 63(1):137–152, 2010.

[13] Lili Dong, Bin Wang, Ming Zhao, and Wenhai Xu. Robust infrared maritime target detection based on visual attention and spatiotemporal filtering. *IEEE Transactions on Geoscience and Remote Sensing*, 55(5):3037–3050, 2017.

[14] Eleftheria Eliopoulou, Apostolos Papanikolaou, and Markos Voulgarellis. Statistical analysis of ship accidents and review of safety level. *Safety Science*, 85:282–292, 2016.

[15] Radwa ElShawi, Youssef Sherif, Mouaz Al-Mallah, and Sherif Sakr. ILIME: local and global interpretable model-agnostic explainer of black-box decision. In *Proceedings of European Conference on Advances in Databases and Information Systems*, pages 53–68. Springer, 2019.

[16] Qingbo Fan, Fusai Jiang, Quandang Ma, and Yong Ma. PSO-based BP neural network-markov prediction model of ship traffic flow. *Journal of Shanghai Maritime University*, 39(2):22–27, 2018.

[17] Shaojun Gan, Shan Liang, Kang Li, Jing Deng, and Tingli Cheng. Ship trajectory prediction for intelligent traffic management using clustering and ANN. In *Proceedings of 2016 UKACC 11th International Conference on Control (CONTROL)*, pages 1–6. IEEE, 2016.

[18] Floris Goerlandt and Jakub Montewka. Maritime transportation risk analysis: Review and analysis in light of some foundational issues. *Reliability Engineering & System Safety*, 138:115–134, 2015.

[19] Xiaoning Han, Youyu Zhang, and Jun Wang. Development and application of smart vessel traffic management system. *Command Information System and Technology*, 10(4):8–13, 2019.

[20] Fengshu He, Lifeng Miao, Fei Tao, and Cun Zhang. A method of multi-target fusion and tracking for sea surveillance radar based on AIS. *Radar Science and Technology*, 15(2):153–158, 2017.

[21] Wei He, Cheng Zhong, Miguel Angel Sotelo, Xiumin Chu, Xinglong Liu, and Zhixiong Li. Short-term vessel traffic flow forecasting by using an improved kalman model. *Cluster Computing*, 22(4):7907–7916, 2019.

[22] Zishuo Huang, Qinyou Hu, Qiang Mei, Chun Yang, and Zheng Wu. Identity recognition on waterways: a novel ship information tracking method based on multimodal data. *The Journal of Navigation*, 74(6):1336–1352, 2021.

[23] Terry Hughes. Vessel traffic services (VTS): Are we ready for the new millenium? *The Journal of Navigation*, 51(3):404–420, 1998.

[24] Fatima Hussain, Rasheed Hussain, and Ekram Hossain. Explainable artificial intelligence (XAI): An engineering perspective. *arXiv preprint arXiv:2101.03613*, 2021.

[25] Fatima Hussain, Rasheed Hussain, and Ekram Hossain. Explainable artificial intelligence (XAI): An engineering perspective. *arXiv preprint arXiv:2101.03613*, 2021.

[26] Mark Ibrahim, Melissa Louie, Ceena Modarres, and John Paisley. Global explanations of neural networks: Mapping the landscape of predictions. In *Proceedings of the 2019 AAAI/ACM Conference on AI, Ethics, and Society*, pages 279–287. Association for Computing Machinery, 2019.

[27] Sheng-Long Kao, Kuo-Tien Lee, Ki-Yin Chang, and Min-Der Ko. A fuzzy logic method for collision avoidance in vessel traffic service. *The Journal of Navigation*, 60(1):17–31, 2007.

[28] Izhar Ahmed Khan, Nour Moustafa, Imran Razzak, Muhammad Tanveer, Dechang Pi, Yue Pan, and Bakht Sher Ali. XSRU-IoMT: Explainable simple recurrent units for threat detection in internet of medical things networks. *Future Generation Computer Systems*, 127:181–193, 2022.

[29] Kwang-il Kim and Keon Myung Lee. Adaptive information visualization for maritime traffic stream sensor data with parallel context acquisition and machine learning. *Sensors*, 19(23), 2019. Art. no. 5273.

[30] Yun Jip Kim, Yun Koo Chung, and Byung Gil Lee. Vessel tracking vision system using a combination of kaiman filter, Bayesian classification, and adaptive tracking algorithm. In *Proceedings of 16th International Conference on Advanced Communication Technology*, pages 196–201. IEEE, 2014.

[31] Gunwoo Lee, Soo-Yeob Kim, and Min-Kyu Lee. Economic evaluation of vessel traffic service (VTS): A contingent valuation study. *Marine Policy*, 61:149–154, 2015.

[32] Ming-Wei Li, Duan-Feng Han, and Wen-long Wang. Vessel traffic flow forecasting by RSVR with chaotic cloud simulated annealing genetic algorithm and KPCA. *Neurocomputing*, 157:243–255, 2015.

[33] Wei Li, Jing Wang, Xiao Feng Zhao, Zhiyuan Wang, and Qi Jiang Zhang. Target detection in color sonar image based on YOLOV5 network. In *Proceedings of 2021 IEEE International Conference on Signal Processing, Communications and Computing (ICSPCC)*, pages 1–5. IEEE, 2021.

[34] Maohan Liang, Ryan Wen Liu, Yang Zhan, Huanhuan Li, Fenghua Zhu, and Fei-Yue Wang. Fine-grained vessel traffic flow prediction with a spatio-temporal multigraph convolutional network. *IEEE Transactions on Intelligent Transportation Systems*, early access, 2022. doi: 10.1109/TITS.2022.3199160.

[35] Igor Linkov, Stephanie Galaitsi, Benjamin D Trump, Jeffrey M Keisler, and Alexander Kott. Cybertrust: From explainable to actionable and interpretable artificial intelligence. *Computer*, 53(9):91–96, 2020.

[36] Ryan Wen Liu, Maohan Liang, Jiangtian Nie, Yanli Yuan, Zehui Xiong, Han Yu, and Nadra Guizani. STMGCN: Mobile edge computing-empowered vessel trajectory prediction using spatio-temporal multigraph convolutional network. *IEEE Transactions on Industrial Informatics*, 18(11):7977–7987, 2022.

[37] Ryan Wen Liu, Jiangtian Nie, Sahil Garg, Zehui Xiong, Yang Zhang, and M Shamim Hossain. Data-driven trajectory quality improvement for promoting intelligent vessel traffic services in 6G-enabled maritime IoT systems. *IEEE Internet of Things Journal*, 8(7):5374–5385, 2020.

[38] Scott M Lundberg and Su-In Lee. A unified approach to interpreting model predictions. *Advances in Neural Information Processing Systems*, 30, 2017.

[39] Szymon Maksymiuk, Alicja Gosiewska, and Przemyslaw Biecek. Landscape of R packages for explainable artificial intelligence. *arXiv preprint arXiv:2009.13248*, 2020.

[40] Joakim Trygg Mansson, Margareta Lutzhoft, and Ben Brooks. Joint activity in the maritime traffic system: perceptions of ship masters, maritime pilots, tug masters, and vessel traffic service operators. *The Journal of Navigation*, 70(3):547–560, 2017.

[41] Fabio Mazzarella, Virginia Fernandez Arguedas, and Michele Vespe. Knowledge-based vessel position prediction using historical AIS data. In *Proceedings of 2015 Sensor Data Fusion: Trends, Solutions, Applications (SDF)*, pages 1–6. IEEE, 2015.

[42] Volodymyr Mnih, Nicolas Heess, Alex Graves, and koray kavukcuoglu. Recurrent models of visual attention. In *Proceedings of Advances in Neural Information Processing Systems*, volume 27, pages 2204–2212. Curran Associates, Inc., 2014.

[43] Sina Mohseni, Niloofar Zarei, and Eric D Ragan. A multidisciplinary survey and framework for design and evaluation of explainable AI systems. *ACM Transactions on Interactive Intelligent Systems*, 11(3–4):1–45, 2021.

[44] Lokukaluge P Perera, Paulo Oliveira, and C Guedes Soares. Maritime traffic monitoring based on vessel detection, tracking, state estimation, and trajectory prediction. *IEEE Transactions on Intelligent Transportation Systems*, 13(3):1188–1200, 2012.

[45] Gesa Praetorius, Erik Hollnagel, and Joakim Dahlman. Modelling vessel traffic service to understand resilience in everyday operations. *Reliability Engineering & System Safety*, 141:10–21, 2015.

[46] Liang Qi, Bangyu Li, Liankai Chen, Wei Wang, Liang Dong, Xuan Jia, Jing Huang, Chengwei Ge, Ganmin Xue, and Dong Wang. Ship target detection algorithm based on improved faster R-CNN. *Electronics*, 8(9):959, 2019.

[47] Sohanur Rahman. Introduction of Bayesian network in risk analysis of maritime accidents in bangladesh. In *Proceedings of AIP Conference Proceedings*, volume 1919, page 20–24. AIP Publishing, 2017.

[48] Atul Rawal, James Mccoy, Danda B Rawat, Brian Sadler, and Robert Amant. Recent advances in trustworthy explainable artificial intelligence: Status, challenges and perspectives. *IEEE Transactions on Artificial Intelligence*, 1(01):1–1, 2021.

[49] AG Rekha, Loganathan Ponnambalam, and Mohammed Shahid Abdulla. Predicting maritime groundings using support vector data description model. In *Proceedings of International Symposium on Computational Intelligence and Intelligent Systems*, pages 329–334. Springer, 2015.

[50] Marco Tulio Ribeiro, Sameer Singh, and Carlos Guestrin. "why should i trust you?" explaining the predictions of any classifier. In *Proceedings of the 22nd ACM SIGKDD International Conference on Knowledge Discovery and Data Mining*, pages 1135–1144. Association for Computing Machinery, 2016.

[51] Wenhao Shi and Bowen An. Port ship detection method based on multi-structure morphology. *Computer Systems & Applications*, 25(10):283–287, 2016.

[52] Karen Simonyan, Andrea Vedaldi, and Andrew Zisserman. Deep inside convolutional networks: Visualising image classification models and saliency maps. *arXiv preprint arXiv:1312.6034*, 2013.

[53] Binbing Song, Hiroko Itoh, and Yasumi Kawamura. Development of training method for vessel traffic service based on cognitive process. *Cognition, Technology & Work*, 24(2):351–369, 2022.

[54] Chien-Min Su, Ki-Yin Chang, and Chih-Yung Cheng. Fuzzy decision on optimal collision avoidance measures for ships in vessel traffic service. *Journal of Marine Science and Technology*, 20(1):5, 2012.

[55] Yongfeng Suo, Wenke Chen, Christophe Claramunt, and Shenhua Yang. A ship trajectory prediction framework based on a recurrent neural network. *Sensors*, 20(18):5133, 2020.

[56] Yanhua Tian and Jinbiao Chen. Vessel traffic flow prediction based on BP neural network. *Ship & Ocean Engineering*, 39(1):122–125, 2010.

[57] Erico Tjoa and Cuntai Guan. A survey on explainable artificial intelligence (XAI): Toward medical XAI. *IEEE Transactions on Neural Networks and Learning Systems*, 32(11):4793–4813, 2020.

[58] Manjunatha Veerappa, Mathias Anneken, Nadia Burkart, and Marco F Huber. Validation of XAI explanations for multivariate time series classification in the maritime domain. *Journal of Computational Science*, 58:101539, 2022.

[59] Erik Veitch and Ole Andreas Alsos. Human-centered explainable artificial intelligence for marine autonomous surface vehicles. *Journal of Marine Science and Engineering*, 9(11):1227, 2021.

[60] Dangli Wang, Yangran Meng, Shuzhe Chen, Cheng Xie, and Zhao Liu. A hybrid model for vessel traffic flow prediction based on wavelet and prophet. *Journal of Marine Science and Engineering*, 9(11):1231, 2021.

[61] Jingyao Wang and Naigong Yu. UTD-Yolov5: A real-time underwater targets detection method based on attention improved YOLOv5. *arXiv preprint arXiv:2207.00837*, 2022.

[62] Senjie Wang and Zhengwei He. A prediction model of vessel trajectory based on generative adversarial network. *The Journal of Navigation*, 74(5):1161–1171, 2021.

[63] Natalia Wawrzyniak, Tomasz Hyla, and Izabela Bodus-Olkowska. Vessel identification based on automatic hull inscriptions recognition. *Plos One*, 17(7):e0270575, 2022.

[64] Natalia Wawrzyniak and Andrzej Stateczny. Automatic watercraft recognition and identification on water areas covered by video monitoring as extension for sea and river traffic supervision systems. *Polish Maritime Research*, S 1:5–13, 2018.

[65] Zhe Xiao, Xiuju Fu, Liye Zhang, and Rick Siow Mong Goh. Traffic pattern mining and forecasting technologies in maritime traffic service networks: A comprehensive survey. *IEEE Transactions on Intelligent Transportation Systems*, 21(5):1796–1825, 2019.

[66] Zhe Xiao, Xiuju Fu, Liye Zhang, and Rick Siow Mong Goh. Traffic pattern mining and forecasting technologies in maritime traffic service networks: A comprehensive survey. *IEEE Transactions on Intelligent Transportation Systems*, 21(5):1796–1825, 2020.

[67] Zhe Xiao, Xiuju Fu, Liye Zhang, Wanbing Zhang, Ryan Wen Liu, Zhao Liu, and Rick Siow Mong Goh. Big data driven vessel trajectory and navigating state prediction with adaptive learning, motion modeling and particle filtering techniques. *IEEE Transactions on Intelligent Transportation Systems*, 23(4):3696–3709, 2022.

[68] Zhe Xiao, Xiuju Fu, Liangbin Zhao, Liye Zhang, Tze Kern Teo, Ning Li, Wanbing Zhang, and Zheng Qin. Next-generation vessel traffic services systems-from" passive" to" proactive". *IEEE Intelligent Transportation Systems Magazine*, early access, 2022. doi: 10.1109/MITS.2022.3144411.

[69] Zhe Xiao, Liye Zhang, Xiuju Fu, Wanbing Zhang, Joey Tianyi Zhou, and Rick Siow Mong Goh. Concurrent processing cluster design to empower simultaneous prediction for hundreds of vessels' trajectories in near real-time. *IEEE Transactions on Systems, Man, and Cybernetics: Systems*, 51(3):1830–1843, 2019.

[70] Yibing Xie, Nichakorn Pongsakornsathien, Alessandro Gardi, and Roberto Sabatini. Explanation of machine-learning solutions in air-traffic management. *Aerospace*, 8(8):224, 2021.

[71] Zhaoqing Xie and Qing Liu. LSTM networks for vessel traffic flow prediction in inland waterway. In *Proceedings of 2018 IEEE International Conference on Big Data and Smart Computing (BigComp)*, pages 418–425. IEEE, 2018.

[72] Ran Yan and Shuaian Wang. Ship detention prediction using anomaly detection in port state control: model and explanation. *Electronic Research Archive*, 30(10):3679–3691, 2022.

[73] Ying Yang and Xianqiao Chen. Research on the statistical method of ship flow based on deep learning and virtual detection line. In *Proceedings of 2019 IEEE 8th Joint International Information Technology and Artificial Intelligence Conference (ITAIC)*, pages 280–285. IEEE, 2019.

[74] Yanan You, Zezhong Li, Bohao Ran, Jingyi Cao, Sudi Lv, and Fang Liu. Broad area target search system for ship detection via deep convolutional neural network. *Remote Sensing*, 11(17):1965, 2019.

[75] Chunyu Zhang, Bingjie Liu, and Jianlan Ren. Automatic recognition method of moving target in ship transportation intelligent transportation system. *Ship Science and Technology*, 43(14):58–60, 2021.

[76] Xiaolong Zhang, Jiachun Zheng, Changchuan Lin, Xiaorui Hu, Qiaoping Chen, and Shuangshuang Zhu. Target association algorithm and simulation for radar and AIS based on BP network. *Journal of System Simulation*, 27(3):506, 2015.

[77] Ze-guo Zhang, Jian-chuan Yin, Ni-ni Wang, and Zi-gang Hui. Vessel traffic flow analysis and prediction by an improved PSO-BP mechanism based on AIS data. *Evolving Systems*, 10(3):397–407, 2019.

[78] Zhong-Xing ZHANG, Hong-Long LI, Guang-Qian ZHANG, Wen-Ping ZHU, Li-Yuan LIU, Jian LIU, and Nan-Jian WU. CCNet: a high-speed cascaded convolutional neural network for ship detection with multispectral images. *Journal of Infrared and Millimeter Waves*, 38(3), 2019.

[79] Yang Zhou, Winnie Daamen, Tiedo Vellinga, and Serge Hoogendoorn. Review of maritime traffic models from vessel behavior modeling perspective. *Transportation Research Part C: Emerging Technologies*, 105:323–345, 2019.

[80] Maede Zolanvari, Zebo Yang, Khaled Khan, Raj Jain, and Nader Meskin. Trust xai: Model-agnostic explanations for ai with a case study on iiot security. *IEEE Internet of Things Journal*, early access, 2021. doi: 10.1109/JIOT.2021.3122019.

CHAPTER 9

An Explainable Model for Detection and Recognition of Traffic Road Signs

Anass Barodi, Abdelkarim Zemmouri, Abderrahim Bajit, and Mohammed Benbrahim

Laboratory of Advanced Systems Engineering (ISA), National School of Applied Sciences, Ibn Tofail University, Kenitra, 14000, Morocco

Ahmed Tamtaoui

National Institute of Posts and Telecommunications (INPT-Rabat), SC Department, Mohammed V University, Rabat, 10000, Morocco

CONTENTS

DOI: 10.1201/9781003324140-9

THE MAIN OBJECTIVE of this work is to propose an efficient approach applicable to Intelligent Transportation Systems (ITS), particularly embedded vision systems, capable of detecting and tracking objects in a dynamic environment. Many previous works have been conducted in this field to design a powerful and robust ITS system taking into account the notion of optimization. However, in this work, we aim to develop an intelligent system based on Computer Vision (CV) and Artificial Intelligence (AI) that combines two distinct phases respectively; namely detection and recognition. To achieve this objective, the paper focuses on the detection and recognition of traffic road signs. The proposed approach is based on an algorithm named Easy Road Sign Detection and Recognition Algorithm (ERSDRA). To detect traffic signs, we relied on colors (red, blue) and shapes (circle, square, triangle, and rectangle), and for traffic sign recognition, we proposed a CNN architecture (Convolution Neural network). The design of this architecture was inspired by two fundamental architectures, namely LeNet-5 and AlexNet, in order to have an optimistic CNN model after Deep Learning (DL). This CNN model is employed for the recognition of each cropped traffic sign, while Grad-CAM is used as explainable AI techniques to interpret visually the results of the CNN model. Finally, after several tests carried out on our improved approach, for the detection module, the error rate range is between 0,001% and 0,025% for triangular and square shapes respectively, while the recognition module reached the best score of 98,47% in accuracy. Given the performance and interpretability offered by the proposed model, we can say that our model can be useful in the automotive field for road safety, especially in Advanced Driving Assistance Systems (ADAS).

9.1 INTRODUCTION

Road transport generates significant traffic [1], particularly in emerging countries. Unfortunately, this increase has consequences on roads. However, according to accidental experts, the human factor represents the first cause in 80% of accidents [2], because drivers can be unconcentrated which leads to a lack of vigilance on traffic roads. For example, not paying attention to a speed limit or not considering traffic signs creates risky situations that can expose drivers to penalties. This is why car manufacturers are increasingly integrating driver assistance systems into new vehicles [3], to assist the behavior of the driver in real traffic [4]. ITS systems have a great interest in road safety, by assisting and keeping the driver informed of potentially dangerous situations, or changing his behavior to avoid the consequences of an accident [5]. The types of these systems are called the Traffic Sign Recognition system (TSR) [6,7] one part of ADAS systems [8,9].

The TSR system assists the driver by informing him of the signs that exist on the road traveled [10,11]. It is usually based on an embedded camera ahead of the vehicle [12] whose objective is analyzing the video stream provided for recognizing the signs to reduce the risk of accidents. For this purpose, a TSR system can be represented in three stages: detection, recognition, and temporal tracking, specifically the traffic road signs [7]. This paper focuses on two important steps, the detection, and recognition of traffic signs, to identify their visual characteristics based on CV and AI. Using AI, requires two essential elements, the first one is the choice of an image dataset containing all types of road signs; for obtaining an efficient trained model in recognition [13]. The second one is the choice of a CNN architecture [14], which has the same philosophy as the human being for its robustness while extracting the characteristics of the object [15]. Finally, this architecture will be exploited in supervised DL to create a CNN model [16–18].

The proposal ERSDRA algorithm in the phase of detection exploits characteristics such: as shape, color, and pictograms, to locate the traffic signs in the images provided by the camera. This task can be done with some descriptors used during the detection as Canny edge, Green's theorem [19,20], arc length [21–23], Hough Transformation (HT) [24–26]. In the phase of recognition also called classification, the ERSDRA extracts the traffic signs features, these features can then be classified using proposal CNN architecture.

The paper is methodical according to the following schedule, Section 9.1 is the introduction, then Section 9.2 explains the general architecture of the proposed approach. Section 9.3 presents the methodology uses in our approach for the detection and recognition of traffic signs. Section 9.4, and Section 9.5 are devoted to the discussion of the results of traffic sign detection and recognition. Finally, Section 9.6 contains the conclusion.

9.2 THE PROPOSED APPROACH OVERVIEW

The heart of the ITS system is the detection and recognition of traffic signs, which are designed to meet the needs of driver safety and comfort [27], it can also be classified as an embedded vision system. This offers a relevant perception of the environment to develop a mode of driving, by recognizing the road signs present in the scene as shown in Figure 9.1. The first step locates the signs in the images based on visual characteristics, such as the color, shape, and center of the sign, to have a high accuracy in identifying the sign in real time. Indeed, the signs appear in several images, before disappearing from the camera's field of view, which allows tracking. The second step is to crop the images in real time, exactly the road signs, according to the dimensions

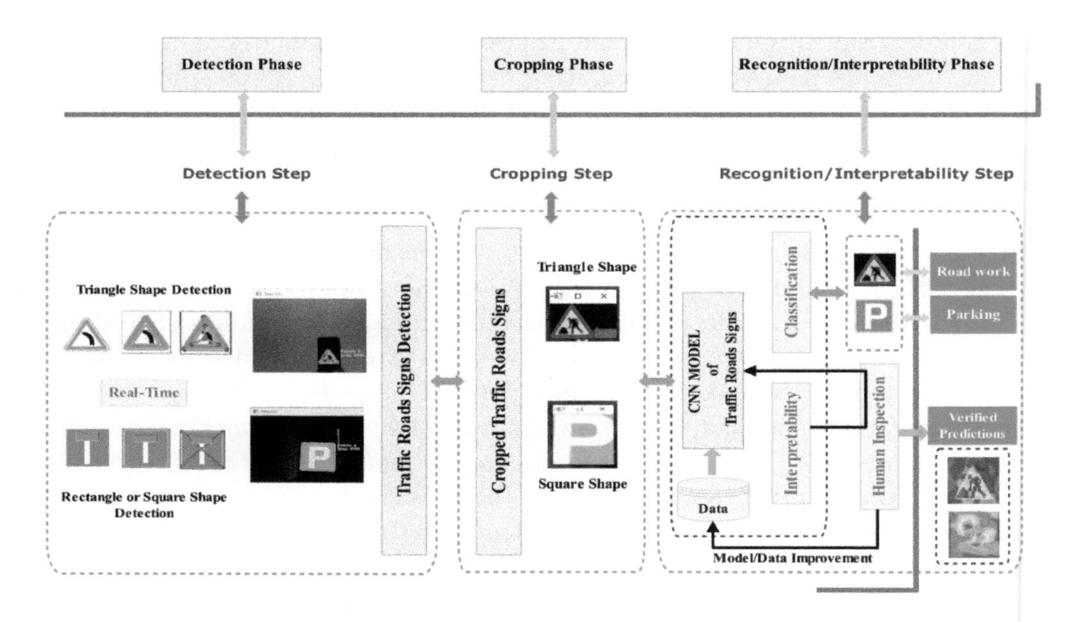

Figure 9.1 Diagram of proposed ITS system for traffic sign detection and recognition.

acceptable by the next phase. The final stage contains the main objective, which is the recognition intelligently based on the DL of the CNN architecture, the type of each sign detected according to its pictogram, and the recognized signs are signaled to the driver or can be transmitted [28, 29] to another ITS system by using V2X communication.

The objective of the paper is to allow a vehicle to recognize the road signs in a field of view (embedded cameras), knowing that it evolves in an urban environment, with very varied backgrounds and in the presence of several fixed and mobile objects: car, pedestrians, bicycle, trees, etc. In addition to these objects, there is a difference in luminance between places due to the shade, for example, climate change, and the time of recognition. To discern road signs in still images or video sequences acquired in real conditions, almost all CV algorithms focus on two steps [30]:

- Sign detection in the road scene.
- Recognition of the sign types that were detected in the detection step.

However, in real-life situations, detection is never perfect [31]. Indeed, the road scene can be quite cluttered and the signs can be obscured or very close to other objects, which does not facilitate the detection task. Detection must take into account different factors such as the diversity of sign categories (prohibition signs, danger signs, mandatory signs) [32]; the change in lighting, and the low contrast which make it difficult to distinguish between signs and background. Some authors based on the colors of signs [33], or others based only on the shape [34] process only a part of the sign contours. The most suitable detector is to combine the color, and shapes for such a case is the optimal corner detector.

This paper demonstrates the supervisory performance of DL for traffic sign recognition. When AI systems are trained on aspects of human visualization (HVS) [35–37] avoid many environmental dangers to reach road safety, such as cloudy or rainy weather that results in obscured view of panels. For DL, we exploit two different architectures of CNN, namely LeNet-5 [36,38], and AlexNet [39,40]. We improve the performance of the DL by creating a CNN model with several CV techniques (histogram equalization and normalization of image datasets) and AI techniques (batch size function and augmentation dataset), to have a robust model and effective [37,38]. This CNN pattern will be subject to traffic sign recognition. Then the evaluation of the effectiveness of all these models according to several criteria, using the confusion matrix, the classification report, and the Grad-CAM technique, with an in-depth analysis of the results obtained by the images taken in the urban world.

More recently, model interpretability has been one biggest trend in the transportation field, this topic has begun to accelerate. One of the reasons for this development: is the growing number of machine learning models in production. On the one hand, this translates into a growing number of end users who need to understand how models make their decisions, as shown in Figure 9.1. On the other hand, a growing number of machine learning developers need to understand why (or why not) a model works in a particular way. For example, it can be used to explainable AI to improve confidence in intelligent modules of autonomous vehicles [41], and can also be used to explain visually Interpretable Machine Learning [42].

9.3 METHODOLOGY

9.3.1 ERSDRA Algorithm

This approach consists of designing an efficacy algorithm for traffic road sign shapes. It thus offers a good gain in robustness against low contrast and poorly illuminated scenes, which is in line with our objective outlined in algorithm 1 called Easy Road Sign Detection and Recognition Algorithm (ERSDRA).

The circular shape (CS) was successfully treated in the detection phase [26] and also in the recognition phase [7]. This approach consists of designing an efficient algorithm for the shapes of road signs. It thus offers a good gain in robustness against low contrast and poorly lit scenes, which our work in this paper focuses on the detection and recognition of RSTS (Rectangle Square Triangle Shape) shapes [43].

9.3.2 Detection Phase

9.3.2.1 Color Extraction

The algorithm requires a fast detector of traffic sign shapes, as shown in Figure 9.2. First of all, in the image, we only need the color of the red or blue panel, which the best solution is to convert the color space from RGB to HSV. To let only the range of the red color or the blue one by a simple condition, that is defining a low filter and a high filter, for each color desired to identify it as indicated in Table 9.2 of

Algorithm 1: Easy Road Sign Detection and Recognition Algorithm (**ERSDRA**)

```
Begin
 Input: RIp: Road Images

Detection
   Input (RIp);
   new frame ← conversion color space (RIp);
   edge ← Canny edges (new frame);

   CS:
   circles ← Hough circles (edge);
   For circles to circles then
        new_1 ← CS draw contour (frame);
        CS sign detected ← CS crop (new_1);
   end;

   RSTS:
   contours ← find contours (edge);
   For contours do
        area ← contour area(contours);
        shape ← a * contour perimeter (contours);
        if shape = N then
             if area > area Min then
                  new_2 ← RSTS draw contour (frame);
                  RSTS sign detected ← RSTS crop (new_2);
             end
        end
   end

Recognition
   predict traffic road sign ← recognition (sign detected);
end
```

this paper [26]. This conversion had the advantages of reducing execution time and facilitate handling. After that, we apply the binarization by thresholding the gray level images, and the pixels that have a higher value will have the max value equal to 255 and those that have a lower value will have the min value equal to 0 or the opposite [44].

9.3.2.2 Contour Tracking

To begin with, most computer vision algorithms use the edge images obtained by a filter such as Sobel, prewet, Canny. Moreover, the Canny filter shows very good performance in traffic scenes [26,43], this one has the result to produce the sharpness of the contours, which will speed up the extraction of the contours. This is why we prefer to opt for the Canny filter. This function returns an image twice as big as the original one, since an outline is not on a pixel but between two pixels, and each

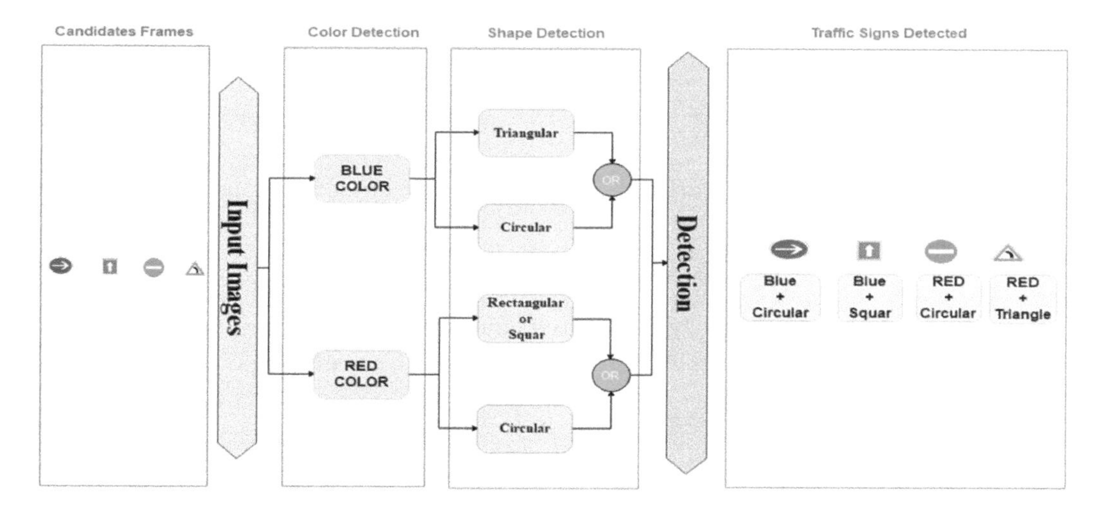

Figure 9.2 Detection steps traffic road signs.

pixel has a value that corresponds to the intensity of the outline. It can therefore be used to eliminate uninteresting contours much more effectively than with a simple Canny detector, this method is very fast which it requires few calculations at each step [7]. It has the advantage of detecting edge pixels in all directions; all pixels will be processed only once.

The contour function thus produced, makes it possible to detect the contours of an object from binary images, these contours can be classified into two categories:

- Those that encompass an object. it is called external; it is a silhouette of the object.
- Those that are integrated into an object are the inner contours.

After the detection of the first contour pixel, the nature of the transition determines its type:

- If this transition is of a background type (0-1), it is an external type contour.
- Otherwise (1-0), the contour is of internal type.

9.3.2.3 *Area Calculation*

Many algorithms have been developed to accelerate the computation of closed contour surfaces. Some of them work for grayscale images, some for binary images, and some for parallel or optical computation. In algorithm 1, we use a discrete version of Green's theorem that applies to binary images [45–48]. This theorem is a simple version of Stokes' theorem [49]. Suppose that R is a region of the XY-plane bounded by a closed curve (sign contour), piecewise smooth C and that the functions $K(x, y)$ and $H(x, y)$ are continuous and also have continuous partial derivatives of the first order in R.

Green's theorem relates a linear integral around a simple closed plane curve C to an ordinary double integral.

$$\vec{C} = K(x,y)\vec{a} + H(x,y)\vec{b} \tag{9.1}$$

If K and H have continuous partial derivatives over an open region including R, then:

$$\int\int(\frac{\partial H}{\partial x} + \frac{\partial K}{\partial y})dxdy = \oint_c Kdx + Hdy \tag{9.2}$$

$\oint_c$ indicate line the integral along C oriented counterclockwise. Using Green's theorem allows us to calculate the area A bounded by a closed parametric curve.

$$A = \int\int_R dA \tag{9.3}$$

With condition

$$\frac{\partial H}{\partial x} - \frac{\partial K}{\partial y} = 1 \tag{9.4}$$

Then the area can be calculated

$$A = \oint_C (Kdx + Hdy) \tag{9.5}$$

For discrete region can be computed the following way:

$$\sum_c (K\Delta x + H\Delta y) \approx \sum_R^m \sum f\Delta x\Delta y \tag{9.6}$$

Denote

$$f = \frac{\partial H}{\partial x} - \frac{\partial K}{\partial y}$$

Green's theorem is applied to calculate the area of road signs in images that have regular shapes. These include triangular, square, and rectangular signs.

9.3.2.4 Contour Perimeter and Approximation of Vertices

After the detection of Canny edges in the image to avoid any noise. Then we use the arc length function to calculate the contour perimeter or the length of a curve. This method has many applications in computer vision in digital images. This function is applied to road sign shapes (triangle, rectangle, and square) [50–52]. After that, the objective is to determine the vertices of the shape detected. Many methods founded the Teh-Chin algorithm, Ramer [29,53], Rosenfeld-Johnson's algorithm, and Douglas-Peucker (DP) approximation [54,55].

The main idea of DP is to move the point with the largest error to the simplified. This operation is repeated until no point has an error that exceeds the given threshold. Another proposed heuristic algorithm maintains both the skeleton of the shape and the semantic meanings of a trajectory [56]. This algorithm outperforms DP under different compression rates, but is not open source and difficult to reproduce [57], and DP is much faster in the identification of shapes like triangles, rectangles in Figure 9.3 from this paper [58].

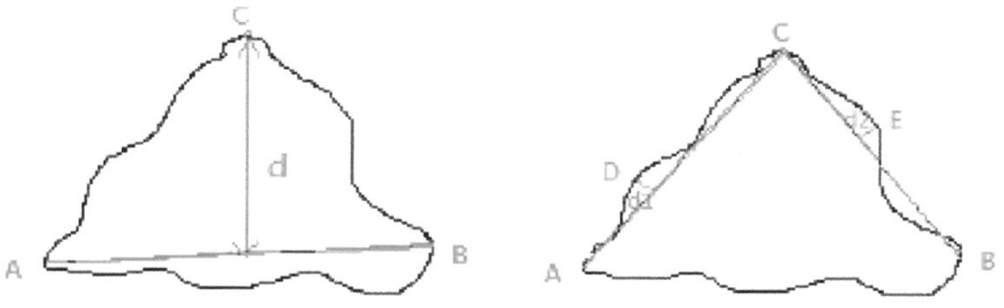

Figure 9.3 Detection steps traffic road signs.

9.3.2.5 *Increase Detection Accuracy*

Hough Transformation: The technique applied to verify the detected traffic signs is the Hough transformation (HT) [59]. The principle of HT is very simple, it is used for the detection of lines which would possibly be present in an analyzed image provided that they are part of the contours. So, the first thing HT does identify all the edge points that construct the geometric shape, using local gradients measured between the pixel values around each point in the image by the Canny filter. These points have the highest gradients in their vicinity, either globally for fixed thresholding, or around the point for dynamic thresholding. In order to leave only the points most likely to belong to the contours. Each point of the identified contours (x, y) will then allow a projection in the HT plane of polar coordinates (ρ, θ) (ρ: radial coordinates, θ: angular coordinates) for all the lines passing through this point [7].

Traffic Sign Center: However, all the methods applied in the previous sections were found to be effective. Assuming that the edges of the signs are well detected, it was now necessary to increase the degree of accuracy, for the second time, before moving to the main step of traffic sign recognition. This step is based on the road sign's center of gravity which makes signs be detected in any environment. The object's gravity center corresponds to a point that would physically hold the object in equilibrium. For this aim, we use the "moments" class that allows the calculation of the order type 3 polygon moments. The moments are essentially based on the object contour vector points, using a function that returns the moment descriptors in a structure that contains the elements sought which are based on the equations 9.7, 9.8, 9.9, and 9.10.

- Spatial Moments

$$M_{ij} = \sum_{x}\sum_{y} x^i y^j I(x, y) \tag{9.7}$$

- Central Moments

$$\mu_{pq} = \sum_{x}\sum_{y} (x - \bar{x})^p (y - \bar{y})^q I(x, y) \tag{9.8}$$

$$\bar{x} = \frac{M_{10}}{M_{00}} \quad \bar{y} = \frac{M_{01}}{M_{00}} \tag{9.9}$$

- Normalized Central Moments

$$n_{ij} = \frac{\mu_{ij}}{\mu_{00}^{1+\frac{i+j}{2}}} \tag{9.10}$$

Thus, the value m_{00} represents the area of an object and the gravity center is calculated as follows: $x = \frac{m_{10}}{m_{00}}$ and $y = \frac{m_{01}}{m_{00}}$, where the values of each parameter x and y represent the object's midpoint.

9.3.3 Recognition Phase

The central point of a convolution neural network is to build an AI system. This one is composed of a set of distributed neurons that are connected each to the other through several layers, to generally solve various statistical and especially classification problems.

9.3.3.1 Convolutional Neural Networks Description

CNNs are the standard form of neural network architecture applied in many fields [60, 61], most layers we can find in CNN 's architecture are convolution layers, pooling layers, fully connected layers, and classification layers. The convolution is very important in CNNs for feature extraction that all neurons detect exactly the same feature [62, 63], and by multiplying the receptive fields, it becomes possible to detect features independently of their position in the visual field. Weight sharing also significantly reduces the number of parameters to be learned, and thus the memory requirements for network operation, which plays a very important role in the speed of the training [64]. The mathematical formula 9.11 describes these operations [65].

$$X_i^L = AF(\sum_{j \in I_j} X_j^{L-1}.W_i^L + b_{ias,j}) \tag{9.11}$$

where $AF()$ is the activation function, X_i^L is the output of the L layer, $X_j^{L-1}$is the feature map of layer $L-1$ multiped by weight matrix W_i^L, $b_{ias,j}$ represent bias value, I_i is the input feature set.

A CNN architecture is formed by a stack of independent processing layers, among these layers, we have the pooling layer that reduces the spatial size of an intermediate image, thereby reducing the number of parameters and computation in the network. It is therefore common practice to periodically insert a pooling layer between two successive convoluted layers of a CNN architecture to control overfitting, it can be represented by a mathematical formula 9.12. The pooling layer works independently on each depth slice of the entry and only resizes it on the surface.

$$pol_j^k = Max\{C_{j:j+r-1}^k\} \tag{9.12}$$

pol_j^k corresponds to the maximum value in the pool region, r represents the size and c^k is the input. It is possible to improve the CNN efficiency of the processing, by interposing between the processing layers a layer which will operate, a mathematical function called activation function 9.13 on the output signals.

$$F(x) = max(0, x) \tag{9.13}$$

After several layers of convolutions and max-pooling, the high-level reasoning in the neural network is done via fully connected layers, the neurons of a fully connected layer are connected to all outputs of the previous layer. Their activation functions can therefore be computed by matrix multiplication followed by a polarization shift. The mathematical formula 9.14 describes these operations.

$$FC^L = \alpha(W^L.FC^{L-1} + b^L) \tag{9.14}$$

$\alpha()$ is the non-linear activation function, b^L represent the bias and W^L is the weight non-linear activation function, $FC^L$is the output characteristics of L-the fully connected layer. Moreover, the FC concept creates an exponential problem of memory, called overfitting, slowing down the processing of information, to prevent this, the dropout layer is used to randomly turn neurons as well as peripheral neurons.

Normally the last layer of the CNN network, various functions adapted to different tasks can be used, the Softmax function is a mathematical function, an N-dimensional vector z of arbitrary real values to an N-dimensional vector $\gamma(z)$ of real values in the range $(0, 1)$ of which the sum is 1, which describe mathematically 9.15.

$$p(z)_i = \frac{e^{z_i}}{\sum_{i=1}^{N_c} e^{z_i}} \quad i \in \{1, ..., N\} \tag{9.15}$$

The Softmax used in the last layer computes the final probabilities for multi-class classification [66]. A loss function is also associated with the final layer to calculate the classification error, this is usually the cross-entropy accuracy [67]. We calculate the weights of each layer by the gradient backpropagation, which progressively computes the parameters to minimize the regularized loss function. Starting from the first layer to the last layer, the optimization is done with an Adam optimizer [27, 28].

9.3.3.2 *Interpretability of Classification Models*

Explainable AI visualization techniques help create explainable models by highlighting image pixels that impact models' decisions, helping to verify whether the model made the prediction based on the correct models of the image. We use the Grad-CAM (Gradient-weighted Class Activation Mapping), one of the most used XAI techniques (Explainable Artificial Intelligence) [70], which applied to an already trained model generates heatmaps for visual interpretation. To obtain the class discriminative localization map Grad-CAM [71] as shown in equation 9.16.

$$L^k_{Grad-CAM} \in \mathbb{R}^{u \times v} \tag{9.16}$$

u is width and v is the height of any class k, the gradient computes the score for class k, y^L (before the Softmax), with respect to feature maps M^c of a convolutional layer,

TABLE 9.1 LeNet-5 Architecture Suitable for Traffic Sign Recognition

Layer	Output Shape	Param
Input	32 × 32 × 1	780
Convolution	(None, 28, 28, 30)	0
Max Pooling	(None, 14, 14, 30)	4065
Activation(ReLU)		
Convolution	(None, 12, 12, 15)	0
Max Pooling	(None, 6, 6, 15)	0
Activation(ReLU)		
Flatten	(None, 540)	0
Dense	(None, 500)	270,500
Dropout	(None, 500)	0
Activation(Softmax)	43	21,543
Total params: 296,888		
Trainable params: 296,888		
Non-trainable params: 0		

i.e. $\frac{\partial y^L}{\partial M^c}$. These gradients flowing back are global average-pooled to obtain the neuron importance weights α_c^L. it performs a weighted combination of forwarding activation maps and follow by a ReLU as shown in equation 9.17, in generally the Grad-CAM show fine-grained importance like pixel-space gradient visualization methods.

$$L^k_{Grad-CAM} = ReLU\left(\sum_c \alpha_c^L M^c\right) \tag{9.17}$$

9.3.3.3 The Adapted LeNet-5 Network Architecture

For our study we have used the LeNet-5 architecture is among the first CNN, developed by LeCun researchers [72]. It has been used for the classification and recognition of handwritten characters in banks and post offices. Table 9.1 illustrates the LeNet-5 architecture description, it includes convolution layers and subsampling layers. These layers are fully connected subsequently from the input layer to output one, and the role of each layer and constituent parameters are explained in detail below.

- **The first layer**: This first layer is applied to a gray level image of size 32 × 32, with six convolution filters of size 5 × 5 and a step (stride) equal to 1. The dimension of the image varies from 32 × 32 × 1 to 28 × 28 × 30.
- **The second layer**: The LeNet-5 applies a subsampling layer with a filter of size 2 × 2 and a step of two. The resulting image dimensions will be reduced to 14 × 14 × 30.
- **The third layer**: is a second convolution layer with 15 feature maps (filters) of size 5 × 5 and a step of 1, the dimensions of the image become 12 × 12 × 15.

- **The fourth layer**: The fourth layer is again a subsampling layer with a filter of size 2×2 and a step of 2. This layer is identical to the second one, the difference is that it has 15 feature maps so that the output is reduced to $6 \times 6 \times 15$.
- **The fifth layer**: The fifth layer is a fully connected convolutional layer with 120 feature maps of size 1×1. Each of the 120 units is connected to all 540 nodes ($6 \times 6 \times 15$) in the fourth layer.
- **The sixth layer**: The sixth layer is applied to connect the outputs of the previous layer having 540 units to obtain 500 units.
- **The last layer**: Finally, the last layer is using the Softmax probability function that output layer 43 possible values which correspond to the classes numbered from 0 to 42.

9.3.3.4 The Adapted AlexNet Network Architecture

AlexNet is a popularized convolutional neural network in computer vision, developed by Geoff Hinton, Ilya Sutskever, and Alex Krizhevsky researchers [73]. This CNN was submitted to the ImageNet base challenge in 2012 and outperformed its competitors [74,75]. This network has a similar concept called the LeNet, it was larger and deeper, it consists of additional convolutional layers previously stacked in LeNet architecture, and each convolutional layer is always followed by a pooling layer.

One of the most commonly used approaches is called Batch Normalization (BN) layer. In CNN, the optimal solution is based on the manipulation of batches (or mini-batches), because the parallelism of the calculations allows for saving time. This layer computes the mean and variance over a batch using the distribution of the sum of the neurons at the input and then repeats it for each training batch. The mathematical operation of BN is described as follows.

$$\tilde{y}^{(l)} = \frac{y^{(l)} - \mathbb{E}[y^{(l)}]}{\sqrt{Var[y^{(l)}]}} \tag{9.18}$$

$$A^{(l)} = \alpha^{(l)} y^{(l)} + \gamma^{(l)} \tag{9.19}$$

$\alpha^{(l)}$ and $\gamma^{(l)}$ represent the parameters to be learned, $y^{(l)}$ and $\gamma^{(l)}$ are the l-th activation of input and output, respectively, $Var[x]$ and $\mathbb{E}[x]$ variance and expectation.

The structure of the AlexNet model is shown in Table 9.2, BN layers are added based on the AlexNet model. The input image for the AlexNet architecture is (32,32,3). This CNN consists of different convolution layers used to extract the image features. The first layer contains 96 filters of size $8 \times 8 \times 3$ and a step of two, and the second one contains 256 filters of size 4×4 and a step of two. These first convolution layers are followed by a similar subsampling window layer of size 2×2 and a step of two. However, the next three convolution layers (3rd, 4th, and 5th layers) are directly connected followed by an under-sampling layer. The output contains two types of fully connected layers and a Softmax probability function layer with 43 classification elements.

TABLE 9.2 AlexNet Architecture Suitable for Traffic Sign Recognition

Layer	Output Shape	Param
Input	32 × 32 × 3	
Convolution	(None, 8, 8, 96)	34,944
Batch normalization	(None, 8, 8, 96)	384
Activation (ReLU)	(None, 8, 8, 96)	0
Max Pooling	(None, 4, 4, 96)	0
Convolution	(None, 4, 4, 256)	614,656
Batch normalization	(None, 4, 4, 256)	1024
Activation (ReLU)	(None, 2, 2, 256)	0
Max Pooling	(None, 4, 4, 256)	0
Convolution	(None, 2, 2, 384)	885,120
Batch normalization	(None, 2, 2, 384)	1536
Activation (ReLU)	(None, 2, 2, 384)	0
Convolution	(None, 2, 2, 384)	1,327,488
Batch normalization	(None, 2, 2, 384)	1536
Activation (ReLU)	(None, 2, 2, 384)	0
Convolution	(None, 2, 2, 256)	884,992
Batch normalization	(None, 2, 2, 256)	1024
Activation (ReLU)	(None, 2, 2, 256)	0
Max Pooling	(None, 1, 1, 256)	0
Flatten	(None, 256)	0
Dense	(None, 4096)	1,052,672
Batch normalization	(None, 4096)	16,384
Activation	(None, 4096)	0
Dropout	(None, 4096)	0
Dense	(None, 4096)	16,781,312
Batch normalization	(None, 4096)	16,384
Activation	(None, 4096)	0
Dropout	(None, 4096)	0
Dense	(None, 1000)	4,097,000
Batch normalization	(None, 1000)	4000
Activation	(None, 1000)	0
Dropout	(None, 1000)	0
Dense	(None, 43)	43,043
Batch normalization	(None, 43)	172
Activation	(None, 43)	0
Total params: 25,763,671		
Trainable params:25,742,449		
Non-trainable params: 21,222		

9.3.3.5 Proposed CNN Architecture

The paper has a great challenge in ITS systems, it is the implementation of the ERSDRA algorithm to effectively recognize the detected traffic signs. For this aim,

TABLE 9.3 The Proposed CNN Architecture for Traffic Sign Recognition

	Layer	Output Shape	Param
BLOCK 1	Conv2D	(None, 32, 32, 8)	608
	ReLU Activation	(None, 32, 32, 8)	0
	Batch normalization	(None, 32, 32, 8)	32
	Max pooling	(None, 16, 16, 8)	0
BLOCK 2	Conv2D	(None, 16, 16, 16)	1168
	ReLU Activation	(None, 16, 16, 16)	0
	Batch normalization	(None, 16, 16, 16)	64
	Conv2D	(None, 16, 16, 16)	2320
	ReLU Activation	(None, 16, 16, 16)	0
	Batch normalization	(None, 16, 16, 16)	64
	Max Pooling	(None, 8, 8, 16)	0
BLOCK 3	Conv2D	(None, 8, 8, 32)	4640
	ReLU Activation	(None, 8, 8, 32)	0
	Batch normalization	(None, 8, 8, 32)	128
	Conv2D	(None, 8, 8, 32)	9248
	ReLU Activation	(None, 8, 8, 32)	0
	Batch normalization	(None, 8, 8, 32)	128
	Max Pooling	(None, 4, 4, 32)	0
BLOCK 4	Flatten	(None, 512)	0
	Dense	(None, 500)	256,500
	Activation	(None, 500)	0
	Batch normalization	(None, 500)	2000
	Dropout	(None, 500)	0
	Dense	(None, 128)	64,128
	Activation	(None, 128)	0
	Batch normalization	(None, 128)	512
	Dropout	(None, 128)	0
	Dense	(None, 43)	5547
	Softmax Activation	(None, 43)	0
Total params: 347,087			
Trainable params: 345,623			
Non-trainable params: 1,464			

we focus our interest on the CNN architectures, we relied on two different CNN architectures, namely LeNet-5 and AlexNet architecture, both of which will be trained based on the GTSRB dataset. For the DL results to make sense, we combine the best performances of the two architectures detailed before. Aiming to obtain an optimized CNN model to be submitted for road sign recognition in automotive systems, the input images for the proposed CNN architecture is (32,32,3). In Table 9.3, we detail the feature map extraction operation experienced in our CNN version. Feature maps

TABLE 9.4 Popular Traffic Road Signs Datasets

Dataset	Train	Test	Validation
GTSRB [76]	34799	12630	4410
DITS [77]	7144	1159	1214
BTSC [78, 79]	4575	2520	—
Rmastif [80]	4044	1784	—
ETS [32]	60546	20442	—

refer to a set of entities created, by applying the same feature extractor to different locations of the input map in a sliding window fixture.

First, we start with the first block that contains a convolution layer, a ReLU activation function, a normalization layer, and finally the MaxPooling layer of the size (2×2) to reduce the dimensionality of the volume. In this block, we implement the convolution layer filter of size (5×5) to distinguish the color and shape of each road sign. Then we continue with the second block which uses the same layers as the first block except for the size of the convolution filter is changed to (3×3) which the third block contains all layers and values of the second block. For the number of filters used in each layer was different from one block to another respectively. For the first we have 8 filters, the second one contains 16 filters, and the third one comprises 32 filters. We increase the number of filters to improve object extraction characteristics and filter precision. The last block of the network is essentially consisted of two fully connected flatten layers and a Softmax classifier. Dropout layers are applied as a form of regularization that aims to prevent overfitting.

9.3.3.6 *Datasets Used*

We used existing public datasets that would allow us to compare our approach to the literature. Therefore, we evaluated the ERSDRA algorithm in the recognition phase, on four new datasets, that have been acquired in the countries concerned: Germany, Belgium, and French. The GTSRB dataset is the primary source of traffic signs and most researchers are limited by this German data. Road sign images vary in height, width, and ratio which is very important, it should be noted that the minimum size of an image in these different datasets corresponds to the dimension (32×32). In addition, some images were acquired under degraded weather conditions, which makes it more accurate to train the CNN architectures in different situations, so Table 9.4 represents a summary of the most popular datasets.

9.4 EXPERIMENTATION AND DISCUSSION RESULTS

9.4.1 Work Environment

The hardware used for the experiments of this approach is composed of a laptop ASUSTek computer core i7-7500 CPU@2.70GHz CPU, 8 Go RAM, 64-bit operating

TABLE 9.5 Road Sign Detection Process

Image	Contour	Shapes	Perimeter Precision (a)	Identification	Perimeter Precision (a)	Vertices Number (N)	Lines Detection using (HT)	Final Image
		Triangle Shapes	0.02		0.1	3		
		Square Shapes				4		

systems, and a digital camera. All deep learning is implemented in GPU processors 920M Nvidia. We employ OpenCV, TensorFlow, and Keras Libraries, for software, that experiments were performed on well-known datasets and on a real environment.

9.4.2 Detection Phase for Traffic Road Signs

Table 9.5 represents the main detection results of the ERSDRA algorithm by exploiting the visual features. The detector applies the corresponding components to traffic signs in red or blue, to search for triangular, square, and rectangular edges. The ERSDRA uses of the OpenCV library is oriented to real-time applications, in the research world as well as in industry, and also optimistic to the identification of objects in images. In a column (1), to distinguish these road signs from other objects in the scene, the HSV space is used because of its low sensitivity to changes in brightness. Relying on the properties related to each color as by defining lower and upper filters of [H, S, V] respectively, for red color [136, 87, 111] and [180, 255, 255], finally for blue color [99, 115, 150] and [110, 255, 255]. After that, we convert the result images from HSV space to gray level so that the following operation applies the Canny edge detection. In this sense, we have opted for a method based on the contours, to detect regions of interest in the image, which involves the elimination of that which requires too much processing. This detector consists of a convolution kernel that provides a maximum response, at the corner in the existence of Gaussian noise smooth it out, the results of this operation, as shown in the column (2).

The contours of these road signs' edges are then highlighted. By calculating the perimeter of the signs that will be multiplied by the accuracy (a), as indicated in the ERSDRA algorithm. The accuracy value of 2% in the column (5), did not exactly identify the triangular shape in the image in a column (6), but the change to 10%, identified the three corners of the object exactly in their relative positions, as shown in the column (9). In contrast, for rectangular or square, no change for square shapes in the identification in columns (5) and (9) of Table 9.5. This calculation of the signs perimeters is done, to have the vertices numbers (N) of each shape detected, precisely the triangle shape which will have 3 vertices and 4 vertices for the square or rectangular shape, as indicated in the column (7). To check the first condition of the ERSDRA algorithm before checking the areas of the shapes. For the verification of the area values, we wanted to verify it to tests in real time shown in the next section.

TABLE 9.6 Detection Shapes Road Signs

Shapes	Shapes Vertices		Shapes Center
Triangle Detected		Precicion	
Square Detected			

The ERSDRA algorithm, in the detection phase, relies on the edges of triangular and square or rectangular signs. The color condition was well validated to identify the road signs, but the related components are not verified sometimes for rectangular or square shapes. Unlike triangular shapes, rectangular shapes do not have well-marked edges. As shown in the column (2) of Table 9.5, the edge of the parking sign is not very sharp. Therefore, the detection of segments is less effective.

Hence, the detection ERSDRA algorithm requires a verification step before drawing the contour of the detected sign, several techniques excite, but must respect the following conditions:

- Find the parallel segments and create the quadrilateral.
- Find the perpendicular segments and create the quadrilateral.

The HT plays on a narrow range of angles around 0° and 60° for the triangular shape and for square or rectangular shapes around 0° and 90°. HT regions are applied to the image segmented by Canny; to work separately on each line detected by the segmentation. Then we will group all the lines found, in order not to complicate the calculation of the vertices for the triangular and rectangular panels which are already discriminating, as shown in the column (8) in Table 9.5. Due to their relative locations in the region of interest. Finally, if the triangle or the square respects the geometric constraints, the corners of the panel are detected around the points of intersection. A tolerance for the effects of the rotation of the signs is introduced at this stage. Table 9.6 shows the high accuracy of detecting the centers of the traffic signs on the images, according to the contour moment equations 9.7, 9.8, 9.9, and 9.10, the center of the sign either triangle or square, it is determined successfully.

9.4.2.1 Test in a Real Environment

Table 9.7 illustrates some cases of traffic sign detection. The contours of the signs are well detected as shown in the column (1), also the vertices of the detected signs are successfully counted, which is three for the triangular shape and four for the rectangular one. The shape identification center, for all signs away from the camera. The example is demonstrated by the triangular shape column (3). It can be concluded

TABLE 9.7 Detection of Traffic Signs in a Real Environment

Contour	Shapes	Detection Precision
	Square Detected 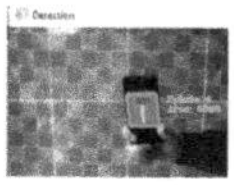 	
	Triangle Detected  	

TABLE 9.8 Area Calculation of Traffic Road Signs Shape

Type of Noise	Shapes	The Reference Area of Shapes (in pixels2) [81]	ACGT [81]	SAGT [81]	ERSDRA
Without any noise	Triangle	5380	5373.65	5379.77	5379.90
	Square	5846	5772.00	5810.79	5844.50
The Percentage Error (PE)		Triangle	0.118%	0.004%	0.001%
		Square	1,260%	0,600%	0,025%

that the approach is robust and aims to make it much more effective with higher real-time accuracy, one can determine the intersection of the coordinates of the vertices points, as shown in the column(4).

To show the effectiveness of using Green's theorem in the area calculation of road signs. We have tried to make comparisons with these methods ACGT and SAGT applied to images in this paper [81]. The idea is to determine if the ERSDRA algorithm can work properly to calculate the area of shapes. The reference or theoretical values were requested for the regular shapes (square and triangle) according to this paper [81].

In order to evaluate the performance of the proposed detection method by the percentage error (PE) formula 9.20, where A_{cal} is calculated area and A_{ref} The theoretical area.

$$PE = \frac{A_{ref} - A_{cal}}{A_{ref}} \tag{9.20}$$

Table 9.8 shows the area calculation of the noise-free shapes. The percentage error range for ERSDRA after several tests, it is between 0.001% and 0.025% for triangular and square shapes respectively. SAGT and ACGT are applied on images [81], but our ERSDRA algorithm is applied in real time, using images in front of a digital camera. The experimental results show that ERSDRA is able to compute the area accurately, as shown in Table 9.7, and that the vertices of the segmented shapes are well profiled.

In the last step, we first limit the candidate panel area from the received video frames that basically are centered and well aligned if their stand is facing the camera and then we crop the detected panels as shown in Table 9.9. These operations are

TABLE 9.9 Detection Shapes Road Signs

Shapes	Shapes Vertices		Cropped Traffic Signs
Triangle Detected	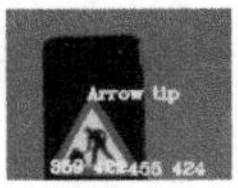	Cropping panels	
Square Detected			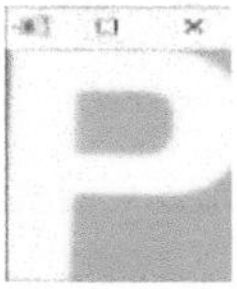

TABLE 9.10 The Hyperparameters for Deep Learning

Optimizer	Activation function	Learning rate	Batch size	Probability function	Loss Function	Processor
ADAM	ReLU	1e-3	64	Softmax	Cross-Entropy	GPU Nvidia Geforce 920M

experienced according to the input of the recognition phase that initially admits images of (32×32) size dimensions.

9.4.3 Recognition Phase

The AI system aims to calculate the class probability scores against the input of the CNN, this is handled by its layers whose data calculation is based on linear combination functions and non-linear transformation functions as shown in Table 9.10 and each function is detailed in the previous section (Section 9.3.2).

9.4.3.1 Preparation of the CNN Recognition Model

In Figure 9.4 we summarize the result obtained by the LeNet-5 model, AlexNet model, and proposed CNN model, applied to the GTSRB dataset. The figures show that the models have the same performance with respect GTSRB dataset in terms of robustness, accuracy, and loss. Note that there is growth during the epochs but their values are different for errors, it stabilizes after epoch number 20. For the same dataset, we have carried out the training only in 150 epochs, because we get the best results at this point, if we add more epochs the model falls into overfitting. Also, Figure 9.3 shows the training process results, which are better when we have more data in the training stage.

The first way to evaluate the proposed classifier is to compare the observed values of the dependent variables to the predicted values provided by the training model. The preferred tool is the confusion matrix which is presented in the form a contingency

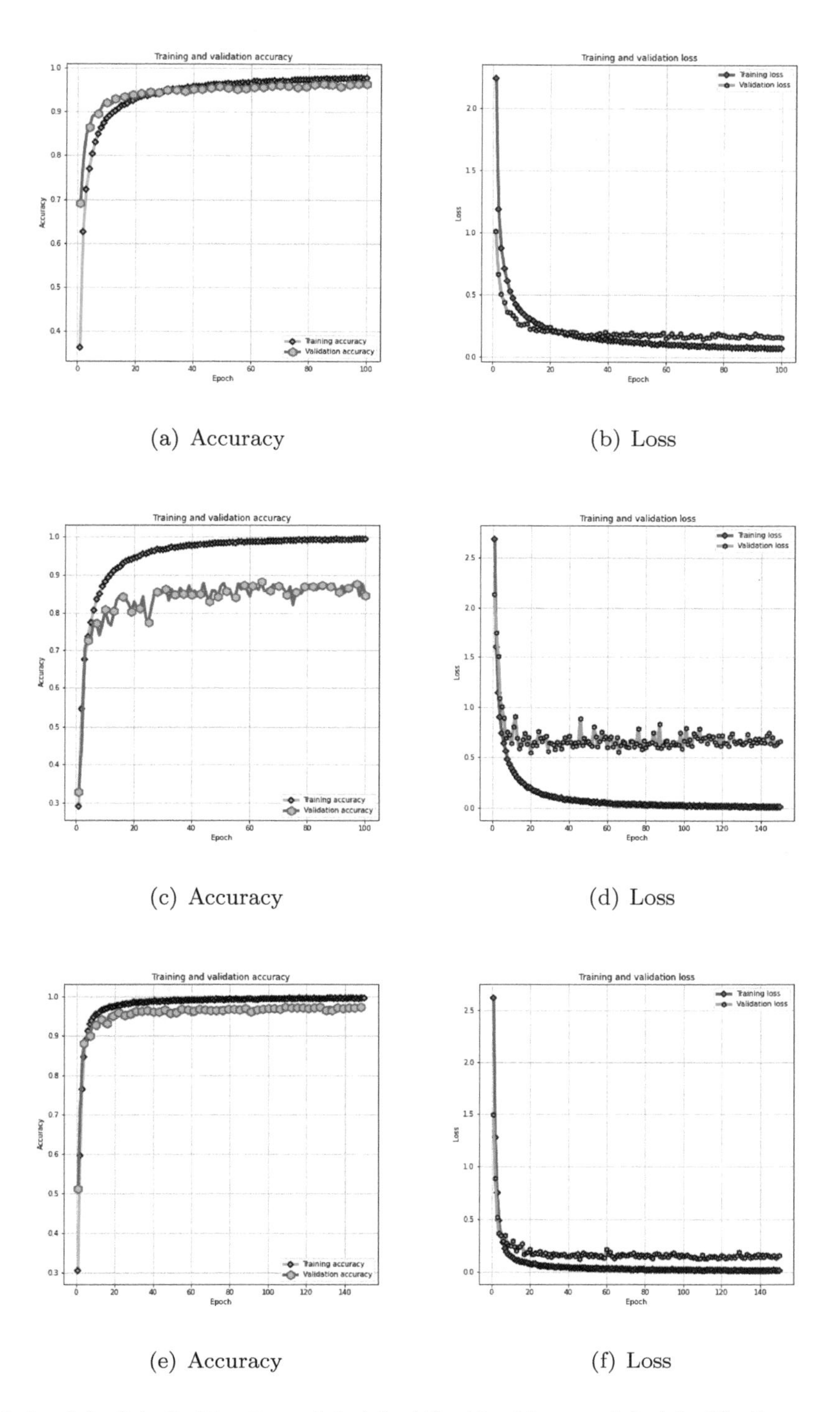

(a) Accuracy (b) Loss

(c) Accuracy (d) Loss

(e) Accuracy (f) Loss

Figure 9.4 (a), (b): LeNet-5 model; (c), (d): AlexNet model; (e), (f): Proposed CNN model.

TABLE 9.11 Performance Prediction for CNN Models using the GTSRB Dataset

Dataset	Architecture	LR	BS	Epochs	Test Loss (%)	Test Accuracy (%)
GTSRB	LeNet-5	1e-3	64	50	16.63	95.64
				100	16.13	95.88
				150	15.53	96.77
	AlexNet			50	63.78	85.58
				100	74.06	84.60
				150	66.20	87.03
	Proposed CNN			50	16.81	96.19
				100	13.99	96.87
				150	15.43	97.15

table that compares the obtained classes (columns) and the desired classes (rows) for the sample. On the main diagonal, we find the well-classified values, outside the diagonal we see the badly classified elements. Table 9.12 contains the confusion matrices that show the percentage of patients of each classified class. When analyzing these matrices, one can observe that the LetNet-5 best performance occurs for 150 epochs, and for AlexNet best performance occurs for 100 epochs.

After the analysis of the obtained results, we notice the following remarks, according to Table 9.11. The learning accuracy of the test increases with the number of epochs, this reflects that at each epoch the model learns much more information. If the accuracy is decreased, then we will need more information to make our model learn, and therefore we must increase the number of epochs and vice versa. Similarly, the learning error of the test decreases with the number of epochs. One can notice that in LeNet-5 architecture the accuracy value reaches a better score of 96.77% than AlexNet. According to the results displayed in Table 9.11, we note that the proposed CNN network offers better results. There are many reasons to approve this. The first one is the homogeneity of the network layers that we experienced managing various combinations aiming to improve the CNN architecture performances. As a result, we find that the CNN model outperforms the LeNet-5 and AlexNet models, especially for small data amounts. In fact, we can observe in Table 9.11 that our proposed CNN model gives reduced training errors than others with 50 epochs, especially for the GTSRB dataset with the best accuracy score is 97.15%.

The confusion matrix allows us to evaluate the performance of our CNN model, as it reflects the True Negative, True Positive, False Negative, and False Positive metrics. Table 9.12 closely illustrates the position of these metrics for each class, the proposed CNN model correctly classified the images of traffic road signs. As result, we observe the figures in Table 9.12, that the LeNet-5 model, AlexNet model, and proposed CNN model are robust with a high degree of accuracy. In a conclusion, if we have a large dataset, we will certainly ensure the DL of the proposed CNN architecture, which for sure will improve performance.

TABLE 9.12 Confusion Matrix of CNN Models

AlexNet	LeNet-5	Proposed CNN Model
50 epochs (*)		
100 epochs (*)		
150 epochs (*)		

* To clearly visualize the names of the classes of road signs, please zoom in on the figures.

9.4.3.2 *Results of the Proposed CNN Architecture on Different Datasets*

In Table 9.13, we represent the test performance values for each dataset, using the proposed CNN architecture. One can observe the influence of the datasets volume after training the CNN model from 100 and 150 epochs for the five datasets. It gives very remarkable results in the ETS dataset. In terms of validation error is equal to 06.60%, while it is equal to 15.43% for the GTSRB dataset. Similarly, this is evident in the values of the other metrics, in terms of test validation, where the best score

TABLE 9.13 Performance Evaluation of Proposed CNN Model on Different Datasets

Dataset	Architecture	LR	BS	Epochs	Test Loss (%)	Test Accuracy (%)
GTSRB				100	13.99	96.87
				150	15.43	97.15
DITS				100	20.57	64.36
				150	(3.34)>100%	52.97
BTSC	**Proposed CNN Model**	1e-3	64	100	18.25	96.07
				150	18.42	96.58
Rmastif				100	28.24	91.47
				150	13.47	96.41
ETS				100	15.40	95.92
				150	**06.60**	**98.47**

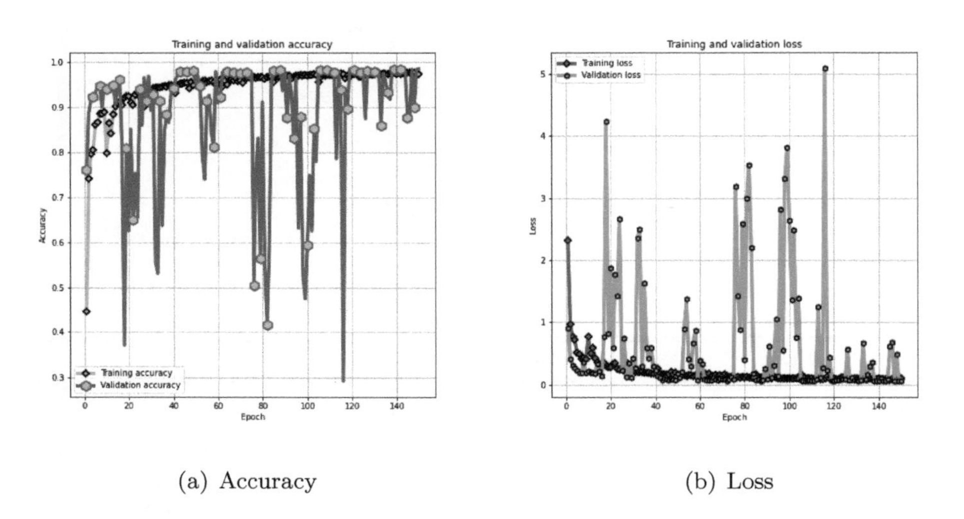

(a) Accuracy (b) Loss

Figure 9.5 Curves of the proposed CNN model trained on the ETS dataset.

is 98.47% in accuracy, while these mentioned metrics are very important to evaluate the performance results. Our decision to choose different datasets based on their sizes was to see how a DL model works in those situations, where we have a large and small volume of datasets.

Figure 9.5 represents the learning curves of the proposed CNN model using the ETS dataset, this validation is only performed on certain classes of traffic signs. This is clearly visible when we have a drop on the validation dataset curve, it is because of lack of data because the ETS dataset sometimes only contains one image for some classes, so the dataset distribution for the training dataset (80%) and in the validation dataset (20%) causes this problem, even the intervention of this problem follows the training is done impeccably. Even with this constraint, the validation curves follow the learning of the model efficiently, and also, we don't have any overfitting, this means that our CNN model is robust and cost-effective.

TABLE 9.14 Accuracy Comparison of Different CNN Models

Model	Type of Image Dataset	Dataset	Type of Activation function	Optimizer	Type of Loss Function	Accuracy (%)
CNN model [82]	Gray-Scale Image Dataset	MNIST	Tanh	SGD	binary cross entropy	97.43
LeNet-5 model	Gray-Scale Image Dataset	GTSRB	ReLU	ADAM	Cross-Entropy	96.77
CNN model [83]	Gray-Scale Image Dataset	GTSDB	ReLU	ADAM	Categorical Cross-Entropy	98.01
AlexNet model	Color Image Dataset	GTSRB	ReLU	ADAM	Cross-Entropy	87.03
Our CNN model	Color Image Dataset	GTSRB	ReLU	ADAM	Cross-Entropy	97.15
Our CNN model	Color Image Dataset	ETS	ReLU	ADAM	Cross-Entropy	98.47

9.5 EVALUATIONS

9.5.1 Compare the Performance of the Proposed CNN Model in the Literature

According to the results obtained, we arrive to obtain an optimized CNN model that can be submitted for road sign recognition in embedded vision systems [82]. However, by deeply analyzing the results of the architectures: LeNet-5, AlexNet, CNN model [82, 83], and our proposed CNN, it was found that using a dataset that contains enough large images can ensure that the model during training has a very high accuracy score, as shown in Table 9.14.

9.5.2 Classification Report for Recognition Traffic Road Signs

The final verdict of our proposed CNN architecture will be given in terms of three evaluation metrics which are, respectively, precision, recall, and f1-score [75, 79]. We took some classes of classification reports, among 146 classes of traffic sign images from the ETS dataset, as shown in Figure 9.5. The formulas for the three-evaluation metrics:

$$Precision = \sum_{i}^{n} \frac{p_i}{n} \tag{9.21}$$

$$Recall = \sum_{i}^{n} \frac{r_i}{n} \tag{9.22}$$

$$F - score = \frac{2 * (Precision * Recall)}{Precision + Recall} \tag{9.23}$$

n: Number of classes. The formulas for the 3-evaluation metrics each class i:

$$p_i = \frac{TP}{TP + FP}, \quad p_i = \frac{TP}{TP + FN}, \quad F - score_i = \frac{2 * (p_i * r_i)}{p_i + r_i} \tag{9.24}$$

TP: (True Positive) Element of the class correctly predicted. FP: (False Positive) Element of the class badly predicted. FN: (False Negative) Element positive classified as negative.

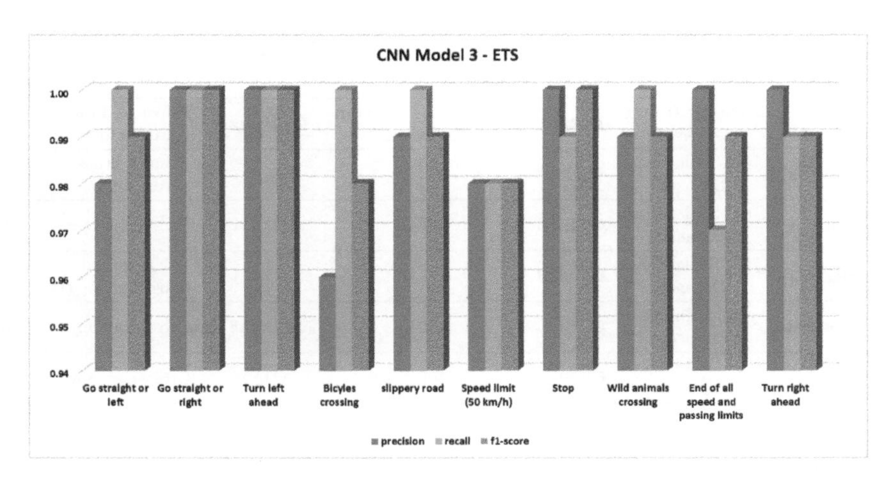

Figure 9.6 Sample CNN model classification report for the ETS dataset.

According to Figure 9.6, the recall allows us to know the percentage of positives well predicted by our model, the recall is very high varying between [97%, 100%], and the model maximizes the number of true positives. That is, it won't miss any positives, but that doesn't mean the CNN model isn't wrong. Nevertheless, this does not give any information on its quality of prediction on the negatives. The precision is quite similar to the recall, the difference is that the precision allows us to know the number of positive predictions well made. As shown in Figure 9.7, the accuracy varies in the interval [96%, 100%], and the accuracy is high. This means that the majority of the positive predictions of the model are well-predicted positives.

While useful, neither precision nor recall can fully evaluate our CNN model, recall and precision metrics fall short, so:

- If the model predicts « positive » all the time, the recall will be high.
- Conversely, if the model never predicts « positive », the precision will be high.

We will therefore based on the f1-score metric which allows us to combine precision and recall. The f1-score allows us to perform a good performance evaluation of our CNN model. After a thorough reading of the classification report, the obtained results are very satisfactory, especially the f1-score parameters which take their values in the interval [98%, 100%], we have the same values as the CNN model of classification report in this paper [83], as shown in Figure 9.7.

9.5.3 Interpretability of Cropped Images

9.5.3.1 XAI Technique

Grad-CAM's visualization generates a heatmap for the "Road Work" and "Parking" classes, showing how well different parts of the image correspond to the Road Work sign, and also a heatmap for the Class Parking sign; indicating how well the parts of the image match for each panel. This XAI produces a heatmap of the input image

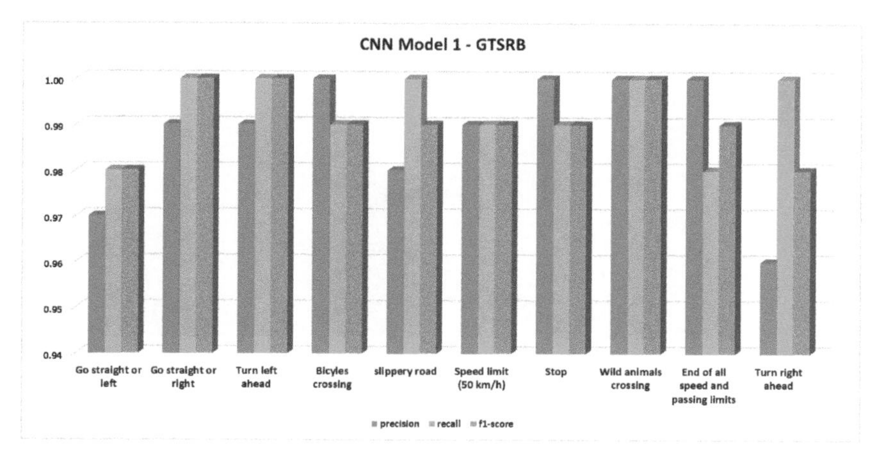

Figure 9.7 Sample CNN model classification report for the GTSRB dataset.

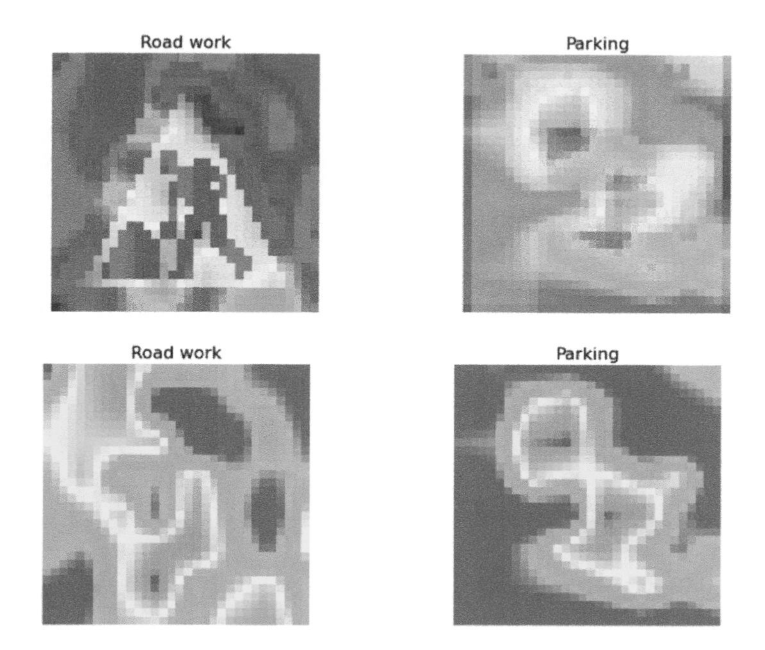

Figure 9.8 Grad-CAM of traffic road signs.

as shown in Figure 9.8, The more a part of the image matches the searched class, it will be colored in red.

9.5.3.2 Saliency Map

A salience map shows us the unique quality of each road sign image as demonstrated in Figure 9.9, saliency shows us how our CNN model, focuses on the pixels that reshape the content of each cropped sign. The map shows the regions which are prominent or noticeable at every location in the visual field and the selection of attended locations, based on the spatial distribution. The resulting plot shows the pixels that had the

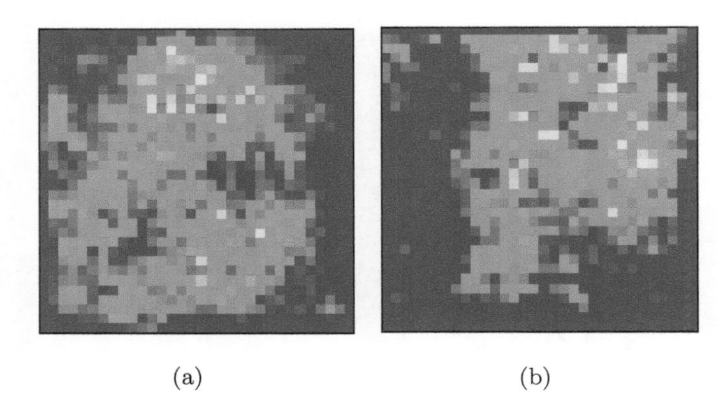

(a) (b)

Figure 9.9 Saliency maps for traffic signs: (a). road work sign and (b). parking signs.

TABLE 9.15 Recognition of Detected Traffic Road Signs

CNN Model	Cropped Signs		Classification
CNN Model on ETS Dataset		Sign Recognition	

greatest influence on the predicted class in colors. Importantly, this technique adds credibility to our CNN model predictions.

9.5.4 Recognition of Cropped Images

In this section, we will experience ERSDRA algorithms on some cropped images given by a digital camera used for an urban environment. The images in Table 9.15 shows some obtained recognition results performed on road work signs and parking signs. We observe that our proposed CNN model performs scores of 91% to correctly identify the traffic signs, and avoids confusion for other classes with a rate of 9%, this confusion can be caused by the training model on the ETS dataset which contains a large number of classes. This dataset includes 146 classes that gather all the European datasets, which execute road sign recognition with very high accuracy.

9.6 CONCLUSION

During the last years, embedded vision systems have been implemented in automotive systems to increase road safety. The majority of applications used in these vision systems are image processing functions, tracking detected objects, and object recognition. This paper presented a very promising ERSDRA algorithm applied to traffic road sign detection and recognition. We have used this algorithm with great success to a large number of images in a particularly difficult context, i.e., urban imagery.

The detection phase cascades two types of discriminated criteria: color and shape. It is particularly efficient in detecting triangular and rectangular signs. Indeed, most detection steps are based on detecting the contour of the signs by the Canny method. We cannot hide the fact that there is still some false detection because in an urban environment many objects can look like signs, these ambiguities were resolved with the introduction of the real-time-based area calculation with Green's theorem and Traffic Sign Center by moments. Also, using a metric arc length for perimeter calculation, with integrating the Hough Transform (HT) technic, consequently allows us to filter out objects that do not have the real forms of a sign.

For the recognition phase, several tests were experienced managing different architectures references like LeNet-5, AlexNet, and a proposed CNN architecture. The proposed CNN model was passed through performance tests and was trained to adopt different traffic sign image datasets like ETS, DITS, BTSC, Rmastif, and GTSRB. With the GTSRB dataset, we reached an accuracy score of 97.15%, and for the ETS dataset, we improved the performance and reached an accuracy score of 98.47%. These encouraging results argued for us to adopt, as a perspective, the proposed CNN model in ITS systems like the TSR Systems. The XAI provides overall information on the decision-making of our CNN model using the technique Grad-CAM by revealing the appropriate level of confidence according to the different types of recognition traffic road signs.

For future work, we have an interest to use in the recognition stage to apply feature extractors by applying residual architectures like ResNet, and DenseNet. Also, the amelioration of the performance of the ERSDRA Algorithm, allowed us to reach to create an optimistic approach to be integrated into embedded vision systems, including ITS system.

Bibliography

[1] H. U. Rashid Khan et al., "The impact of air transportation, railways transportation, and port container traffic on energy demand, customs duty, and economic growth: Evidence from a panel of low-, middle-, and high -income countries," J. Air Transp. Manag., vol. 70, no. February 2017, pp. 18–35, 2018, doi: 10.1016/j.jairtraman.2018.04.013.

[2] T. Vogelpohl, M. Kuhn, T. Hummel, and M. Vollrath, "Asleep at the automated wheel—Sleepiness and fatigue during highly automated driving," Accid. Anal. Prev., vol. 126, no. July 2017, pp. 70–84, 2019, doi: 10.1016/j.aap.2018.03.013.

[3] A. Barodi, A. Bajit, M. Benbrahim, and A. Tamtaoui, "Improving the transfer learning performances in the classification of the automotive traffic roads signs," E3S Web Conf., vol. 234, p. 00064, 2021, doi: 10.1051/e3sconf/202123400064.

[4] H. Gao, B. Cheng, J. Wang, K. Li, J. Zhao, and D. Li, "Object Classification Using CNN -Based Fusion of Vision and LIDAR in Autonomous Vehicle Environment," IEEE Trans. Ind. Informatics, vol. 14, no. 9, pp. 4224–4230, 2018, doi: 10.1109/TII.2018.2822828.

[5] N. Kalra and S. M. Paddock, "Driving to safety: How many miles of driving would it take to demonstrate autonomous vehicle reliability?," Transp. Res. Part A Policy Pract., vol. 94, pp. 182–193, 2016, doi: 10.1016/j.tra.2016.09.010.

[6] A. Arcos-Garcia, J. A. Alvarez-Garcia, and L. M. Soria-Morillo, "Deep neural network for traffic sign recognition systems: An analysis of spatial transformers and stochastic optimisation methods," Neural Networks, vol. 99, pp. 158–165, 2018, doi: 10.1016/j.neunet.2018.01.005.

[7] A. Barodi, A. Bajit, A. Tamtaoui, and M. Benbrahim, "An Enhanced Artificial Intelligence -Based Approach Applied to Vehicular Traffic Signs Detection and Road Safety Enhancement," Adv. Sci. Technol. Eng. Syst. J., vol. 6, no. 1, pp. 672–683, 2021, doi: 10.25046/aj060173.

[8] F. Zaklouta and B. Stanciulescu, "Real-time traffic sign recognition in three stages," Rob. Auton. Syst., vol. 62, no. 1, pp. 16–24, 2014, doi: 10.1016/j.robot.2012.07.019.

[9] S. Bouguezzi, H. Ben Fredj, H. Faiedh, and C. Souani, "Improved architecture for traffic sign recognition using a self-regularized activation function: SigmaH," Vis. Comput., no. 0123456789, 2021, doi: 10.1007/s00371-021-02211-5.

[10] S. Zhou, C. Deng, Z. Piao, and B. Zhao, "Few-shot traffic sign recognition with clustering inductive bias and random neural network," Pattern Recognit., vol. 100, p. 107160, 2020, doi: 10.1016/j.patcog.2019.107160.

[11] W. A. Haque, S. Arefin, A. S. M. Shihavuddin, and M. A. Hasan, "DeepThin: A novel lightweight CNN architecture for traffic sign recognition without GPU requirements," Expert Syst. Appl., vol. 168, no. November 2020, p. 114481, 2021, doi: 10.1016/j.eswa.2020.114481.

[12] E. Marti, M. A. De Miguel, F. Garcia, and J. Perez, "A Review of Sensor Technologies for Perception in Automated Driving," IEEE Intell. Transp. Syst. Mag., vol. 11, no. 4, pp. 94–108, 2019, doi: 10.1109/MITS.2019.2907630.

[13] K. T. Islam and R. G. Raj, "Real-time (Vision-based) road sign recognition using an artificial neural network," Sensors (Switzerland), vol. 17, no. 4, pp. 14–16, 2017, doi: 10.3390/s17040853.

[14] K. Ho, A. Gilbert, H. Jin, and J. Collomosse, "Neural architecture search for deep image prior," Comput. Graph., vol. 98, pp. 188–196, 2021, doi: 10.1016/j.cag.2021.05.013.

[15] G. Anusha and P. Deepa, "Design of approximate adders and multipliers for error tolerant image processing," Microprocess. Microsyst., vol. 72, p. 102940, 2020, doi: 10.1016/j.micpro.2019.102940.

[16] S. Sengupta et al., "A review of deep learning with special emphasis on architectures, applications and recent trends," Knowledge-Based Syst., vol. 194, p. 105596, 2020, doi: 10.1016/j.knosys.2020.105596.

[17] M. A. Islas et al., "A fuzzy logic model for hourly electrical power demand modeling," Electron., vol. 10, no. 4, pp. 1–12, 2021, doi: 10.3390/electronics10040448.

[18] J. de J. Rubio, "Stability Analysis of the Modified Levenberg – Marquardt Algorithm," Ieee Trans. Neural Networks Learn. Syst., vol. 32, no. 8, pp. 1–15, 2020.

[19] Y. Le Cozler et al., "Volume and surface area of Holstein dairy cows calculated from complete 3D shapes acquired using a high-precision scanning system: Interest for body weight estimation," Comput. Electron. Agric., vol. 165, no. July, p. 104977, 2019, doi: 10.1016/j.compag.2019.104977.

[20] K. F. Mulchrone and K. R. Choudhury, "Fitting an ellipse to an arbitrary shape: Implications for strain analysis," J. Struct. Geol., vol. 26, no. 1, pp. 143–153, 2004, doi: 10.1016/S0191-8141(03)00093-2.

[21] J. Yang, H. Wang, J. Yuan, Y. Li, and J. Liu, "Invariant multi-scale descriptor for shape representation, matching and retrieval," Comput. Vis. Image Underst., vol. 145, pp. 43–58, 2016, doi: 10.1016/j.cviu.2016.01.005.

[22] A. Suhadolnik, J. Petrisic, and F. Kosel, "An anchored discrete convolution algorithm for measuring length in digital images," Meas. J. Int. Meas. Confed., vol. 42, no. 7, pp. 1112–1117, 2009, doi: 10.1016/j.measurement.2009.04.005.

[23] K. Chen, P. Zhao, and R. Yang, "A linear approach to metric circumference computation for digitized convex shapes," J. Electron., vol. 25, no. 4, pp. 572–575, 2008, doi: 10.1007/s11767-008-0013-z.

[24] J. Chen, H. Qiang, J. Wu, G. Xu, and Z. Wang, "Navigation path extraction for greenhouse cucumber-picking robots using the prediction-point Hough transform," Comput. Electron. Agric., vol. 180, no. July 2020, p. 105911, 2021, doi: 10.1016/j.compag.2020.105911.

[25] W. Lin, X. Ren, T. Zhou, X. Cheng, and M. Tong, "A novel robust algorithm for position and orientation detection based on cascaded deep neural network," Neurocomputing, vol. 308, pp. 138–146, 2018, doi: 10.1016/j.neucom.2018.04.061.

[26] A. Barodi, A. Bajit, M. Benbrahim, and A. Tamtaoui, "An Enhanced Approach in Detecting Object Applied to Automotive Traffic Roads Signs," in 6th International Conference on Optimization and Applications, ICOA 2020 - Proceedings, Apr. 2020, pp. 1–6, doi: 10.1109/ICOA49421.2020.9094457.

[27] A. Gudigar, S. Chokkadi, and Raghavendra U, "A review on automatic detection and recognition of traffic sign," Multimed. Tools Appl., vol. 75, no. 1, pp. 333–364, 2016, https://link.springer.com/article/10.1007/s11042-014-2293-7.

[28] A. Kumar et al., "Black hole attack detection in vehicular ad-hoc network using secure AODV routing algorithm," Microprocess. Microsyst., vol. 80, no. October 2020, p. 103352, 2021, doi: 10.1016/j.micpro.2020.103352.

[29] A. Zemmouri, M. Alareqi, R. Elgouri, M. Benbrahim, and L. Hlou, "Integration and implimentation system-on-aprogrammable-chip (SOPC) in FPGA," J. Theor. Appl. Inf. Technol., vol. 76, no. 1, pp. 127–133, 2015.

[30] J. Stallkamp, M. Schlipsing, J. Salmen, and C. Igel, "Man vs. computer: Benchmarking machine learning algorithms for traffic sign recognition," Neural Networks, vol. 32, pp. 323–332, 2012, doi: 10.1016/j.neunet.2012.02.016.

[31] H. Jing and X. Xiaoqiong, "Sports image detection based on FPGA hardware system and particle swarm algorithm," Microprocess. Microsyst., vol. 80, no. September 2020, p. 103348, 2021, doi: 10.1016/j.micpro.2020.103348.

[32] C. G. Serna and Y. Ruichek, "Classification of Traffic Signs: The European Dataset," IEEE Access, vol. 6, pp. 78136–78148, 2018, doi: 10.1109/ACCESS.2018.2884826.

[33] X. Xu, J. Jin, S. Zhang, L. Zhang, S. Pu, and Z. Chen, "Smart data driven traffic sign detection method based on adaptive color threshold and shape symmetry," Futur. Gener. Comput. Syst., vol. 94, pp. 381–391, 2019, doi: 10.1016/j.future.2018.11.027.

[34] N. Barnes, A. Zelinsky, and L. S. Fletcher, "Real-time speed sign detection using the radial symmetry detector," IEEE Trans. Intell. Transp. Syst., vol. 9, no. 2, pp. 322–332, 2008, doi: 10.1109/TITS.2008.922935.

[35] J. Yang, F. Wei, Y. Bai, M. Zuo, X. Sun, and Y. Chen, "An effective multi-task two-stage network with the cross-scale training strategy for multi-scale image super resolution," Electron., vol. 10, no. 19, 2021, doi: 10.3390/electronics10192434.

[36] Z. Abdelkarim, R. Elgouri, and M. A. Alareqi, "Design and implementation of pulse width modulation using hardware / software microblaze soft-core Design and Implementation of Pulse Width Modulation Using Hardware / Software MicroBlaze Soft - Core," no. January 2018, pp. 167–175, 2017, http://doi.org/10.11591/ijpeds.v8.i1.pp167-175.

[37] Q. He et al., "Predictive models for daylight performance of general floorplans based on CNN and GAN: A proof-of-concept study," Build. Environ., vol. 206, no. September, p. 108346, 2021, doi: 10.1016/j.buildenv.2021.108346.

[38] A. Zemmouri, R. Elgouri, M. Alareqi, H. Dahou, M. Benbrahim, and L. Hlou, "A comparison analysis of PWM circuit with arduino and FPGA," ARPN J. Eng. Appl. Sci., vol. 12, no. 16, pp. 4679–4683, 2017.

[39] C. C. J. Kuo, "Understanding convolutional neural networks with a mathematical model," J. Vis. Commun. Image Represent., vol. 41, pp. 406–413, 2016, doi: 10.1016/j.jvcir.2016.11.003.

[40] R. Karthik, M. Hariharan, S. Anand, P. Mathikshara, A. Johnson, and R. Menaka, "Attention embedded residual CNN for disease detection in tomato leaves," Appl. Soft Comput. J., vol. 86, p. 105933, 2020, doi: 10.1016/j.asoc.2019.105933.

[41] K. Aslansefat, S. Kabir, A. Abdullatif, V. Vasudevan, and Y. Papadopoulos, "Toward Improving Confidence in Autonomous Vehicle Software: A Study on Traffic Sign Recognition Systems," Computer (Long. Beach. Calif)., vol. 54, no. 8, pp. 66–76, 2021, doi: 10.1109/MC.2021.3075054.

[42] G. Yadam, A. K. Moharir, and I. Srivastava, "Explainable and Visually Interpretable Machine Learning for Flight Sciences," Proc. CONECCT 2020 - 6th IEEE Int. Conf. Electron. Comput. Commun. Technol., 2020, doi: 10.1109/CONECCT50063.2020.9198505.

[43] A. Barodi, A. Bajit, M. Benbrahim, and A. Tamtaoui, "Applying Real-Time Object Shapes Detection to Automotive Traffic Roads Signs," 2020, doi: 10.1109/ISAECT50560.2020.9523673.

[44] Y. Zhong and L. Zhang, "Sub-pixel mapping based on artificial immune systems for remote sensing imagery," Pattern Recognit., vol. 46, no. 11, pp. 2902–2926, 2013, doi: 10.1016/j.patcog.2013.04.009.

[45] P. Davis and S. Raianu, "Computing areas using Green's theorem and a software planimeter," Teach. Math. its Appl., vol. 26, no. 2, pp. 103–108, 2007, doi: 10.1093/teamat/hrl017.

[46] L. Yang and F. Albregtsen, "Fast and exact computation of Cartesian geometric moments using discrete Green's theorem," Pattern Recognit., vol. 29, no. 7, pp. 1061–1073, 1996, doi: 10.1016/0031-3203(95)00147-6.

[47] J. Hoshino, A. Mori, H. Kudo, and M. Kawai, "A new fast algorithm for moment computation," Pattern Recognit., vol. 26, no. 11, pp. 1619–1621, 1993, https://doi.org/10.1016/0031-3203(93)90017-Q

[48] A. Zemmouri, R. Elgouri, M. Alareqi, M. Benbrahim, and L. Hlou, "Design and implementation of pulse width modulation using hardware/software microblaze soft-core," Int. J. Power Electron. Drive Syst., vol. 8, no. 1, pp. 167–175, 2017, http://doi.org/10.11591/ijpeds.v8.i1.pp167-175.

[49] M. Ruzhansky and D. Suragan, "Layer potentials, Kac's problem, and refined Hardy inequality on homogeneous Carnot groups," Adv. Math. (N. Y)., vol. 308, pp. 483–528, 2017, doi: 10.1016/j.aim.2016.12.013.

[50] D. Soendoro and I. Supriana, "Traffic sign recognition with Color-based Method, shape-arc estimation and SVM," Proc. 2011 Int. Conf. Electr. Eng. Informatics, ICEEI 2011, no. July, 2011, doi: 10.1109/ICEEI.2011.6021584.

[51] L. A. Elrefaei, M. Omar Al-musawa, and N. Abdullah Al-gohany, "Development of an Android Application for Object Detection Based on Color, Shape or Local Features," Int. J. Multimed. Its Appl., vol. 9, no. 1, pp. 21–30, 2017, doi: 10.5121/ijma.2017.9103.

[52] H. Alt and L. J. Guibas, Discrete Geometric Shapes: Matching, Interpolation, and Approximation**Partially supported by Deutsche Forschungsgemeinschaft (DFG), Grant No. A1 253/4-2. Woodhead Publishing Limited, 2000.

[53] N. L. Fernandez Garcia, L. D. M. Martinez, A. C. Poyato, F. J. Madrid Cuevas, and R. M. Carnicer, "Unsupervised generation of polygonal approximations based on the convex hull," Pattern Recognit. Lett., vol. 135, pp. 138–145, 2020, doi: 10.1016/j.patrec.2020.04.014.

[54] E. Practice, "a Digitized Line or its Caricature," Class. Cartogr., 2011.

[55] J. L. G. Pallero, "Robust line simplification on the plane," Comput. Geosci., vol. 61, pp. 152–159, 2013, doi: 10.1016/j.cageo.2013.08.011.

[56] Y. Chen, K. Jiang, Y. Zheng, C. Li, and N. Yu, "Trajectory simplification method for location-based social networking services," GIS Proc. ACM Int. Symp. Adv. Geogr. Inf. Syst., no. c, pp. 33–40, 2009, doi: 10.1145/1629890.1629898.

[57] W. Cao and Y. Li, "DOTS: An online and near-optimal trajectory simplification algorithm," J. Syst. Softw., vol. 126, no. M, pp. 34–44, 2017, doi: 10.1016/j.jss.2017.01.003.

[58] Z. Zheng, H. Zhang, B. Wang, and Z. Gao, "Robust traffic sign recognition and tracking for Advanced Driver Assistance Systems," IEEE Conf. Intell. Transp. Syst. Proceedings, ITSC, pp. 704–709, 2012, doi: 10.1109/ITSC.2012.6338799.

[59] P. Mukhopadhyay and B. B. Chaudhuri, "A survey of Hough Transform," Pattern Recognit., vol. 48, no. 3, pp. 993–1010, 2015, doi: 10.1016/j.patcog.2014.08.027.

[60] B. B. Traore, B. Kamsu-Foguem, and F. Tangara, "Deep convolution neural network for image recognition," Ecol. Inform., vol. 48, no. September, pp. 257–268, 2018, doi: 10.1016/j.ecoinf.2018.10.002.

[61] H. M. Song, J. Woo, and H. K. Kim, "In-vehicle network intrusion detection using deep convolutional neural network," Veh. Commun., vol. 21, p. 100198, 2020, doi: 10.1016/j.vehcom.2019.100198.

[62] L. Eren, T. Ince, and S. Kiranyaz, "A Generic Intelligent Bearing Fault Diagnosis System Using Compact Adaptive 1D CNN Classifier," J. Signal Process. Syst., vol. 91, no. 2, pp. 179–189, 2019, doi: 10.1007/s11265-018-1378-3.

[63] Z. Tang, C. Li, and S. Sun, "Single-trial EEG classification of motor imagery using deep convolutional neural networks," Optik (Stuttg)., vol. 130, pp. 11–18, 2017, doi: 10.1016/j.ijleo.2016.10.117.

[64] S. Han, H. Mao, and W. J. Dally, "Deep compression: Compressing deep neural networks with pruning, trained quantization and Huffman coding," 4th Int. Conf. Learn. Represent. ICLR 2016 - Conf. Track Proc., pp. 1–14, 2016.

[65] Y. Zhu, G. Li, R. Wang, S. Tang, H. Su, and K. Cao, "Intelligent fault diagnosis of hydraulic piston pump combining improved LeNet-5 and PSO hyperparameter optimization," Appl. Acoust., vol. 183, p. 108336, 2021, doi: 10.1016/j.apacoust.2021.108336.

[66] M. K. Titsias, "One-vs-each approximation to softmax for scalable estimation of probabilities," Adv. Neural Inf. Process. Syst., no. 1, pp. 4168–4176, 2016.

[67] Z. Zhang and M. R. Sabuncu, "Generalized cross entropy loss for training deep neural networks with noisy labels," Adv. Neural Inf. Process. Syst., vol. 2018-Decem, no. NeurIPS, pp. 8778–8788, 2018.

[68] A. Barodi, A. Bajit, M. Benbrahim, and A. Tamtaoui, "Improving the transfer learning performances in the classification of the automotive traffic roads signs," E3S Web Conf., vol. 234, p. 00064, 2021, doi: 10.1051/e3sconf/202123400064.

[69] D. O. Melinte and L. Vladareanu, "Facial expressions recognition for human–robot interaction using deep convolutional neural networks with rectified adam optimizer," Sensors (Switzerland), vol. 20, no. 8, 2020, doi: 10.3390/s20082393.

[70] N.Altini et al.,"NDG-CAM: Nuclei Detection in Histopathology Images with Semantic Segmentation Networks and Grad-CAM," pp. 1–19, 2022.

[71] R. R. Selvaraju, M. Cogswell, A. Das, R. Vedantam, D. Parikh, and D. Batra, "Grad-CAM : Why did you say that? Visual Explanations from deep networks via gradient-based localization," Rev. do Hosp. das Clínicas, vol. 17, pp. 331–336, 2016, [Online]. Available: http://arxiv.org/abs/1610.02391.

[72] W. Cui, Q. Lu, A. M. Qureshi, W. Li, and K. Wu, "An adaptive LeNet-5 model for anomaly detection," Inf. Secur. J., vol. 30, no. 1, pp. 19–29, 2021, doi: 10.1080/19393555.2020.1797248.

[73] F. N. Iandola, S. Han, M. W. Moskewicz, K. Ashraf, W. J. Dally, and K. Keutzer, "SqueezeNet: AlexNet-level accuracy with 50x fewer parameters and ¡0.5MB model size," pp. 1–13, 2016, [Online]. Available: http://arxiv.org/abs/1602.07360.

[74] S. Wan, Y. Liang, and Y. Zhang, "Deep convolutional neural networks for diabetic retinopathy detection by image classification," Comput. Electr. Eng., vol. 72, pp. 274–282, 2018, doi: 10.1016/j.compeleceng.2018.07.042.

[75] A. Zemmouri, A. Barodi, A. Satif, M. Alareqi, R. Elgouri, L. Hlou, and M. Benbrahim., "Proposal of a reliable embedded circuit to control a stepper motor using microblaze soft-core processor," Int. J. Reconfigurable Embed. Syst., vol. 11, no. 3, p. 215, 2022, doi: 10.11591/ijres.v11.i3.pp215-225.

[76] J. Stallkamp, M. Schlipsing, J. Salmen, and C. Igel, "The German Traffic Sign Recognition Benchmark for the IJCNN'11 Competition," Proc. Int. Jt. Conf. Neural Networks, pp. 1453–1460, 2011, [Online]. Available: http://ieeexplore.ieee.org/xpls/abs_all.jsp?arnumber=6033395.

[77] J. Blanc-Talon, A. Kasinski, W. Philips, D. Popescu, and P. Scheunders, Erratum to: Advanced Concepts for Intelligent Vision Systems. 2017.

[78] A. Arcos-Garcia, M. Soilan, J. A. Alvarez-Garcia, and B. Riveiro, "Exploiting synergies of mobile mapping sensors and deep learning for traffic sign recognition systems," Expert Syst. Appl., vol. 89, pp. 286–295, 2017, doi: 10.1016/j.eswa.2017.07.042.

[79] A. Zemmouri, A. Barodi, H. Dahou, M. Alareqi, R. Elgouri, L. Hlou and M. Benbrahim, "a microsystem design for controlling a Dc motor by pulse width modulation using Microblaze soft-core," Int. J. Electr. Comput. Eng., vol. 13, no. 2, pp. 1–11, 2023, doi: 10.11591/ijece.v13i2.pp1437-1448.

[80] A. Jain, A. Mishra, A. Shukla, and R. Tiwari, "A Novel Genetically Optimized Convolutional Neural Network for Traffic Sign Recognition: A New Benchmark on Belgium and Chinese Traffic Sign Datasets," Neural Process. Lett., vol. 50, no. 3, pp. 3019–3043, 2019, doi: 10.1007/s11063-019-09991-x.

[81] M. F. Esmaile, M. H. Marhaban, R. Mahmud, and M. I. Saripan, "Cross-sectional area calculation for arbitrary shape in the image using star algorithm with Green's theorem," IEEJ Trans. Electr. Electron. Eng., vol. 8, no. 5, pp. 497–504, 2013, doi: 10.1002/tee.21886.

[82] R. Udendhran, M. Balamurugan, A. Suresh, and R. Varatharajan, "Enhancing image processing architecture using deep learning for embedded vision systems," Microprocess. Microsyst., vol. 76, p. 103094, 2020, doi: 10.1016/j.micpro.2020.103094.

[83] A. BARODI, A. Bajit, A. ZEMMOURI, M. Benbrahim, and A. Tamtaoui, "Improved Deep Learning Performance for Real-Time Traffic Sign Detection and Recognition Applicable to Intelligent Transportation Systems," Int. J. Adv. Comput. Sci. Appl., vol. 13, no. 5, pp. 712–723, 2022, doi: 10.14569/IJACSA.2022.0130582.

CHAPTER 10

An Interpretable Detection of Transportation Mode Considering GPS, Spatial, and Contextual Data Based on Ensemble Machine Learning

Sajjad Sowlati and Rahim Ali Abbaspour

School of Surveying and Geospatial Engineering, College of Engineering, University of Tehran, Tehran, Iran

Alireza Chehreghan

Faculty of Mining Engineering, Sahand University of Technology, Tabriz, Iran

CONTENTS

DOI: 10.1201/9781003324140-10

CONSTRUCTING AND DEVELOPING intelligent transportation is one of the requirements for smart cities. Transportation Mode detection is essential for various transportation structures and planning. Nowadays, with the advancements in GPS technology and its integration into tools like smartphones, coupled with the availability of open-source data, vast sources of data can be accessed. This study aims to predict Transportation Modes of the bike, bus, car, train, and walking based on GPS-recorded points, spatial data, and contextual data. Until now, we have found no study that has utilized kinematic, spatial, and contextual group features together while handling the interpretability of Transportation Modes based on the feature group and checking each important feature from the separate classification models. In the proposed approach of this paper, after extracting all three group features, a hybrid feature selection is implemented to reduce the complexity. Then, RF, GB, XGBoost, CatBoost, LightGBoost, SVM, and KNN classifications are analyzed by F-Score evaluation and the SHAP method. Furthermore, for the first time in Transportation Mode detection, supervised machine learning classification models are applied along with ensemble learning techniques, including Stacking, MVE, and WAPVE. This ensemble learning techniques help prevent biased prediction and enhance reliability. After the implementation of the proposed methods, it is found that using spatial features alongside kinematic features improves the accuracy of forecasting all Transportation Modes. Moreover, using contextual and kinematic features together have the most significant impact on car, bus, and bike modes. Employing a feature selection algorithm reduces training time in all classification models and increases the F-score in the KNN, SVM, and CatBoost models relative to all the features. Among the models, the stacking algorithm for the bike, train, and walking modes, in addition the LightGBoost algorithm for the bus and car modes, shows a higher F-score. A stacking algorithm with an F-score weighted average of 91.32% and the micro average of 91.42% is the best classification model for predicting all Transportation Modes.

10.1 INTRODUCTION

Identifying Transportation Modes is a fundamental components of Intelligent Transportation Systems in Smart Cities [1, 2]. Furthermore, identifying modes of transportation for each trip of citizens is of great significance at both the macro and micro levels of urban management. Presently, traffic congestion and air pollution are two extremely important issues in big cities. Understanding how to distribute Transportation Modes based on location, time, and the relationship between the use of Transportation Modes and environmental issues [3] at the macro level provides the ground for the government to design intelligent transportation development strategies and assist traffic management and urban planning by implementing the policies [4]. At the micro-level, service providers provide personalized advertisements according to the mode of transportation, location of citizens, and traffic information on the roads. Citizens can use this information to reduce the cost and time of their travel [5] and benefit from targeted advertising in accordance with the mode of transportation they use [6]. After having information on Transportation Modes, health applications can be used for health monitoring and exercising goals [7, 8].

Previously, information on the use of Transportation Modes was prepared and completed via surveying citizens through questionnaires or phone calls [9, 10]. Being time-consuming and costly, having low response rates as well as gathering incomplete and inaccurate information are the main disadvantages of using traditional surveys [10]. Nowadays, advances in technology and the spread of GPS and smartphones have led to the production and use of vast amounts of data and information by researchers [11]. GPS data has many benefits in the construction and development of smart transportation, including behavior analysis [12, 13], traffic congestion estimation [14, 15], transportation monitoring [15, 16], activity detection [17, 18], and more.

In some studies, only GPS data has been used to predict Transportation Modes [10, 19]. In some others, various sensors such as the accelerometer and the gyroscope are used in addition to GPS [20, 21] while in some of them, sensors other than GPS are employed [22, 23]. The use of information from different sensors improves the prediction accuracy; however, as the number of sensors increases, the complexity of the calculations increases, and the battery life of the gadget used is reduced [24]. In some studies, citizens' personal information is used [9, 17]. In this case, in addition to the difficulty of accessing information, there is a possibility of endangering the privacy of citizens.

Spatial data such as road network lines and stations for buses, trains, and subways are publicly available worldwide using open-source volunteered geospatial information projects such as OpenStreetMap [25]. In some studies, this information is used alongside the information obtained from the sensors [4, 6, 26, 27]. Contextual data is available at a low cost. For example, low-cost weather conditions have a direct impact on citizens' decisions to use Transportation Modes [28, 29]. Using this information along with GPS data can improve prediction accuracy.

In several articles, many machine learning [19, 26, 30–32] and deep learning classification algorithms [1, 10, 33, 34] have been used to predict Transportation Modes. The use of deep learning algorithms does not require feature extraction manually but needs large data, constant length of input information, and high complexity in training. It does not take into account, the potential features extracted in these algorithms, which reduces the performance of the prediction. For these reasons, in this study, a machine learning algorithm is used.

In previous studies, the effectiveness of extracted features based on the feature set has not been investigated and the importance of feature selection on final accuracy has often been overlooked. In this work, 128 features from three groups of kinematic, spatial, and contextual travel were extracted and seven machine learning classification algorithms including Random Forest (RF), Extreme Gradient Boosting (XGBoost), Categorical Boosting (CatBoost), Light Gradient Boosting (LightGBoost), Gradient Boosting (GB), Support Vector Machine (SVM), and K-Nearest Neighbor (KNN) were implemented. The effectiveness of each of these subsets of the three groups of features in predicting Transportation Modes was investigated. Moreover, a hybrid method was implemented for feature selection to reduce training costs and increase processing speed in each of the classification models to achieve effective features. In addition, a new approach for implementing three ensemble learning models including Majority Voting Ensemble Classification (MVE), Weighted Average Probabilities Voting Ensemble Classification (WAPVE), and Stacked Generalization (Stacking) was used to increase the level of predictive reliability, combining the answers of the classification models, and prevent biased predicting.

This paper is organized as follows. In Section 10.2, we review previous studies. Section 10.3 is devoted to expressing the methodology of the problem. In Section 10.4, we will explain the implementation and evaluation of results and, finally, Section 10.5 contains the conclusions.

10.2 RELATED WORK

Many studies have examined the predictions of walking, bike, bus, car, and train modes. Xiao, et al. [32], a Bayesian network model is implemented using kinematic features and GPS data recorded in Shanghai city, which can predict transport modes with an additional mode of E-Bike except for train mode. This study predicts Transportation Mode by recall and precision of each mode above 80% and walks mode above 97%. In Xiao, et al. [19], 111 features were extracted from GPS data. KNN,

Decision Tree (DT), SVM, RF, GB, and XGBoost models were implemented, among which the XGBoost model was the best predictor with 90.77% accuracy. In Mäenpää, et al. [25], an F-score of 90.70% was achieved using GPS, feature selection data regardless of train mode, and implementation of Bayes classier, RF, and Artificial Neural Network (ANN). In Liang, et al. [4], 29 features were employed using GPS and spatial data to train DT, RF, Adaptive Boosting (AdaBoost), Bagging, GB, and XGBoost models. The RF model was able to predict Transportation Modes plus airplane mode with 86.50% accuracy. In Wang, et al. [30], 8 basic and 3 advanced features were extracted and two models, i.e., XGBoost and LightGBoost, were implemented. The LightGBoost model was able to achieve the best accuracy with 90.61%. Dabiri and Heaslip [10] used a Convolutional Neural Network (CNN) model considering the characteristics of kinematic points including velocity, acceleration, jerk, and bearing rate. This model could predict Transportation Modes with 84.8% accuracy. In Dabiri, et al. [33] a semi-supervised model was employed to use unlabeled trips alongside the labeled trips. In Nawaz, et al. [31], GPS data is used to extract kinematic features. The grid system method was employed to divide the study area and weekday was also utilized which results in the extraction of 37 features. They could predict Transportation Modes, except for train mode, with 93.99% accuracy. Huang, et al. [35] used five features obtained from GPS data and a spatial feature similar to the route line with the bus network to implement a noise-resistant collaborative isolation forest model, which could predict Transportation Modes with 81.02% accuracy. In Li, et al. [36], a CNN model with Generative Adversarial Network (GAN) was used due to the low volume of labeled trips. This achieved 86.70% accuracy. In another work, four kinematic point features including velocity, acceleration, jerk, and bearing rate as well as four additional features of geographical region, weather attributes, day of the week, and time slice were employed to learn a Convolutional Long Short-Term Memory (ConvLSTM) deep learning model. This model predicts Transportation Modes, except for trains, with an F-score of 83.97% [1]. Yu, et al. [34] extracted 25 features, including location, speed, acceleration, direction change, rate of directional change, distance from each basic transportation facility, the average speed of the related location set, and the average distance between the related locations and each basic transportation facility. Then, 18 features remain for training multi-layer neural networks and post-processing based on a moving window by filtering features based on information entropy. This method can predict all users' Transportation Modes and divide train mode into train and subway mode by 80.61% F-measure value. In Li, et al. [26], 15 features were extracted using GPS and spatial data. DT, RF, AdaBoost, XGBoost, LightGBoost, and ANN classification models were implemented, among which RF was able to obtain the best predicting accuracy of 91.1%. Despite these studies, there will be a need for additional research to examine the models and impact of features on the prediction of Transportation Modes. In most of these work, only GPS data has been used. In this study, more features is extracted and evaluated due to consideration of the availability of spatial and contextual data. Moreover, in most of the research, several classification models have been used to select the best model, while in our study, the combination of answers from all the models is implemented

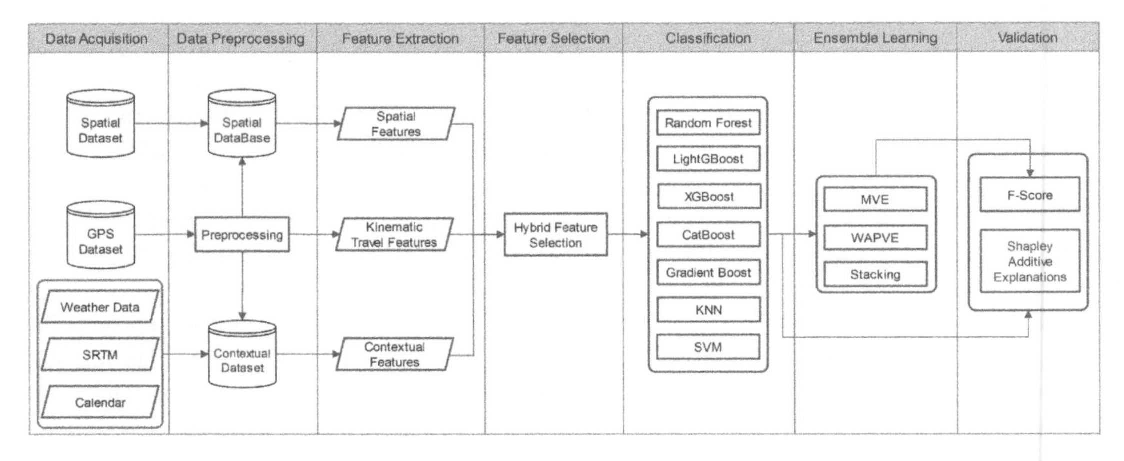

Figure 10.1 Proposed Method.

TABLE 10.1 Velocity and Acceleration Threshold Values Based on Transportation Modes

Transportation Mode	Velocity Threshold (m/s)	Acceleration Threshold (m/s^2)
Walk	7	3
Bike	12	3
Bus	34	2
Car	50	10
Train	34	3

in addition to selecting the best model in order to increase the confidence level of the final answer.

10.3 METHODOLOGY

The approach to identifying Transportation Modes in this study is illustrated as a flowchart in Figure 10.1. First, the data is gathered and the features are extracted after pre-processing. Through the feature selection method, effective features are selected and classification models are implemented. Ensemble learning models are implemented using classification models and the evaluations are performed at the end. These steps are explained in the following subsections.

10.3.1 Pre-Processing

Because incorrect measurements and recordings are possible due to the limitations of positioning techniques, the systematic error of point recorders, and human operator error, GPS data pre-processing is required [37]. For this purpose, the following steps are conducted to prepare the data.

- Points with the same time recording in a trip and out-of-range coordinates are deleted.
- To eliminate noises, the acceleration and velocity thresholds are considered according to Table 10.1 and the natural characteristics of each Transportation Mode [10].
- Due to the need for at least three GPS logs for kinematic travel features, trips of less than three GPS logs are eliminated.
- Trips, during which several Transportation Modes are used, are divided into one mode of transportation due to the nature and need for classification.
- Due to the possibility of turning off the positioning receivers or lack of signals, trips where the time interval between two points is more than 15 minutes are divided into two trips [38].

To process context data, elevation and weather data are also stored for the study area. Moreover, the date and time of the recorded data are converted into the time zone of the studied area. Spatial data of the study area is stored in a Spatial Database.

10.3.2 Feature Extraction

In this study, three groups of features including kinematic travel features, spatial features, and contextual features are extracted. Each of them is discussed as follows.

10.3.2.1 Kinematic Travel Features

Each GPS spatiotemporal log is stored as $P_i = [Latitude_{(i)}, Longitude_{(i)}, Time_{(i)}]$. To extract the features, the features of kinematic points must be extracted first. In this study, the features extracted based on each log of GPS coordinates are called point kinematic features. Point kinematic features including velocity, acceleration, jerk, and bearing rate are extracted between two successive GPS logs, i.e., P_i and P_{i+1}, in the time interval ΔT [10]. Eq. (1) show how these features are calculated.

$$V_{p_i} = \frac{Distance(P_i, P_{i+1})}{\Delta T}, \ A_{p_i} = \frac{V_{P_{i+1}} - V_{P_i}}{\Delta T}, \ J_{p_i} = \frac{A_{P_{i+1}} - A_{P_i}}{\Delta T} \tag{10.1}$$

$$x = cos[p_i(lat)] \times sin[p_{i+1}(lat)] - sin[p_i(lat)] \times cos[p_{i+1}(lat)] \times cos[P_{i+1}(long) - P_i(long)]$$

$$y = sin[p_{i+1}(long) - p_i(long)] \times cos[p_{i+1}(lat)]$$

$$B_{(p_i)} = arctan(y, x)$$

$$BR_{(p_i)} = |B_{(p_{i+1})} - B_{(p_i)}|$$

where $V(p_i)$, $A(p_i)$, and $J(p_i)$ are velocity, acceleration, and jerk at point P_i, respectively. The Distance between the points in this equation is calculated using the inverse formula in Vincenty and the reference ellipsoid WGS84 [39]. In these equations, x and y represent the change of direction in two directions while lat and long represent the latitude and longitude of the point, respectively. B and BR are, respectively, bearing and bearing rate calculated by three log points.

To extract features from point kinematic features, two methods of extracting statistical features during the trip are used. Some advanced travel features are used to handle traffic congestion and different weather conditions, which increase the possibility of error prediction, [19,40]. In the following, each of these features is explained.

Statistical Travel Features: This group of features is evaluated by calculating the kinematic feature statistics of points extracted from GPS logs during the trip. Fourteen statistics including average(avg), median(med), maximum(max), minimum(min), standard deviation(std), 10th percentile(10th), 25th percentile(25th), 75th percentile(75th), 90th percentile(90th), value range, interquartile, coefficient of variation(cv), kurtosis, and skewness(skew) are extracted and calculated for each point of kinematic features.

Advanced Travel Features: In certain traffic and weather conditions, Transportation Modes may be predicted with error. For example, the movement behavior of a car in heavy traffic conditions is very similar to that of walking. For this reason, advanced features are used along with setting the appropriate threshold. Three features including bearing change rate, stop rate, and velocity change rate are extracted and introduced as follows [40].

- Bearing Change Rate (BCR): It indicates the number of GPS logs, the bearing rate of which is greater than the specified threshold (A_{br}) relative to the distance traveled. It is calculated according to Equation 10.2.

$$BCR = \frac{|P_c|}{Distance}, \quad P_c = \{p_i | p_i \in P, BR(p_i) > A_{br}\} \tag{10.2}$$

  where $|P_c|$ is the frequency of the bearing rate value of GPS points that are greater than the threshold.

- Stop Rate (SR): It indicates the number of GPS logs, the velocity of which is less than the threshold (V_s) relative to the distance traveled. It is calculated according to Equation 10.3.

$$SR = \frac{|P_s|}{Distance}, \quad P_s = \{p_i | p_i \in P, V(p_i) < V_s\} \tag{10.3}$$

  where $|P_s|$ is the frequency of the velocity value of GPS points that are lower than the threshold.

- Velocity Change Rate (VCR): It indicates the number of GPS logs, the velocity rate of which is greater than the specified threshold (V_r) relative to the distance traveled. It is calculated based on Equation (4).

$$VCR = \frac{|P_v|}{Distance}, \quad P_v = \{p_i | p_i \in P, VR_i > V_r\}, \ VR_i = \frac{V_{p_{i+1}} - V_{p_i}}{V_{p_i}} \tag{10.4}$$

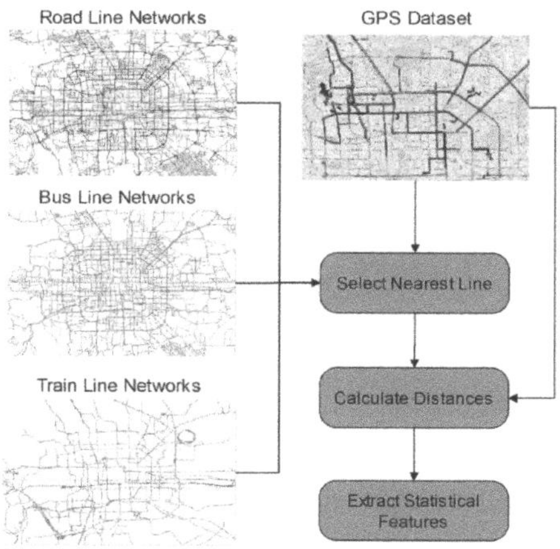

Figure 10.2 Calculation of Distance of Points from the Lines Network.

where VR_i is the Velocity rate of the GPS point p_i and $|P_v|$ is the frequency of the velocity rate of GPS points that are greater than the set threshold.

10.3.2.2 Spatial Features

Spatial features for linear and point networks include road routes, bus routes, railroad tracks, and locations for bus stations and train stations. The extracted spatial features will be described as follows.

Distance of Points from the Lines Network: To extract this feature, first, three road networks including road (Droad), bus routes (Dbus), and railway networks including train, subway, and light rails (Dtrain) are considered. For all points in a trip and each of the considered networks, one of the nearest neighbor line networks is selected and the distance of points to the line network is calculated. Statistical features such as kinematic travel statistical features for the obtained distances are then extracted. Figure 10.2 shows how to calculate these features.

Bus/train Station Indicator: According to Equation 10.5, this feature represents the ratio of number of logs with a velocity less than the threshold at the buffer distance specified by the bus/train station to all logs with a velocity of less than the threshold during the trip [26].

$$SI = \frac{P_B}{P_S},\ P_S = \{p_i | p_i \in P, V(p_i) < V_s\},\ P_B = \{p_i | p_i \in P_S, p_i \cap B_S\} \qquad (10.5)$$

where V_s is the velocity threshold calculated according to the stop rate threshold and B_S represents the intended buffer size of the stations. $|P_S|$ is the frequency of the velocity of GPS points that are lower than V_s. $|P_B|$ is the frequency of the GPS points that are intersection by bus/train stations buffer. The calculation of threshold value for this equation is in described in Section 10.4.

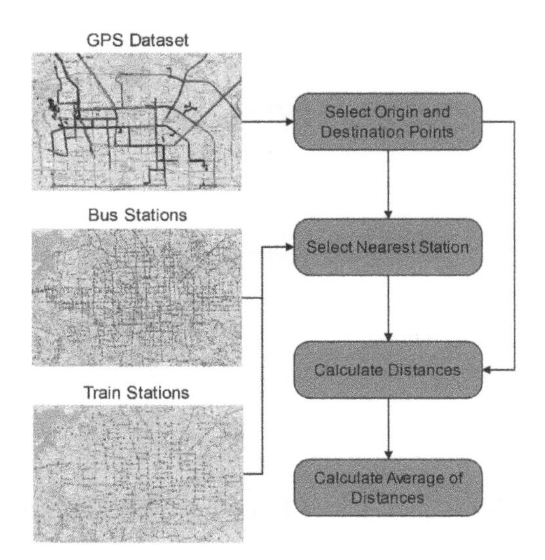

Figure 10.3 Calculation of the distance from the bus/train stations.

Distance from Bus/train Station: To calculate this feature, first, one of the nearest bus/train stations along with two origin and destination points of the trip is determined [9, 27]. Then, the distance of these two points from the nearest stations (Dorg and Ddes) in addition the average of two extraction distances (Dorg_des_avg) is extracted. Figure 10.3 shows how to calculate these features.

10.3.2.3 Context Features

To increase the accuracy, some other easily accessible features are extracted as contextual features in this study. The way to extract these features is discussed below.

Elevation Features: In the real world, it is almost impossible to use walking and bike modes on high-elevation routes where the elevation difference is large. For this reason, Shuttle Radar Topography Mission (SRTM) dataset is used to achieve the height of each GPS log. Then, the statistical characteristics of the value range, coefficient of variation, kurtosis, skewness, and standard deviation for the elevation obtained based on the GPS logs of each trip are extracted.

Holiday Feature: On holidays, the movement pattern of citizens may be changed from working to leisure conditions. Hence, the holidays or workdays feature is extracted using the calendar of the studied area.

Path Geometry Changes (based on Distance) Feature: Studying the geometry of the route traveled is one of the features that distinguishes Transportation Modes. One of the least expensive ways is to calculate the distance geometry changes based on distance [26]. In this feature, the cumulative distance relative to the direct distance of the path is calculated according to Equation 10.6. Calculating this feature can provide general information about the geometry of the path:

$$DC = \frac{\sum_{i=1}^{n-1} Distance(P_i, P_{i+1})}{Distance(P_1, P_n)} \tag{10.6}$$

In this equation, n is equal to the number of points per trip.

Weather Features: Decisions in choosing Transportation Modes change according to the weather conditions; hence, weather conditions are directly related to identifying Transportation Modes [28, 29]. Thus, 11 daily weather features including average temperature (T°C), maximum temperature (TM°C), minimum temperature (Tm°C), average relative humidity (%H), average visibility (VVkm), average wind speed (Vkm/h), maximum sustained wind speed (VMkm/h), rain or drizzle (RA), snow indicator (SN), storm indicator (TS) and fog indicator (FG) were extracted in this study.

10.3.3 Feature Selection

Some of the extracted features may be irrelevant or additional. These types of features can be overlooked without significant information loss in model training [41]. In this study, using the hybrid method, a subset of features that will cause the best accuracy was selected according to the classification model and used as input in the model.

10.3.3.1 Hybrid Feature Selection

Features that have the highest correlation with the objective function and minimum correlation with other features will cause high accuracy [42]. For this purpose, initially, due to a large number of extractive features and the possibility of high correlation between features, Correlation-based Feature Selection (CFS) was used to calculate the linear correlation coefficient and filter them into pairs according to Equation (7) when the correlation coefficient is over 90%. Sequential Forward Floating Selection (SFFS) was then used to calculate and select the effective feature subset according to each of the classification models [43]. During the implementation, three-fold cross-validation was used to prevent overfitting [44]:

$$r = \frac{\sum_{i=1}^{n}(X_i - \bar{X}_i)(Y_i - \bar{Y}_i)}{\sqrt{\sum_{i=1}^{n}(X_i - \bar{X}_i)^2}\sqrt{\sum_{i=1}^{n}(Y_i - \bar{Y}_i)^2}} \tag{10.7}$$

where r, n, X and Y, $\bar{X}$, $\bar{Y}$ represent the correlation coefficient, the number of trips studied features, and the average values of the features, respectively.

10.3.4 Classification of Modes

To predict Transportation Modes using the extracted features, seven models of machine learning classification, which have been used in most of the previous studies and resulted in high accuracy after classification, were implemented in this study.

10.3.4.1 Support Vector Machine (SVM)

By selecting a support vector, this model examines and selects the best line and margin, so that classes can be separated in the best way. The SVM model performs well in terms of memory and works well for high-dimensional data. In this study, the soft margin method is used. Classification is performed according to Equation 10.8 and its solution by the Lagrange method [45]:

$$min\frac{1}{2}\left\|\theta\right\|^2 + C\sum_{t=1}^{m}\epsilon^t \tag{10.8}$$

$$s.t\ y^t\theta^T\varphi(x^t) \geq 1-\epsilon^t\ ,\ \epsilon^t \geq 0$$

where the x data distance from the decision boundary is equal to $|\theta^t x + \theta_0| \geq \rho\left\|\theta\right\|$ and $\left\|\theta\right\|$ is vector norm. According to Equation 10.8, to maximize the margin of the line, $\left\|\theta\right\|$ must be minimized. In this equation, C is equal to the penalty parameter, ϵ^t is a soft margin error and y^t represents the segment class label, the value of which for the upper line data is +1 and for the lower line data is equal to -1.

10.3.4.2 Gradient Boosting (GB)

This model is used in regression and classification. In this method, weak models are constructed through decision trees and boosting and calculating the loss function repeatedly to make predictions. This model tries to reduce bias and could adjust parameters, which will make the model more flexible. Equation 10.9 shows how to calculate the gradient amplification model [46]:

$$F_0 = argmin\sum_{i=1}^{n} L(y_i,\gamma) \tag{10.9}$$

$$L(y_i,\gamma) = -log(likelihood) = -[y_i log(P) + (1 - y_i log(1-P))]$$

$$\frac{d}{dlog(odds)} = -y_i + \frac{e^{odds}}{1+e^{log(odds)}} = -observed + Predicted$$

$$F_m(x) = F_{m-1}(x) + v\sum_{j=1}^{j_m}\gamma_{jm}I(x \in R_{jm})$$

$$\gamma = \frac{Residual}{P(1-P)}$$

where L represents the cost function, y_i shows the observation label, x represents features, P represents the predicted probability, v is the learning rate, γ converts several probabilities of a leaf into one probability, F_0 the first stage probability estimator and $F_m(x)$ is the estimator probability of other stages, except for the first one.

10.3.4.3 Light Gradient Boosting (LightGBoost)

This model is one of the Gradient Boosting (GB) types in which the prediction speed can be increased compared to the traditional model using a data histogram. As the dimensions of features and data increase, it is difficult to calculate the information gain to find the split point, which leads to reduced effectiveness and scalability. Using the two techniques of Gradient-based One-Side Sampling (GOSS) and Exclusive Feature Bundling (EFB) alongside the GB model is called *LightGBoost*. In this model, data with higher gradients play an important role in constructing the decision tree. In GOSS, a significant proportion of samples with high gradients, along with a random sampling of small gradients, estimates information gain. This model can control a large amount of data due to low memory usage. For this purpose, the data is first sorted based on the absolute value of the calculated gradient; a×100% of the highest value is sampled and stored in subset b×100% of the remaining data is randomly sampled and stored in subset B. Finally, randomly sampled samples are amplified with constant (1-a)/b. The estimated value of variance gain is calculated according to Equation 10.10 [47]:

$$\tilde{V}_j(d) = \frac{1}{n}\left(\frac{(\sum_{x_i \in A_l} g_i + \frac{1-a}{b}\sum_{x_i \in B_l} g_i)^2}{n_l^j(d)} + \frac{(\sum_{x_i \in A_r} g_i + \frac{1-a}{b}\sum_{x_i \in B_r} g_i)^2}{n_r^j(d)}\right) \quad (10.10)$$

$$B_l = \{x_i \in B : x_{ij} \leq d\}, A_r = \{x_i \in A : x_{ij} > d\},$$
$$B_r = \{x_i \in B : x_{ij} > d\}, A_l = \{x_i \in A : x_{ij} \leq d\}$$

where x_i is the data vector and g_i is the value of the negative gradient calculated from the cost function. Using the greedy EFB algorithm, the number of features can be reduced accurately without any damage. High-dimensional data is usually scattered and does not take non-zero values along the rows; therefore, the EFB method is used to reduce the size of the properties and bundle build features by combining them. In EFB, color graphs are used to find properties that can be combined according to the histogram of the features [47].

10.3.4.4 Categorical Boosting (CatBoost)

In this study, CatBoost classification has been used for the first time in detecting Transportation Modes. In the innovative algorithm, the categorical feature is automatically converted into a number during training. In this model, the decision trees are made symmetrically, which will lead to quick prediction [48]. One way to convert a categorical feature is to use greedy TBS. In this method, the average value of the label is used to find the split point of the decision tree. In the CatBoost model, a priority value is added to greedy TBS and using a random permutation in the data, the average value of the label for each data p is replaced according to Equation 10.11,

$$x_{\sigma_p,k} = \frac{\sum_{j=1}^{p-1}(x_{\sigma_j,k} = x_{\sigma_p,k}) \times y_{\sigma_j} + a \times P}{\sum_{j=1}^{p-1}(x_{\sigma_j,k} = x_{\sigma_p,k}) + a} \quad (10.11)$$

The dataset are $D = [(X_i, Y_i)]_{i=1 \cdots n}$, where $X_i = (x_{i,1}), \ldots, (x_{i,m})$ is a vector with m features and $Y_i \in R$ is the data label. The permutations are shown as $\sigma = (\sigma_1, \ldots, \sigma_n)$, Categorical feature with K, initial value with P, and parameter $a > 0$ are the weight of the initial value. To obtain the optimal result, P can be changed at any stage. Also, if the models use the same training data for prediction, it causes a prediction shift. To avoid it, we used ordered boosting [49].

10.3.4.5 Random Forest (RF)

The sum of several Decision Trees constitutes the forest. If the decision trees are formed randomly, a random forest can be created. This model is an ensemble classification that builds decision trees by bagging and, finally, estimates the most popular class by voting. This model tries to reduce the variance, can train decision trees in parallel, and gains acceptable accuracy among the machine learning classification functions [50].

10.3.4.6 Extreme Gradient Boosting (XGBoost)

This model is an improved GB method that reduces overfitting by adding a regularization value; also, due to the inheritance of the RF method, it will have the ability to be trained in parallel. Equation 10.12 shows how to calculate the cost function in XGBoost [51]. The first term of this equation represents cost function and the second term represents regularization. T represents the number of leaves of the decision tree and the user-defined penalty gamma for tree pruning. By solving the Taylor expansion to the second step, the cost function is calculated according to Equation 10.13.

$$L(y_i, P_i) = [\sum_{i=1}^{n} L(y_i, P_i^m)] + \sum_{k} \Omega(f_k) \tag{10.12}$$

$$\Omega(f) = \gamma T + \frac{1}{2}\lambda \left\| w \right\|^2$$

$$L^t(q) = -\frac{1}{2}\sum_{j=1}^{T} \frac{(\sum_{i \in I_j} g_i)^2}{\sum_{i \in I_j h_j} + \lambda} + \gamma T \tag{10.13}$$

In Equation 10.13, g_i represents the first derivative of the cost function and h_i represents the second one.

10.3.4.7 K-Nearest Neighbors (KNN)

KNN is a simple and efficient algorithm that maintains the class of the instruction set during learning and, for prediction, each class in the k-nearest neighbors of the instruction set is assigned as the input class [52]. In this algorithm, the 'nearest' represents the distance between the extracted features of the trips. In this study, the 2nd-degree Minkowski distance was used considering the Inverse Distance Weighted (IDW), so that close neighbors had a greater impact on the prediction [53].

10.3.5 Ensemble Learning (EL)

Using a set of distinct solutions has better and more reliable results than using just one solution. The approach of the article, after implementing the classification models, is to combine the predictions of the classification models. For this purpose, three methods of combination are considered, which are described below.

10.3.5.1 Majority Voting Ensemble Classification (MVE)

Majority voting is one of the simplest and most popular ensemble approaches that consider the number of classes of each of the base learners' predictions to vote and the maximum number of predictions as predictions. Each classifier has only one vote in this method. If j is the number of classifiers and Ct represents a classifier, then, $t = 1, 2, \ldots, j$ and the set of classifiers is represented as $E = [C_1, C_2, \ldots, C_j]$, the output is similar to Equation 10.14 [54].

$$MVE_{trip} = max(\sum_{t=1}^{j} d_{t,n}), \quad 1 \leq n \leq i \tag{10.14}$$

In this case, the decision of each classifier, Ct, is equal to $d_{t,n} \in 0, 1$, where $n = 1, 2, \ldots, i$, and i is the number of classes. If the output of the classifier t prediction belongs to class n, the value $d_{t,n}$ will be equal to 1; otherwise, it will be equal to 0.

10.3.5.2 Weighted Average Probabilities Voting Ensemble Classification (WAPVE)

Not all base learners can properly predict and considering the same weight for each of them may not be optimal. For this reason, in this method, a weight is selected for each classifier. The final class selection in this method is to construct the sum of all the votes and consider the maximum weight [54]. This algorithm consists of three parts. Initially, the weight of each base learner is determined by considering five-fold cross-validation. Then, base learners are trained in probabilities and multiclass objectives using train data; finally, method evaluation is performed using testing data [54, 55]. The prediction of the Transportation Mode of a trip can be calculated using Equations 10.15,

$$\begin{aligned} WAPV_{trip_{i,1}} &= prob_{i,j} \times Classifiers_{weight_{j,1}}, \\ Prediction_{label} &= argMax(WAPV_{i,1}) \end{aligned} \tag{10.15}$$

$$\begin{bmatrix} prob_{1,1} & \ldots & prob_{1,j} \\ \vdots & \ddots & \vdots \\ prob_{i,1} & \ldots & prob_{i,j} \end{bmatrix}, Classifiers_{weight} = \begin{bmatrix} w_1 \\ \vdots \\ w_j \end{bmatrix}$$

$prob$ indicates the probability of occurrence of any of the Transportation Modes, $Classifiers_{weight}$ indicates the weight of each classification and $argMax$ is the maximum value selected as the Transportation Mode label of the desired trip. $Prediction_{label}$ represents the output label that is predicted after executing the WAPVE method. The total probability in each classification model is 1, where i

is equal to the number of classes in the model and j is the number of considered classifiers. The weight selection of each model is reviewed in the implementation section.

10.3.5.3 Stacked Generalization (Stacking)

The stacking method can discover how to combine the output of basic learners regardless of their weight. This method uses the meta-learner algorithm to learn which classifications are reliable. In this study, to prevent overfitting, the stacking method is combined with cross-validation, which includes three steps [56].

Step (1): According to the available data set and classification models, the input data are divided into K subsets using K-fold cross-validation. Each time, the K-1 subset is used for classification to predict the remaining subset and to estimate the learning accuracy.

Step (2): A new set is created based on the outputs of the first step. In this stage, the output labels predicted in the first stage are considered new features, and the main labels are stored in the new collection.

Step (3): Based on the new data set, a second-level classifier is trained and implemented, which is the result of predicting the base models.

10.3.6 Evaluation

10.3.6.1 F-Score

After implementing the classification models, there is a need to evaluate the models. To do so, the F-score according to Equation 10.16 is used [57]. Since the number of classes to be predicted is more than two, the results of each class prediction should be averaged. For this reason, two averages, micro according to Equation 10.17 and weighted based on the number of samples per class according to Equation 10.18 are used [58, 59]:

$$P = \frac{TP}{TP + FP}, \quad R = \frac{TP}{TP + FN}, \quad F_{score} = \frac{2PR}{P + R} \tag{10.16}$$

$$\begin{aligned} P_{micro} &= \frac{\sum_{i=1}^{n} TP_i}{\sum_{i=1}^{n} TP_i + \sum_{i=1}^{n} FP_i}, \quad R_{micro} = \frac{\sum_{i=1}^{n} TP_i}{\sum_{i=1}^{n} TP_i + \sum_{i=1}^{n} FN_i}, \\ F_{score(micro)} &= \frac{2P_{micro}R_{micro}}{P_{micro} + R_{micro}} \end{aligned} \tag{10.17}$$

$$\begin{aligned} P_{weighted} &= \frac{\sum_{i=1}^{n} P_i \times W_i}{\sum_{i=1}^{n} W_i}, \quad R_{weighted} = \frac{\sum_{i=1}^{n} R_i \times W_i}{\sum_{i=1}^{n} W_i}, \\ F_{score(weighted)} &= \frac{2P_{weighted}R_{weighted}}{P_{weighted} + R_{weighted}} \end{aligned} \tag{10.18}$$

where P, R are precision and recall, respectively, which lead to F_{Score} calculation. TP is a positive sample number of class members and is correctly identified as a member of the same class. FN is the number of positive class member samples that are mistaken for a negative class member. FP is the number of instances, in which a sample is a negative class member and is mistaken for a positive class member. W stands for weight, which is calculated using the number of samples belonging to that class.

10.3.6.2 Shapley Additive Explanations (SHAP)

The process of implementing powerful classification algorithms is in the form of a black box that humans cannot understand, so it is necessary to add interpretability to these algorithms [60]. One of these methods is Shapley Additive Explanations [61]. This algorithm uses the game theory method, and its goal is to distribute each player's points fairly and collectively to reach a specific result. Understanding why a model makes a particular prediction can be as critical as prediction accuracy in many applications [61]. The importance of each feature used in model training can be calculated by implementing the SHAP algorithm in classification algorithms. Shapley's value calculation for one feature is based on Equation 10.19 [61].

$$\emptyset_i = \sum_{S \subseteq F \setminus \{i\}} \frac{|S|!(|F| - |S| - 1)!}{|F|!} [f_{S \cup \{i\}}(x_{S \cup \{i\}}) - f_S(x_S)] \tag{10.19}$$

It assigns an importance value to each feature, indicating the effect on the model's prediction of including that feature. Where F is the set of all features and $F \setminus \{i\}$ is the set of all possible combinations of features excluding the i. $f_{S \cup \{i\}})$ is the result of training the model with i, and f_S is a model without using the desired feature, and x_S shows the values of the input features in the set S. Since the effect of not considering a feature depends on the rest of the model features, this value is calculated for all possible subsets of $S \subseteq F \setminus \{i\}$. Until today, we can use the Kernel-SHAP method in addition to all classification models, Tree-SHAP in addition to tree base classification models and especially tree-based ensemble models, and Deep-SHAP in deep learning models to describe the features [61–63].

10.4 IMPLEMENTATION

In this study, a processing system with 16GB RAM, CPU of 10870 16core, GPU of 1660ti 6GB, and 500GB SSD memory was used. Kinematic travel and contextual features were extracted using the Python programming language and spatial features were extracted using the PostgreSQL spatial database (PostGIS). All the classification models and SHAP method were implemented using the Python programming language and the Scikit-learn, LightGBM, CatBoost, XGboost, and SHAP libraries.

10.4.1 Dataset

The GPS trajectory considered in this study was the Geolife database [40, 64, 65]. This was recorded as spatiotemporal data over five years by 182 people; 69 users

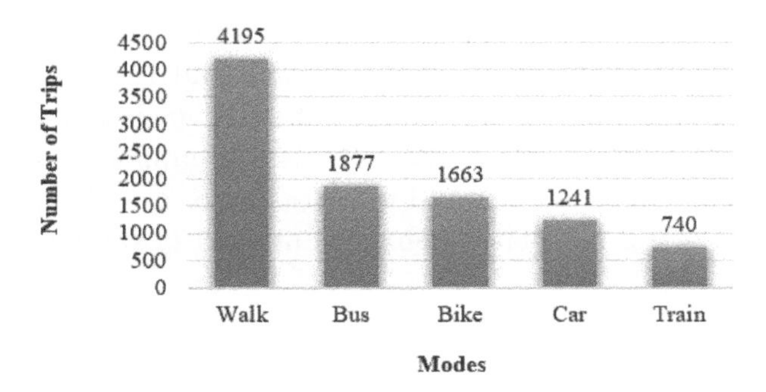

Figure 10.4 Distribution of Trips Based on Transportation Modes.

have labeled their trips, which include 17,621 trips with a total length of 1,292,951 km and a total of 50,176 hours, which are publicly available. The studied area of this database included 30 cities in China and several cities in the United States and Europe, more than 70% of the data was stored in Beijing city. Elevation data was obtained from GeoTIFF images, US Geological Survey (USGS), and SRTM Digital Elevation Models (DEMs) [66]. Weather information for the contextual data subset was stored from the tutiempo website [67] and spatial data was stored through the Beijing City Lab (BCL) site [68] and OpenStreetMap.

10.4.2 Pre-Processing

After implementing the items mentioned in the pre-processing, to avoid the interference of transportation networks in different cities, the available data was limited to the city of Beijing. Moreover, among the travel modes in the dataset, only ground Transportation Modes including walking, bike, car, taxi, train, and subway were considered. According to the data guidance, taxi and car modes were considered car modes. Due to the possibility of light rails being labeled as train or subway by users and owing to seamlessly connected light rails and subway lines in Beijing city, the train and subway modes were also considered as train modes. This led to combining network lines and train, subway, and light rail stations and considering them as network lines and the location of train stations. All of the time frames in the dataset are in GMT, but, the raw data time was converted into the time zone of the study area for use in contextual features. In total, 9716 trips were made and Figure 10.4 shows how the trips are distributed based on Transportation Modes.

10.4.3 Determining Hyperparameters

Classification models and several extracted features need thresholds before making the prediction. For this purpose, first, the whole data was divided into two parts: 70% of the data as the training data and 30% of the data as the testing data in a stratified form [69]. Determining each of the hyperparameters was explained as follows.

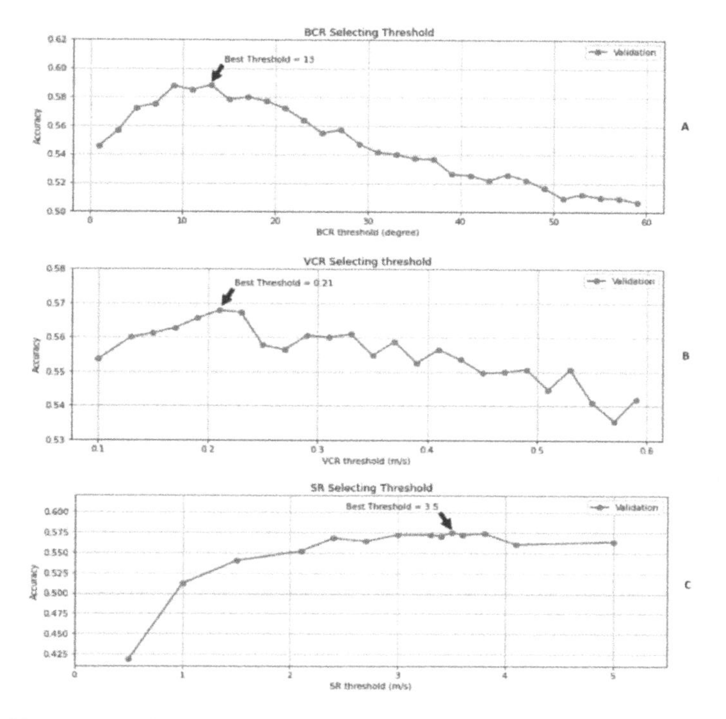

Figure 10.5 Kinematic features thresholds.

10.4.3.1 Threshold of Kinematic Travel Features

The bearing change rate, velocity change rate, and stop rate features require thresholds. For this purpose, first, the threshold setting range for these features is assumed. Then, the best threshold is selected using the training data, performing fivefold cross-validation on the LightGBoost model, and calculating the F1-score weighted. According to Figure 10.5, the bearing change rate, velocity change rate, and stop rate thresholds are selected as 13 degrees, 0.21 m/s, and 3.5 m/s, respectively.

10.4.3.2 Threshold of Spatial Features

The features of bus/train station indicator need to set a buffer size of the distance from the stations. For this purpose, an interval of distances is considered for each of the bus and train stations, which are calculated as determining the threshold of kinematic properties. According to Figure 10.6, the buffer size is 50m for the distance from the bus stations and 230m for the distance from the train stations. Velocity limit (Vs) is equal to the value of the stop rate threshold, i.e., 3.5 m/s.

10.4.3.3 Determining Hyperparameters of Classification Models

Since Transportation Modes are multiclass, all the classifications are objectively implemented as multiclass. To create a comparative condition in tree-based models including RF, XGBoost, LightGBoost, GB, and CatBoost, the number of trees in all of them manually is considered equal to 1000, and in the LightGBoost model, GOSS

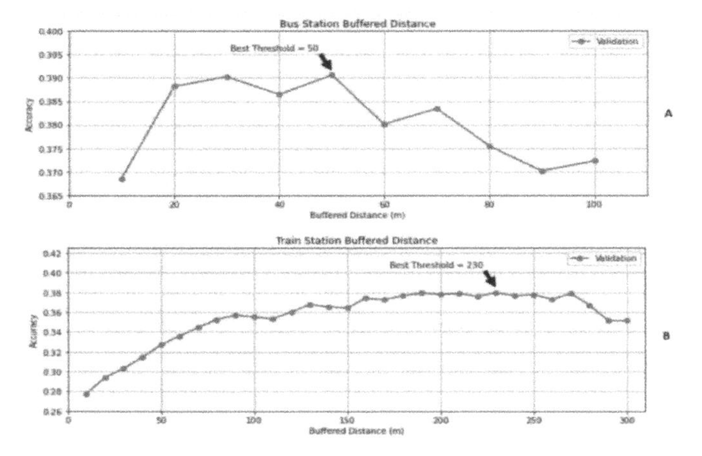

Figure 10.6 Spatial features thresholds.

boosting type is used. Other parameters in these models are determined according to the default value of the python library and no tune is considered. In SVM, the C value is manually set to 1. In KNN, the value of $K = 5$ has achieved the highest accuracy by fivefold cross-validation according to the interval 1, 3, 5, 7, 9 and the remainder of the parameters are the default value in the Scikit-learn library.

10.4.3.4 Determining Hyperparameters of Ensemble Learning Models

The weight of KNN, LightGBoost, RF, CatBoost, SVM, XGBoost, and GB base learners in the WAPVE model by considering F-score weighted and five-fold cross-validation, was calculated as 86.92, 91.87, 90.90, 91.51, 86.91, 91.62, and 90.96, respectively. For the meta-learner, the stacking method used RF with five-fold cross-validation.

10.4.4 Examining Features Group

In this section, a subset of the feature group extracted according to Table 10.2 is considered. In each model, the estimated F-score and the effectiveness of each feature group are different. According to Table 10.2 in the single-group subset, kinematic travel features for bike mode; kinematic travel features and then spatial features for the bus and car modes; kinematic travel features and spatial features approximately to the same extent for the train mode; and kinematic travel features and then contextual features for walking mode were able to estimate the best F-score. In the subgroups of dual groups, in bike and walking modes, spatial-kinematic travel features and contextual-kinematic travel features, almost equally, in bus and car modes, first, spatial-kinematic travel features and, then, contextual-kinematic travel features, and in train mode, spatial-kinematic travel features were able to predict with a higher F-score than other groups of features. In general, the use of spatial features group along with kinematic travel features has increased the predictive F-score in all modes of transportation, especially motor modes (bus, car, train). The use of contextual

TABLE 10.2 F-Score Value of Each of the Classifiers Based on Extracted Features Group

XGBoost	Bike	Bus	Car	Train	Walk	Average
Kinematic	90.99	86.47	85.93	83.11	94.34	90.31
Spatial	74.82	76.38	79.24	82.20	86.22	81.17
Contextual	67.12	64.90	66.49	61.82	87.29	74.91
Kinematic and Spatial	91.84	88.48	90.29	87.22	94.37	91.73
Spatial and Contextual	78.20	80.39	82.56	84.83	89.21	84.44
Kinematic and Contextual	91.20	86.62	87.17	82.64	94.39	90.52
All	91.49	88.60	90.75	87.62	94.30	91.75
LightGBoost	**Bike**	**Bus**	**Car**	**Train**	**Walk**	**Average**
Kinematic	91.58	87.07	86.10	83.18	94.43	90.60
Spatial	74.96	76.20	79.20	81.75	86.20	81.11
Contextual	67.21	63.71	65.80	58.89	87.25	74.37
Kinematic and Spatial	91.99	88.66	89.93	87.87	94.42	91.82
Spatial and Contextual	79.22	80.90	83.43	84.40	89.67	84.99
Kinematic and Contextual	91.36	87.29	87.39	83.28	94.43	90.77
All	91.79	89.04	90.96	87.89	94.50	92.02
SVM	**Bike**	**Bus**	**Car**	**Train**	**Walk**	**Average**
Kinematic	81.2	69.90	67.01	51.56	88.07	77.91
Spatial	52.1	57.07	55.40	51.13	59.97	56.80
Contextual	57.37	42.11	52.06	34.19	87.35	64.92
Kinematic and Spatial	82.89	72.88	74.17	58.65	88.4	80.37
Spatial and Contextual	67.29	67.87	66.95	64.75	86.09	75.28
Kinematic and Contextual	80.89	71.11	67.64	56.60	86.94	78.06
All	80.9	70.51	70.38	65.26	86.64	78.83
KNN	**Bike**	**Bus**	**Car**	**Train**	**Walk**	**Average**
Kinematic	53.51	47.48	53	39.10	82.22	63.57
Spatial	51.92	40.89	44.16	49.19	72.59	57.51
Contextual	51.22	42.93	49.51	29.24	77.01	58.86
Kinematic and Spatial	53.58	47.68	53.12	38.75	82.23	63.62
Spatial and Contextual	51.30	42.72	50.27	43.50	77.53	60.24
Kinematic and Contextual	53.50	47.75	53.23	38.88	82.50	63.76
All	53.72	47.87	53.45	38.14	82.45	63.77
GB	**Bike**	**Bus**	**Car**	**Train**	**Walk**	**Average**
Kinematic	90.53	85.87	85.57	81.23	94	89.78
Spatial	73.57	75.31	77.29	80.15	85.28	79.94
Contextual	65.95	63.22	64.05	58.30	87.21	73.77
Kinematic and Spatial	91.42	87.63	89.58	86.57	94.46	91.39
Spatial and Contextual	77.44	79.90	82.07	84.39	88.93	84
Kinematic and Contextual	90.69	86.38	86.36	81.20	94.08	90.04
All	91.30	88.21	90.35	86.88	94.40	91.58
Catboost	**Bike**	**Bus**	**Car**	**Train**	**Walk**	**Average**
Kinematic	90.02	85.89	85.14	82.20	94.22	89.81
Spatial	74.50	77.22	79.99	82.49	86.32	81.44
Contextual	67.18	64.95	65.49	59.66	87.75	74.84
Kinematic and Spatial	91.21	87.61	89.46	87.22	94.35	91.34
Spatial and Contextual	77.95	80.79	82.49	83.57	89.38	84.44
Kinematic and Contextual	90.83	86.03	85.33	82.05	94.25	90.01
All	90.59	87.88	89.60	87.83	94.23	91.30
Catboost	**Bike**	**Bus**	**Car**	**Train**	**Walk**	**Average**
Kinematic	90.11	84.45	83.95	81.78	93.90	89.23
Spatial	72.42	73.80	78.40	81.87	84.96	79.58
Contextual	67.40	63.69	65.28	55.93	87.97	74.42
Kinematic and Spatial	90.13	86.20	88.07	87.20	93.75	90.45
Spatial and Contextual	75.74	79.04	81.41	84.52	88.88	83.44
Kinematic and Contextual	90.32	84.07	83.58	81.28	93.90	89.11
All	90.43	86.31	88.03	87.02	93.95	90.59

TABLE 10.3 F-Score Evaluation of Each Transportation Mode After Feature Selection

Classification Models	Bike	Bus	Car	Train	Walk	Average
XGBoost	91.11	89.03	90.53	87.66	94.02	91.62
LightGBoost	91.52	89.19	90.56	87.49	94.37	91.87
RF	90.31	87.65	89.11	86.20	93.95	90.90
GB	90.52	87.48	89.31	86.47	93.97	90.96
CatBoost	91.05	88.35	89.63	87.69	94.34	91.51
KNN	87.70	82.46	79.33	77.45	92.53	86.92
SVM	87.09	80.77	81.51	77.34	92.88	86.91

features group along with kinematic travel features has also had the greatest impact on the Transportation Modes of bus, bike, and car.

10.4.5 Examining Feature Selection

To increase the processing speed and avoid additional processing costs, feature selection from the union of all features was implemented according to each of the classification models, the results of which are based on all Transportation Modes in Table 10.3. Table 10.4 shows the number of features, F-score, and training speed of each of the classification models after feature selection and its differences according to the consideration of all features in parentheses. The feature selection method in SVM, KNN, and CatBoost models increased the F-score by 8.08%, 23.15%, and 0.21% in all modes of transportation, respectively. Obviously, due to the embedded tree-based models, considering some of the features by feature selection, according to the third column of Table 10.4, could cause a limited reduction in F-score compared to considering all the features. However, due to the acceptable reduction in the size of the input features, which was between 36 and 51 features out of 128 features (second column of Table 10.4) as well as the increase in training speed, this limited amount of F-score can be neglected.

10.4.6 Examining Classification Models

MVE, WAPVE, and Stacking models were implemented using training data, base classifiers, and selected features through the feature selection method and evaluated by the testing data. Table 10.5 shows a comparison of the implemented models. In EL models, Stacking, MVE and WAPVE methods estimated the best prediction with F-score with the weighted average of 91.32%, 91.03%, 90.93%, and the micro average of 91.42%, 91.18%, and 91.08%, respectively. Compared to the base models, all three models improved bike mode. The order of base models based on the average F-score was LightGBoost, XGBoost, CatBoost, RF, GB, KNN, and SVM, respectively. In addition to gaining the best F-score, the stacking method predicted bike mode with 91.36%, and train and walking modes with F-scores of 82.35% and 94.13%, respectively, with the highest values. In bus and car modes, the LightGBoost method

TABLE 10.4 Investigating the Effectiveness of Feature Selection Based on Classification Models

Classification Models	Number of features by feature selection of 128 feature	F-score weighted obtained from the feature selection (Minus All features groups)	Training Time per second (Minus All features groups)
XGBoost	51	% 91.62 (- % 0.13)	14.21 (-12.4)
LightGBoost	45	% 91.87 (- % 0.05)	2.99 (-1.47)
RF	36	% 90.90 (- % 0.31)	3.21 (-2.51)
GB	43	% 90.96 (-% 0.62)	258.56 (-539.20)
CatBoost	40	% 91.51 (+ % 0.21)	12.68 (-23.06)
KNN	36	% 86.92 (+ % 23.15)	0.01 (0)
SVM	45	% 86.91 (+ % 8.08)	1.42 (-60.04)

TABLE 10.5 F-Score Evaluation of Each Transportation Mode after Feature Selection

Classification Models	Bike	Bus	Car	Train	Walk	Weighted Average	Micro Average
RF	90.48	86.45	87.70	80.62	93.49	89.90	90.08
GB	90.47	86.45	87.86	80.60	93.44	89.89	90.02
CatBoost	90.60	88.13	90.42	82	94.07	90.94	91.08
KNN	87.93	81.19	79.09	76.41	82.71	86.69	86.93
SVM	87.44	81.71	81.32	72.09	82.75	86.68	86.83
XGBoost	91.20	88.79	91.18	81.30	93.62	91.02	91.15
LightGBoost	91.09	89.27	91.33	82	93.91	91.30	91.41
MVE	91.28	88.46	90.05	81.93	93.98	91.03	91.18
WAPVE	91.29	88.05	90.12	81.72	93.93	90.93	91.08
Stacking	91.36	88.83	90.86	82.35	94.13	91.32	91.42

predicted better than the other models with F-scores of 89.27% and 91.33%, respectively. Considering the prediction of all the Transportation Modes in EL mode, the stacking method obtained the F-score weighted and micro average of 91.32% and 91.42%, respectively. In the case of using only one model, the LightGBoost method obtaining the F-score weighted and micro average of 91.30% and 91.41% were able to best predict Transportation Modes.

10.4.7 Interpretability of Classification Models

Ensemble learning in this research calculates a combination of implemented classification methods that improve performance. But this decreases the model's interpretability. To solve this problem, it is necessary to check the effectiveness of the features in each classification algorithm to check the level of interpretability. At this stage, the selected features should be examined. Based on this scenario, the intersection of the

TABLE 10.6 F-Score Evaluation of Each Transportation Mode after Feature Selection

Number of Interaction Models	Kinematic Features	Spatial Features	Contexual Features
7	V_avg, A_min	Dtrain_avg, Dtrain_max, Bus_SI	-
6	V_std, A_75th, J_avg, J_10th, BCR	Droad_avg, Dtrain_std, Train_SI, Ddes_bus	-
5	A_avg, A_90th, J _std, BR_10th, V_max, V_cv	Droad_max, Droad_min, Droad_std, Droad_interquantile, Dbus_avg, Dbus_min, Dorg_des_avg_bus, Dorg_train, Ddes_train	Holiday, T, DC
4	V_skew, V_min, BR_med, BR_max	Dtrain_cv, Dbus_interquantile	SN, FG
3	A_10th, J _med, J _skew, A_med, A_skew, J_kurtosis, A_kurtosis	Droad_90th, Droad_cv, Dbus_max, Dbus_std, Dorg_bus	TS, Elv_cv, H, Elv_valuerange
2	V_25th, V_10th, BR_avg, Dist, BR_std, BR_skew	Dbus_cv, V_interquantile, Droad_kurtosis, Dbus_skew	VM, RA, VV
1	A_std, J_cv, BR_min, V_kurtosis, BR_75th, BR_cv, A_cv, A_max	Droad_skew, Dtrain_kurtosis, Dbus_kurtosis	-

selected features in the seven implemented classification models according to Table 10.6 has been calculated. Then, SHAP Values are calculated based on the selected features and separately from the classification models. For this purpose, Tree-SHAP is calculated in tree-based classification models, including XGB, LightGBoost, RF, GB, and CatBoost. In SVM and KNN models, Kernel-SHAP is calculated and 1000 trips are considered due to the high processing time in the kernel. Because of the limitations in the number of pages, SHAP values have been calculated for LightGBoost classification as one of the best models separating all the transport modes shown by the bee swarm plot in Figure 10.7. In this figure, a single dot on each feature represents trips; the y-axis represents SHAP values; and pile-up dots represent density. The color of the points indicates the original value of the feature whereas the red and blue points, respectively, indicate the higher and lower selected feature values. The mean of |SHAP Values| for each classification model and selected features according to Figure 10.8 has been calculated. According to bee swarm plots, such as Figures 10.7 and mean of |SHAP Values| in Figures 10.8, the best features with the most impact in Bike, Car, and Walk modes are velocity average and velocity standard deviation. These important features in bus mode are acceleration minimum and velocity standard deviation. In train mode, spatial features such as train station indicators are important in addition to the kinematic features of velocity. Based on the features repeated in Table 10.6 and Figure 10.8, velocity average in the kinematic features group, bus station indicator in the spatial features group, and path geometry changes feature based on distance in the contextual features group are important features.

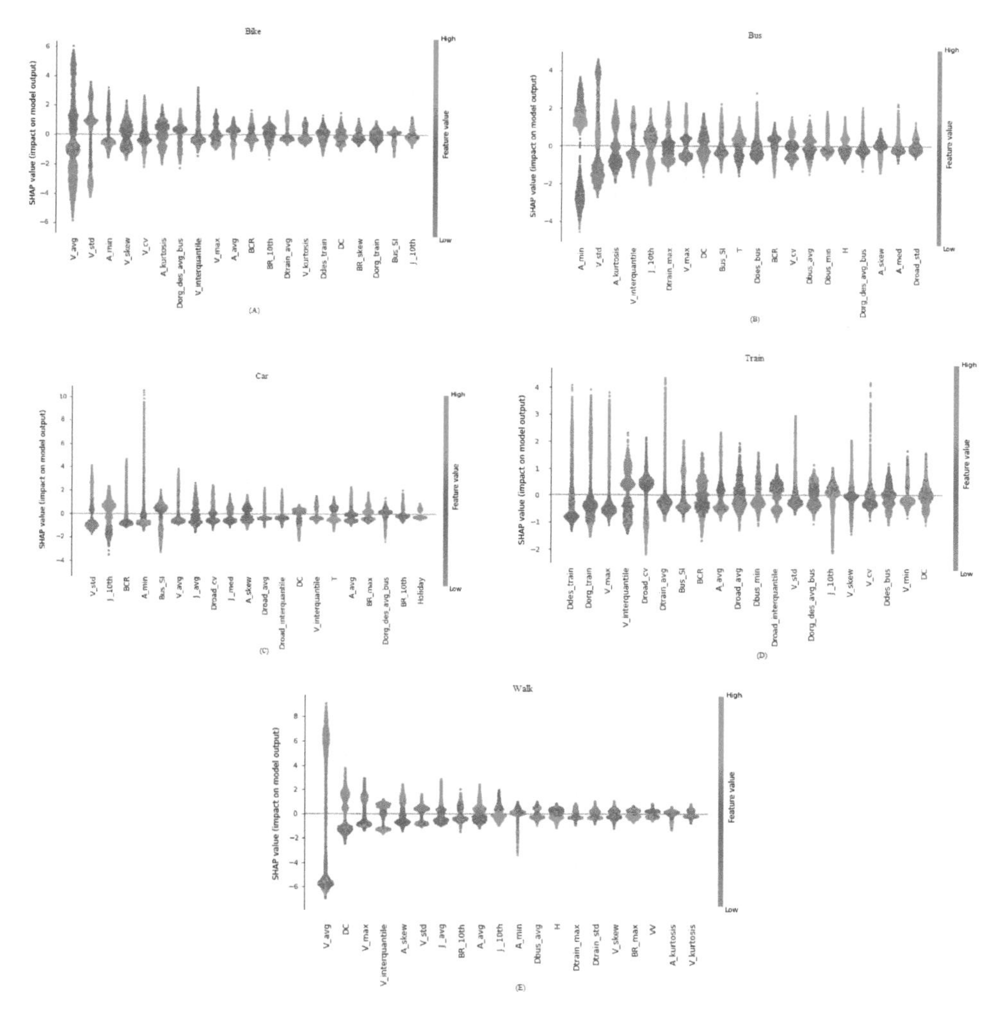

Figure 10.7 SHAP values for LightGBoost Classification.

10.4.8 Comparison with Other Studies

In the final evaluation phase, the proposed model was compared with other studies that have predicted Transportation Modes in Geolife data. The interpretability of features using all three groups of them separately from classification models as well as separate investigation of each feature has not been studied. The interpretability of features has only been based on the accuracy obtained based on the classification. In most of them, only the accuracy of different classification models has been investigated. In Table 10.7, the results of the F-score weighted average in ten studies are reviewed, Interpretability classification has been done in all of them and only in some features interpretability has been investigated based on the accuracy of the classification model. Different modes have been predicted in the studies, in which the ground modes of the bike, bus, car, train, and walking were the most common. In the approach of our article, these modes were also used as predictable modes. As

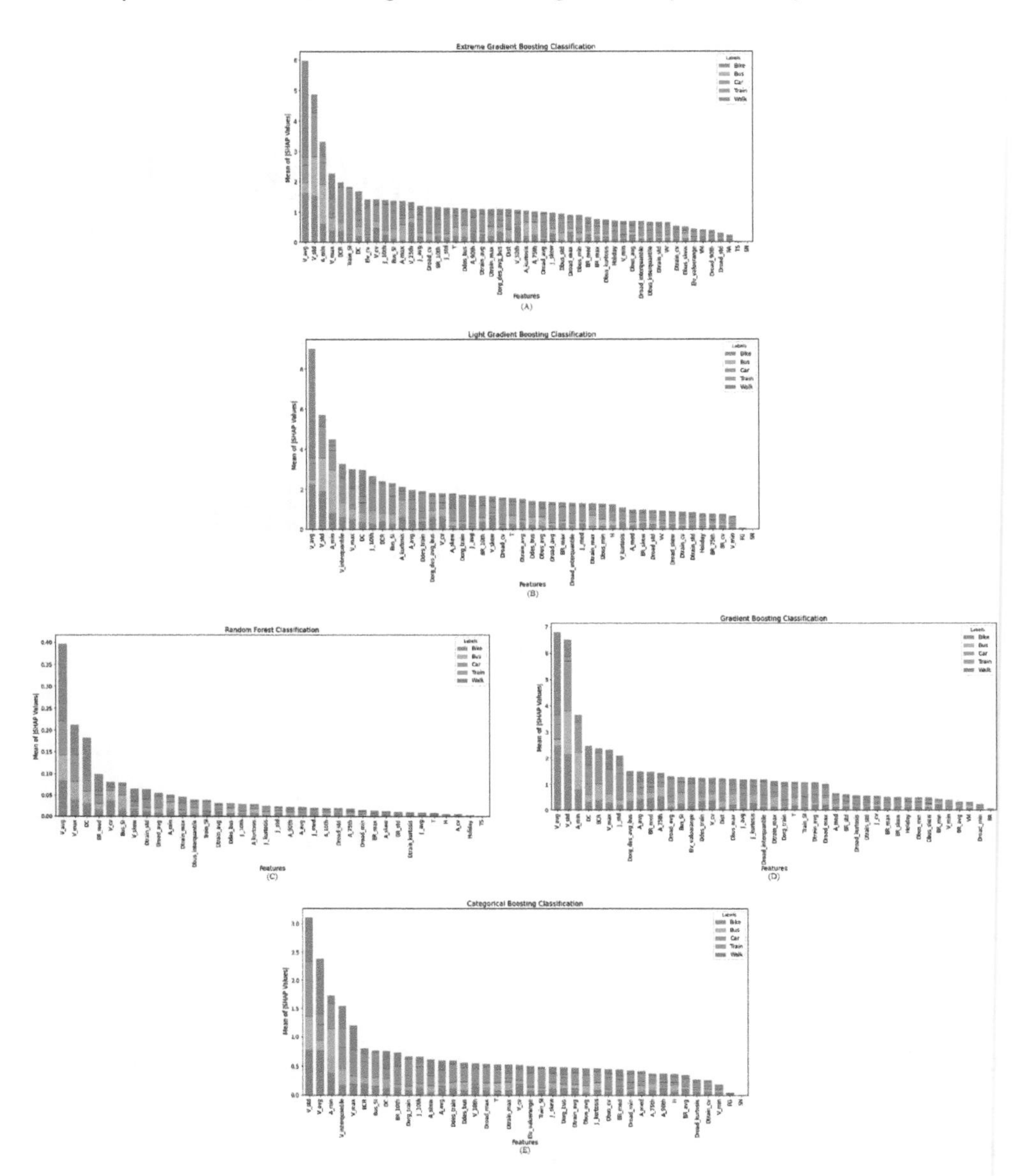

Figure 10.8 Mean of |SHAP value| for each classification model.

mentioned, studies that have been used in deep learning methods have scored lower on the F-score than those using machine learning and feature extraction. Studies have used various data, all of which share GPS data. In some studies, spatial data are used while in others contextual data is utilized. All three datasets were used in our paper approach. Among these studies, the proposed model in our study was able to provide the best prediction and obtain a weighted average F-score of 91.32% in the stacking ensemble models, and in the case of using only one classifier, the LightGBoost classification method can achieve a weighted average F-score of 91.30%.

TABLE 10.7 Comparison of F-Score of Proposed Models Than Related Works

Methods	Mode	Classification model	Feature Interpretability	Classification Interpretability	Data	F-Score (%)
Dabiri and Heaslip [10]	Bike, Bus, Car, Train, Walk	CNN, KNN, SVM, DT, RT, MLP			GPS	84.63
Xiao, et al. [19]	Bike, Bus, Car, Train, Subway, Walk	DT, KNN, SVM, RF, GB, XGBoost			GPS	90.92
Wang, et al. [30]	Bike, Bus, Car, Taxi, Train, Subway, Walk	XGBoost, LightGBoost			GPS	90.51
Dabiri, et al. [33]	Bike, Bus, Car, Train, Walk	Semi-Supervised Convolutional Autoencoder, KNN, SVM, DT, MLP, CNN, LSTM, Semi-Two-Steps, Semi-Pseudo-Label			GPS	77.49
Huang, et al. [35]	Bike, Bus, Car, Subway, Walk	Collaborative isolation forest, LR, KNN, SVM, RF			GPS, Spatial	81.43
Nawaz, et al. [1]	Bike, Bus, Car, Walk	Convolutional LSTM, SVM, RF, MLP, LSTM, CNN			GPS, Contextual	83.79
Li, et al. [26]	Bike, Bus, Car, Subway, Walk	ANN, LightGBoost, RF, XGBoost, DT, AdaBoost			GPS, Spatial	91.05
Li, et al. [36]	Bike, Bus, Car, Train, Walk	GAN (generative adversarial networks) + CNN, ANN, RF, SVM			GPS	86.84
Liang, et al. [4]	Bike, Bus, Car, Train, Walk, Plane	DT, RF, AdaBoost, Bagging, GBDT, XGBoost			GPS, Spatial	86.19
Yu, et al. [34]	Bike, Bus, Car, Train, Subway, Walk	Multi-layer neural network and post-processing, SVM, CNN, Back Propagation Neural Network, Multi-layer neural network			GPS, Spatial	80.61
Proposed (LightG-Boost)	Bike, Bus, Car, Train, Walk	LightGBoost			GPS, Spatial, Contextual	91.30
Proposed (Stacking)	Bike, Bus, Car, Train, Walk	Stacking (KNN, RF, CatBoost, LightGBoost, XGBoost, GB, SVM)			GPS, Spatial, Contextual	91.32

10.5 CONCLUSION

Nowadays, with the expansion of cities, the development and implementation of intelligent transportation play an essential role in the daily lives of citizens. Identifying Transportation Modes is one of the prerequisites for the development of intelligent transportation. With the spread of GPS and the use of this technology by citizens on their smartphones, a massive amount of location data can be accessed. This data is free of any Transportation Mode labels and requires an approach to predict the Transportation Modes used in each citizen's trip. This study aims to present an approach that can be used to extract Transportation Modes with high accuracy via extracting important features from the available data. Due to the availability of spatial data such as road networks, contextual data such as weather data, and elevation maps as open-source, more features can be used to improve predicting accuracy. In this study, using GPS, spatial and contextual data, and several machine learning classification algorithms including GB, XGBoost, LightGBoost, CatBoost, RF, KNN, and SVM, we can accurately predict Transportation Modes. This article aims to study the effect of travel kinematics and spatial and contextual feature groups on walking, bike, bus, car, and train modes. Moreover, the next objective is to select the classification model that could best predict. Using three groups of features improved the prediction compared to using the kinematic travel feature by itself. Using all the features could reduce the speed of training of all models and reduce the F-score of KNN, SVM, and CatBoost models. For this reason, the hybrid feature selection method was implemented, by which effective features were selected and used in model training. It uses the intersection of features and the SHAP method to determine which features have influenced classification models and to find important features. To prevent biased prediction and achieve more reliable results using a new approach, three methods of combining implemented machine learning models, including MVE, WAPVE, and stacking, were used. After implementation, in the case of using only one model, the LightGBoost method and, in the case of combining models, the stacking method was able to predict the Transportation Modes in the best way by obtaining the highest F-score. In future studies, more classification methods, such as deep learning algorithms, will be implemented, so that the prediction combination of more models can be used for the final prediction. In addition, more features are extracted, so that after implementing feature selection, more important features can be used in implementing the classifiers.

Bibliography

[1] A. Nawaz, H. Zhiqiu, W. Senzhang, Y. Hussain, I. Khan, and Z. Khan, "Convolutional LSTM based Transportation Mode learning from raw GPS trajectories," IET Intelligent Transport Systems, vol. 14, no. 6, pp. 570–577, 2020. DOI: 10.1049/iet-its.2019.0017

[2] C. Cong, Y. Kwak, and B. Deal, "Incorporating active Transportation Modes in large scale urban modeling to inform sustainable urban development," Computers, Environment and Urban Systems, vol. 91, p. 101726, 2022. https://doi.org/10.1016/j.compenvurbsys.2021.101726

[3] H. M. Mir, K. Behrang, M. T. Isaai, and P. Nejat, "The impact of outcome framing and psychological distance of air pollution consequences on Transportation Mode choice," Transportation Research Part D: Transport and Environment, vol. 46, pp. 328–338, 2016. https://doi.org/10.1016/j.trd.2016.04.012

[4] J. Liang et al., "An enhanced Transportation Mode detection method based on GPS data," in International Conference of Pioneering Computer Scientists, Engineers and Educators, 2017: Springer, pp. 605–620. DOI: 10.1007/978-981-10-6385-5_51

[5] Y. Zheng, Y. Chen, Q. Li, X. Xie, and W.-Y. Ma, "Understanding Transportation Modes based on GPS data for web applications," ACM Transactions on the Web (TWEB), vol. 4, no. 1, pp. 1–36, 2010. DOI: 10.1145/1658373.1658374

[6] X. Zhu, J. Li, Z. Liu, S. Wang, and F. Yang, "Learning transportation annotated mobility profiles from GPS data for context-aware mobile services," in 2016 IEEE International Conference on Services Computing (SCC), 2016: IEEE, pp. 475–482. DOI: 10.1109/SCC.2016.68

[7] J. Parkka, M. Ermes, P. Korpipaa, J. Mantyjarvi, J. Peltola, and I. Korhonen, "Activity classification using realistic data from wearable sensors," IEEE Transactions on Information Technology in Biomedicine, vol. 10, no. 1, pp. 119–128, 2006. DOI: 10.1109/TITB.2005.856863

[8] H. F. Nweke, Y. W. Teh, G. Mujtaba, and M. A. Al-Garadi, "Data fusion and multiple classifier systems for human activity detection and health monitoring: Review and open research directions," Information Fusion, vol. 46, pp. 147–170, 2019. DOI: 10.1016/j.inffus.2018.06.002

[9] B. Wang, L. Gao, and Z. Juan, "Travel mode detection using GPS data and socioeconomic attributes based on a random forest classifier," IEEE Transactions on Intelligent Transportation Systems, vol. 19, no. 5, pp. 1547-1558, 2017. DOI: 10.1109/TITS.2017.2723523

[10] S. Dabiri and K. Heaslip, "Inferring Transportation Modes from GPS trajectories using a convolutional neural network," Transportation Research Part C: Emerging Technologies, vol. 86, pp. 360–371, 2018. https://doi.org/10.1016/j.trc.2017.11.021

[11] J. Zhang, Y. Zheng, D. Qi, R. Li, and X. Yi, "DNN-based prediction model for spatio-temporal data," in Proceedings of the 24th ACM SIGSPATIAL International Conference on Advances in Geographic Information Systems, 2016, pp. 1–4. https://doi.org/10.1145/2996913.2997016

[12] A. Khosroshahi, E. Ohn-Bar, and M. M. Trivedi, "Surround vehicles trajectory analysis with recurrent neural networks," in 2016 IEEE 19th International Conference on Intelligent Transportation Systems (ITSC), 2016: IEEE, pp. 2267–2272. DOI: 10.1109/ITSC.2016.7795922

[13] A. Bolbol, T. Cheng, I. Tsapakis, and J. Haworth, "Inferring hybrid Transportation Modes from sparse GPS data using a moving window SVM classification," Computers, Environment and Urban Systems, vol. 36, no. 6, pp. 526–537, 2012. https://doi.org/10.1016/j.compenvurbsys.2012.06.001

[14] S. Wang, L. He, L. Stenneth, S. Y. Philip, Z. Li, and Z. Huang, "Estimating urban traffic congestions with multi-sourced data," in 2016 17th IEEE International Conference on Mobile Data Management (MDM), 2016, vol. 1: IEEE, pp. 82–91. DOI: 10.1109/MDM.2016.25

[15] S. Wang, X. Zhang, F. Li, S. Y. Philip, and Z. Huang, "Efficient traffic estimation with multi-sourced data by parallel coupled hidden markov model," IEEE Transactions on Intelligent Transportation Systems, vol. 20, no. 8, pp. 3010–3023, 2018. DOI: 10.1109/TITS.2018.2870948

[16] S. Wang et al., "Computing urban traffic congestions by incorporating sparse GPS probe data and social media data," ACM Transactions on Information Systems (TOIS), vol. 35, no. 4, pp. 1–30, 2017. https://doi.org/10.1145/3057281

[17] A. Yazdizadeh, Z. Patterson, and B. Farooq, "An automated approach from GPS traces to complete trip information," International Journal of Transportation Science and Technology, vol. 8, no. 1, pp. 82–100, 2019. https://doi.org/10.1016/j.ijtst.2018.08.003

[18] K. Sila-Nowicka, J. Vandrol, T. Oshan, J. A. Long, U. Demsar, and A. S. Fotheringham, "Analysis of human mobility patterns from GPS trajectories and contextual information," International Journal of Geographical Information Science, vol. 30, no. 5, pp. 881–906, 2016. https://doi.org/10.1080/13658816.2015.1100731

[19] Z. Xiao, Y. Wang, K. Fu, and F. Wu, "Identifying different Transportation Modes from trajectory data using tree-based ensemble classifiers," ISPRS International Journal of Geo-Information, vol. 6, no. 2, p. 57, 2017. https://doi.org/10.3390/ijgi6020057

[20] R. E. Guinness, "Beyond where to how: A machine learning approach for sensing mobility contexts using smartphone sensors," Sensors, vol. 15, no. 5, pp. 9962–9985, 2015. https://doi.org/10.3390/s150509962

[21] P. Nirmal, I. Disanayaka, D. Haputhanthri, and A. Wijayasiri, "Transportation Mode Detection Using Crowdsourced Smartphone Data," in 2021 28th Conference of Open Innovations Association (FRUCT), 2021: IEEE, pp. 341–349. DOI: 10.23919/FRUCT50888.2021.9347625

[22] M. Gjoreski et al., "Classical and deep learning methods for recognizing human activities and modes of transportation with smartphone sensors," Information Fusion, vol. 62, pp. 47–62, 2020. https://doi.org/10.1016/j.inffus.2020.04.004

[23] D. Shin et al., "Urban sensing: Using smartphones for Transportation Mode classification," Computers, Environment and Urban Systems, vol. 53, pp. 76-86, 2015. https://doi.org/10.1016/j.compenvurbsys.2014.07.011

[24] G. Xiao, Z. Juan, and J. Gao, "Travel mode detection based on neural networks and particle swarm optimization," Information, vol. 6, no. 3, pp. 522–535, 2015. https://doi.org/10.3390/info6030522

[25] H. Maenpaa, A. Lobov, and J. L. M. Lastra, "Travel mode estimation for multimodal journey planner," Transportation Research Part C: Emerging Technologies, vol. 82, pp. 273–289, 2017. https://doi.org/10.1016/j.trc.2017.06.021

[26] J. Li, X. Pei, X. Wang, D. Yao, Y. Zhang, and Y. Yue, "Transportation Mode identification with GPS trajectory data and GIS information," Tsinghua Science and Technology, vol. 26, no. 4, pp. 403–416, 2021. DOI: 10.26599/TST.2020.9010014

[27] H. Gong, C. Chen, E. Bialostozky, and C. T. Lawson, "A GPS/GIS method for travel mode detection in New York City," Computers, Environment and Urban Systems, vol. 36, no. 2, pp. 131–139, 2012. https://doi.org/10.1016/j.compenvurbsys.2011.05.003

[28] V. S. Brum-Bastos, J. A. Long, and U. Demsar, "Weather effects on human mobility: a study using multi-channel sequence analysis," Computers, Environment and Urban Systems, vol. 71, pp. 131–152, 2018. https://doi.org/10.1016/j.compenvurbsys.2018.05.004

[29] L. Ma, H. Xiong, Z. Wang, and K. Xie, "Impact of weather conditions on middle school students' commute mode choices: Empirical findings from Beijing, China," Transportation Research Part D: Transport and Environment, vol. 68, pp. 39–51, 2019. https://doi.org/10.1016/j.trd.2018.05.008

[30] B. Wang, Y. Wang, K. Qin, and Q. Xia, "Detecting Transportation Modes based on LightGBM classifier from GPS trajectory data," in 2018 26th International Conference on Geoinformatics, 2018: IEEE, pp. 1–7. DOI: 10.1109/GEOINFORMATICS.2018.8557149

[31] A. Nawaz et al., "Mode Inference using enhanced Segmentation and Pre-processing on raw Global Positioning System data," Measurement and Control, vol. 53, no. 7–8, pp. 1144-1158, 2020. https://doi.org/10.1177/0020294020918324

[32] G. Xiao, Z. Juan, and C. Zhang, "Travel mode detection based on GPS track data and Bayesian networks," Computers, Environment and Urban Systems, vol. 54, pp. 14–22, 2015. https://doi.org/10.1016/j.compenvurbsys.2015.05.005

[33] S. Dabiri, C.-T. Lu, K. Heaslip, and C. K. Reddy, "Semi-supervised deep learning approach for Transportation Mode identification using GPS trajectory data," IEEE Transactions on Knowledge and Data Engineering, vol. 32, no. 5, pp. 1010–1023, 2019. DOI: 10.1109/TKDE.2019.2896985

[34] Q. Yu, Y. Luo, D. Wang, C. Chen, L. Sun, and Y. Zhang, "Using information entropy and a multi-layer neural network with trajectory data to identify Transportation Modes," International Journal of Geographical Information Science, vol. 35, no. 7, pp. 1346–1373, 2021. https://doi.org/10.1080/13658816.2021.1901904

[35] Z. Huang, P. Wang, and Y. Liu, "Statistical characteristics and Transportation Mode identification of individual trajectories," International Journal of Modern Physics B, vol. 34, no. 10, p. 2050092, 2020. https://doi.org/10.1142/S0217979220500927

[36] L. Li, J. Zhu, H. Zhang, H. Tan, B. Du, and B. Ran, "Coupled application of generative adversarial networks and conventional neural networks for travel mode detection using GPS data," Transportation Research Part A: Policy and Practice, vol. 136, pp. 282–292, 2020. https://doi.org/10.1016/j.tra.2020.04.005

[37] R. B. Langley, "Innovation: the GPS error budget," GPS World, vol. 8, no. 3, pp. 51-56, 1997.

[38] M. Guo, S. Liang, L. Zhao, and P. Wang, "Transportation Mode recognition with deep forest based on GPS data," IEEE Access, vol. 8, pp. 150891–150901, 2020. DOI: 10.1109/ACCESS.2020.3015242

[39] T. Vincenty, "Direct and inverse solutions of geodesics on the ellipsoid with application of nested equations," Survey Review, vol. 23, no. 176, pp. 88–93, 1975. https://doi.org/10.1179/sre.1975.23.176.88

[40] Y. Zheng, Q. Li, Y. Chen, X. Xie, and W.-Y. Ma, "Understanding mobility based on GPS data," in Proceedings of the 10th International Conference on Ubiquitous Computing, 2008, pp. 312–321. https://doi.org/10.1145/1409635.1409677

[41] G. Xiao, Q. Cheng, and C. Zhang, "Detecting travel modes from smartphone-based travel surveys with continuous hidden Markov models," International Journal of Distributed Sensor Networks, vol. 15, no. 4, p. 1550147719844156, 2019. https://doi.org/10.1177/1550147719844156

[42] L. Yu and H. Liu, "Feature selection for high-dimensional data: A fast correlation-based filter solution," in Proceedings of the 20th International Conference on Machine Learning (ICML-03), 2003, pp. 856–863.

[43] P. Pudil, J. Novovicova, and J. Kittler, "Floating search methods in feature selection," Pattern Recognition Letters, vol. 15, no. 11, pp. 1119–1125, 1994. https://doi.org/10.1016/0167-8655(94)90127-9

[44] J. Han, M. Kamber, and J. Pei, "Data Mining: Concepts and Techniques Third Edition [M]," The Morgan Kaufmann Series in Data Management Systems, vol. 5, no. 4, 2012.

[45] R. S. Shah, "Support vector machines for classification and regression," 2007. https://doi.org/10.1039/B918972F

[46] J. H. Friedman, "Greedy function approximation: a gradient boosting machine," Annals of Statistics, pp. 1189–1232, 2001. DOI: 10.1214/aos/1013203451

[47] G. Ke et al., "Lightgbm: A highly efficient gradient boosting decision tree," Advances in Neural Information Processing Systems, vol. 30, pp. 3146–3154, 2017.

[48] L. Prokhorenkova, G. Gusev, A. Vorobev, A. V. Dorogush, and A. Gulin, "CatBoost: unbiased boosting with categorical features," Advances in Neural Information Processing Systems, vol. 31, 2018. https://doi.org/10.48550/arXiv.1706.09516

[49] A. V. Dorogush, V. Ershov, and A. Gulin, "CatBoost: gradient boosting with categorical features support," arXiv preprint arXiv:1810.11363, 2018. https://doi.org/10.48550/arXiv.1810.11363

[50] A. Liaw and M. Wiener, "Classification and regression by randomForest," R News, vol. 2, no. 3, pp. 18–22, 2002.

[51] T. Chen and C. Guestrin, "Xgboost: A scalable tree boosting system," in Proceedings of the 22nd ACM SIGKDD International Conference on Knowledge Discovery and Data Mining, 2016, pp. 785–794. https://doi.org/10.1145/2939672.2939785

[52] P. Cunningham and S. J. Delany, "k-Nearest neighbour classifiers: (with Python examples)," arXiv preprint arXiv:2004.04523, 2020. https://doi.org/10.1145/3459665

[53] S. A. Dudani, "The distance-weighted k-nearest-neighbor rule," IEEE Transactions on Systems, Man, and Cybernetics, no. 4, pp. 325–327, 1976. DOI: 10.1109/TSMC.1976.5408784

[54] A. Dogan and D. Birant, "A weighted majority voting ensemble approach for classification," in 2019 4th International Conference on Computer Science and Engineering (UBMK), 2019: IEEE, pp. 1–6. DOI: 10.1109/UBMK.2019.8907028

[55] A. Rojarath and W. Songpan, "Probability-Weighted Voting Ensemble Learning for Classification Model," Journal of Advances in Information Technology Vol, vol. 11, no. 4, 2020.

[56] C. C. Aggarwal, "Data Classification: Algorithms and Applications," in Chapman & Hall/CRC Data Mining and Knowledge Discovery Series: Chapman and Hall/CRC, 2014, pp. 498–501. https://doi.org/10.1201/b17320

[57] J. D. Novakovic, A. Veljovic, S. S. Ilic, Z. Papic, and T. Milica, "Evaluation of classification models in machine learning," Theory and Applications of Mathematics & Computer Science, vol. 7, no. 1, pp. 39–46, 2017.

[58] M. Grandini, E. Bagli, and G. Visani, "Metrics for multi-class classification: an overview," arXiv preprint arXiv:2008.05756, 2020. https://doi.org/10.48550/arXiv.2008.05756

[59] Z. C. Lipton, C. Elkan, and B. Naryanaswamy, "Optimal thresholding of classifiers to maximize F1 measure," in Joint European Conference on Machine Learning and Knowledge Discovery in Databases, 2014: Springer, pp. 225–239. DOI: 10.1007/978-3-662-44851-9_15

[60] Z. Li, "Extracting spatial effects from machine learning model using local interpretation method: An example of SHAP and XGBoost," Computers, Environment and Urban Systems, vol. 96, p. 101845, 2022. https://doi.org/10.1016/j.compenvurbsys.2022.101845

[61] S. M. Lundberg and S.-I. Lee, "A unified approach to interpreting model predictions," Advances in Neural Information Processing Systems, vol. 30, 2017.

[62] S. M. Lundberg et al., "From local explanations to global understanding with explainable AI for trees," Nature Machine Intelligence, vol. 2, no. 1, pp. 56–67, 2020. https://doi.org/10.48550/arXiv.1905.04610

[63] S. M. Lundberg, G. G. Erion, and S.-I. Lee, "Consistent individualized feature attribution for tree ensembles," arXiv preprint arXiv:1802.03888, 2018. https://doi.org/10.48550/arXiv.1802.03888

[64] Y. Zheng, L. Zhang, X. Xie, and W.-Y. Ma, "Mining interesting locations and travel sequences from GPS trajectories," in Proceedings of the 18th International Conference on World Wide Web, 2009, pp. 791–800. https://doi.org/10.1145/1526709.1526816

[65] Y. Zheng, X. Xie, and W.-Y. Ma, "Geolife: A collaborative social networking service among user, location and trajectory," IEEE Data Eng. Bull., vol. 33, no. 2, pp. 32–39, 2010.

[66] "SRTM digital elevation model data." https://earthexplorer.usgs.gov.

[67] "Global climate data." https://en.tutiempo.net/climate.

[68] "Beijing City Lab." https://www.beijingcitylab.com.

[69] P. Refaeilzadeh, L. Tang, and H. Liu, "Cross-validation," Encyclopedia of database systems, vol. 5, pp. 532–538, 2009.

CHAPTER 11

Blockchain and Explainable AI for Trustworthy Autonomous Vehicles

Ouassima Markouh
Laboratory of Artificial Intelligence, Data Science and Emerging Systems, ENSA Fez, USMBA, Morocco

Amina Adadi
ISIC Research Team, L2ISEI Laboratory, Moulay Ismail University, Meknes, Morocco

Mohammed Berrada
Laboratory of Artificial Intelligence, Data Science and Emerging Systems, ENSA Fez, USMBA, Morocco

CONTENTS

Autonomous Vehicles, (AVs) have become subject to trust and ethical concerns. If AVs are to be accepted and largely adopted, we need to create designs that build trust. At the heart of technologies of trust lies Explainable Artificial Intelligence and Blockchain. The aim of the chapter is to present an exploratory analysis of how Explainable Artificial Intelligence (XAI) and Blockchain technology could leverage trust in autonomous vehicles industry. First, we present a preliminary overview of the literature related to the use of blockchain architectures and XAI models in AV. Then, we open discussion for the possibility of coupling the two technologies (Blockchain and XAI) for more robust and trustworthy AVs.

DOI: 10.1201/9781003324140-11

11.1 INTRODUCTION

AVs are considered a big disruptive technology in the smart transportation domain. However, in spite of the improvements in the infrastructure, the noticeable increase in the number of driverless vehicles opened the door for more traffic issues. The existing autonomous transportation solutions are struggling to overcome the traffic problems resulted by this continued evolution. In this regard, the Internet of Vehicles (IoV) was found to ensure the communication between different driverless vehicles to avoid certain traffic-related issues [13]. Yet, the IoV has introduced other forms of problems regarding data sharing, privacy, security, interoperability and much more [13]. The blockchain technology has proved its efficiency in finance, which is illustrated by the success the cryptocurrencies have gained. A blockchain consists of a chain of blocks linked together via cryptographic hash, each block contains the cryptographic hash of the previous block. the power of the blockchain is seen in this linkage, hacking into one block will cause the malfunction of all the previous ones, thus ensuring high level of security. Blockchain being decentralized means the ledger is shared between different parties and they can all publicly view the data displayed on the ledger thus eliminating the single authority problem. A key feature of the blockchain technology is smart contracts referring to self-executing piece of code containing a set of agreements and terms stored on the blockchain behaving like legal documents managing the transactions between different parties. The code and the conditions in the contracts are publicly available on the ledger, the smart contracts is automatically executed by a distributed ledger system. The trust that the scientific communities put into the Distributed Ledger Technology (DLT) has exponentially increased over the years. Although the primary use of blockchain was in finance, the researchers from different fields take advantage of the opportunities this technology held. The investigations in the IoV filed have explored the potential use of decentralized-based solution in AVs to ensure trust, and how would it help with the concerns that couldn't be completely solved using centralized-based solution.

On the other end of the spectrum, deep learning models are dominating the Artificial Intelligence (AI) approaches used in AVs. Yet, the deep architectures used in an AV context are not always trusted to be accurate especially when arguing about the scene perception feature. The more the advances in AI-empowered driving systems accrue, the greater the number of autonomous vehicles we see on the road. Thus, the skeptical opinion on the use of smart cars on roads raises. For instance, in [18] the authors investigated the ethical issues of adopting AVs by analysis the Trolley Problem as a core example. The issue is how to blame when an accident accrues, the autonomous car or the driver which can refer to as the responsibility gap issue. If the algorithms output a non-ethical decision could the driver change the course of events? For example, when being in a situation where both the passengers and the driver are in danger, the automotive system here is in a lost point where no decision could be made. The system is not designed to deal with such a situation because of the non-sufficient data which is referred to as semantic gap issue. Another issue is the AV regulation, the growing concern caused by the advances in AI algorithms rises the need for an AV standards and regulation system to monitor the data circulating in an intelligent transportation system network. The main objective is to protect stakeholders' privacy by securing their personal data. The aforementioned issues lead the stakeholders to ask the "why" question to understand the causes leading to the consequences of these challenges. Thus, we notice the need for explanations about the decisions of driverless cars. Explainable AI is a set of methods designed to help understand and interpret predictions made by AI models without impacting their performance. This technology holds the potential to enhance trust and public acceptance of AV.

Against this background, we propose in this chapter to investigate through the literature lens the use of XAI and Blockchain technologies in the AVs context. The aim is to promote more synergy between these two technologies of trust. The rest of the chapter is organized as follows. Section 11.2 provides a review of the application of Blockchain in AV and IoV. Section 11.3 explores the importance of the explainability in AV industry. Section 11.4 opens up the discussion on the potential of converging XAI and Blockchain technologies for more robust AV, and finally, Section 11.5 concludes the paper.

11.2 BLOCKCHAIN IN AUTONOMOUS VEHICLES

A literature review on the use of blockchain in autonomous vehicles was introduced in this section. The purpose is to identify and examine the added value of DLT in IoV field based on existing works in the literature. To achieve this purpose, the analysis was focused on a scanning of articles related to the subject published in the Scopus database. Scopus was selected due to its broad coverage of indexed publications. To obtain relevant articles, a specific search query was conducted. We limited the search to articles providing a blockchain -based solution in autonomous vehicles. The main criteria for selecting relevant literature are: (i) the article is written in English, (ii) the full-text is available, (iii) review, meta-analysis, survey, or commentary articles were excluded from the results, (iv) the study is conducted in an autonomous transportation context, finally, (v) the article should be tackling at least one autonomous vehicles -related problem. For each chosen article a full-text analysis was conducted to identify the transportation-related problematic solved using blockchain.

By examining the findings of the studied papers, we tried to compile them into a unified classification. The current autonomous transportation problems that have been addressed in the literature using blockchain /DLT can be categorized into four main classes, scilicet, (i) Trust Management, and (ii) Security Management, (iii) Safety Management, and (iv) Sustainability management. Figure 11.1 illustrates the proposed taxonomy of autonomous vehicles applications discussed in the selected articles. Each of these classes can be subcategorized into other subclasses. The occurrence frequency calculation of each class shows that trust management approaches are the most dominant application in literature. Trust management approaches could be divided into two subcategories, (a) trust management systems related to decision-making algorithms and (b) trust management systems related to data-managing. The second class of application is security management. The security mentioned in vehicular network depends primarily on (a) identity protection and (b) network scanning to prevent malicious participants from interrupting the sensitive data circulating in

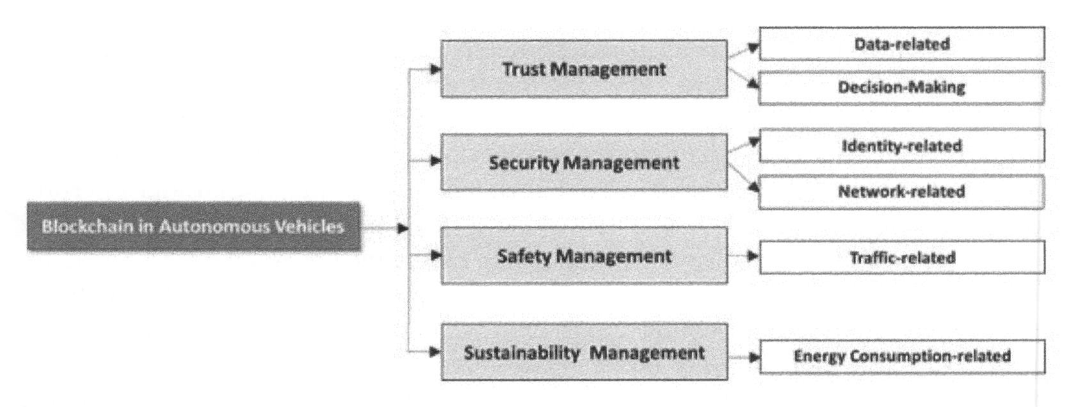

Figure 11.1 Application of blockchain in AV.

the network. The safety management approaches focus on applications related to traffic safety management. As for sustainability management solutions, they mainly address the energy consumption issue.

11.2.1 Trust Management

The AVs industry has known a huge growth in terms of investments and researches. Yet, the technology behind driverless vehicles cannot be fully trusted by communities because of the malicious attacks on the network that could result in unwanted manipulation over the vehicle or feeding the systems with inaccurate data because of a wrong classification that may lead unreasonable decision-making that do not meet the needs of a certain situation. In a transportation context each small mistake can cost priceless lives. The blockchain technology introduced a system based on decentralized distributed trustless ledger. The data carried by transactions are verified by the network nodes thus they are always trusted to be accurate. To take advantage of the innumerous advantages of intelligent cars, trust related problems should be overcame. The blockchain based approaches for trust management aimed to regain the trust in autonomous vehicles by focusing on two main areas: data related trust and decision-making related trust.

1. ***Data-related trust***
   Implementing trust mechanisms in autonomous vehicles depends primarily on the data circulating on the IoV network. In [12] and [11] the authors proposed a system based on a blockchain network to keep stakeholders' record of reputation. The aim is to allow reputation data to propagate at a local and interlocal levels. The reputation score of each vehicle determines either the data provided by that vehicle is trusted or not. The vehicles use the trusted data shared on the ledger to make predication about their environment. Each vehicle on the network has a public and private key where the transactions made by a certain vehicle is consistently verified and proved by stakeholders existing on the network to ensure its trustfulness. Regularly, autonomous vehicular networks depend on decentralized resources, thus utilizing centralized systems in IoV does not

answer the trust concern in vehicular networks. This issue was addressed in [10], where the authors utilized a decentralized protocol with no central authority to provide a secure and trusted way for vehicular registration. The primary users on this decentralized network are registered driverless vehicles using a 20-byte address to ensure secure communication between autonomous vehicles. To ensure trust in the data provided, a set of predefined transactions are defined. The transaction will oversee adding a new entity to the network after verifying the reliability of the data provided by the autonomous vehicle. Autonomous vehicle platoon can be a promising solution for traffic congestion issues. Yet, sometimes the benefits of platooning cannot be fully developed. Highway Electronic Toll Collection (ETC) charging problem is one of the platooning paradigm concerns that was addressed in [23] and [15]. The authors propose a blockchain -based efficient highway toll paradigm for the opportunistic platoon to ensure that the autonomous vehicles trust each other and to also detect the autonomous vehicles trying to escape the ETC charging by following the platoon. To overcome these concerns, a blockchain based solution will record and verify the driving history and the credential information of every registered vehicle. The authors in [2] suggested a new model called Witness of Things based on keeping decision data of autonomous vehicles in distributed ledgers using IoV networks. By providing a single version of the truth, the proposed model will help related industry and governmental institutions learn the true causes of traffic problems and cars accidents.

2. ***Decision-making related trust***
   Situation awareness data in autonomous vehicles consist of the road conditions, traffic control, incident and congestion. It helps the algorithms running the autonomous vehicle to make the right decisions. The authors in [18] and [12] provided a solution based on an edge platform running on a distributed network to enable participants to share information to build the big picture about traffic and environmental conditions, and to help the system making better decision depending on the road factors. In [24], an AI-enabled trust management system using blockchain was addressed to solve the problems related to decision-making by increasing the vehicular network security using blockchain to prevent any malicious autonomous vehicles from interrupting the network and manipulation the data which can affect driving decision, any wrong decision can have disastrous consequences.

11.2.2 Security Management

As the autonomous transportation technology evolves, the security challenges also become more and more complex. The centralized vehicular networks make the whole system vulnerable and easy to attack. The need for a distributed solution increased over the years. The cryptographic algorithms used in the blockchain networks can be utilized in autonomous transportation to increase security and prevent network attack.

1. ***Network Security Management***
   AVs make use of different machine learning (ML) algorithms to preform predictive tasks. The collaboration of decentralized systems could improve the robustness of the ML models. The authors in [9] suggested a novel consensus protocol named PoVS-BFT to assure vehicular communication in a safe way without interruption from outsiders, assuming that the autonomous vehicles utilize conventional algorithms for model optimization. The protocol is based on two-steps transaction verification mechanism to eliminate malicious actors. Ensuring privacy is one of the key characteristics in blockchain networks, with the privacy and trust being a big concern in automation transportation systems, the blockchain technology could be utilized to protect the vehicular network users' privacy by applying a network intrusion detection system [14]. Requesting a ride share in autonomous vehicles is no easy task to preform since the responding autonomous vehicle should verify and process the data related to the vehicle. The networking between vehicles should be fully secure. Relaying on trusted third party in this equation is not the optimal choice, since if the trusted authority rogues or loses credibility, the whole system will collapse. To address this challenge, the authors in [7] introduced Chained of Things framework which is a blockchain -based secure ride-sharing service between passengers and autonomous vehicles. When a new transaction accrues the network peers share the transaction information between each other using the blockchain network. In this case the information of the users requesting ride shares are verified by the peers of the network to ensure the credibility of the data to prevent susceptible entities from integrating the network. In [20] the authors investigated the feasibility of a regional blockchain while achieving a low 51% attack success probability. The approach suggests driving a condition that grantees a low 51% attack success probability in terms of the numbers of good nodes and malicious nodes, the message delivery time, and the puzzle computation time.

2. ***Identity Management***
   Protecting the users' privacy in an IoV network is a priority. This issue was addressed in a number of articles trying to integrate the blockchain technology to ensure the network peers' privacy. Specifically, the Platooning technologies enable autonomous vehicles to drive collectively. The registration process of a vehicle requires the driver's personal information which rises the privacy concern since the vehicle's information will be verified before guaranteeing access to join the platoon. Answering this issue, in reference [15] the authors suggested a privacy-preserving verification protocol based on zero-knowledge proof and permissioned blockchain. A location-aware verification protocol that validates the identity of the autonomous vehicle in a privacy-preserving manner was introduced. The IoV ecosystem requires consistent updates to ensure the efficiency of the system, yet these updates might cause breaches in the network and identity information loss when relying on a centralized solution. The

authors in [21] provided a blockchain based identity management framework for the IoV ecosystem, to ensure a secure software update.

11.2.3 Safety Management

The ultimate aim of AVs industry is to provide high levels of safety. The decision-making process in AV should be enhanced to ensure safe traffic. Real-time decision-making based on vehicle sensor data, traffic data, and environmental and situational data becomes imperative. Thus, the collaboration between different stakeholders should be facilitated to provide a safe traffic for all road users. The data about the road and traffic conditions provide a situation awareness for the algorithm running the AVs for better decision making. To ensure the traffic safety, the blockchain based solutions were integrated to provide more insights on the traffic. A decentralized framework that enables smart contracts between traffic data producers and consumers based on blockchain was introduced in [18]. The system acts as a data collector and analyzer. The smart contracts get the situational and environmental information carried by transactions as inputs then they output the statue of the traffic ensuring safety and minimizing road accidents.

11.2.4 Sustainability Management

At the level of energy consumption, the AVs industry aims to provide a low energy consuming system to overcome the environmental concerns caused by regular cars. Majority of autonomous electric vehicles (AEVs) are furnished with bidirectional chargers. Yet, during the energy trading process, the centralized systems are subject to many flaws like the high energy transmission loss and cannot resist the single-point failure. As consequence, the use of a decentralized system is important to keep the safety of energy trading between AEVs. In [26] the authors proposed a decentralized energy trading architecture. The AEVs can broadcast their requirements to other AEVs and negotiate with them on the transmission parameters. The solution is based on smart contracts that are deployed among RSUs which are automatically executed after the negotiation process. Following the same approach, the authors in [5] proposed a blockchain based system for AEV charging in a decentralized network of charging stations. The solution main objective is to minimize electric energy consumption.

11.3 EXPLAINABILITY IN AUTONOMOUS VEHICLES

Presently, the AI breakthroughs, especially in ML algorithms, made the tremendous advances seen in AVs possible. ML has become an essential enabler in self-driving cars. The ML algorithms are trained on a set of selected data before being deployed into action which makes it a strong aspect of any autonomous operation. With the help of AI models, the automotive companies promise a safer and more ecologically friendly transportation systems. Despite the fact that AI technology behind AV is so advanced, the dream of reaching a fully autonomous car without human intervention is still beyond reaching. The semi-automotive driving is the dominant solution developed by most autonomous vehicles ' companies. This issue rises the concern of having a collaborative control between the human driver and the intelligent system. For instance, an AV might be in a situation where it must handover the full control to the driver, such transition might cause sever damages if not thought of in advance. The driver must be aware of the contextual and environmental factors causing the handover, meaning the driver is in need of explanations for such a behavior. Another key concern is the social trust, people cannot trust what they cannot understand. The technology behind AVs still not understandable by common users, which makes them question the decisions of a driverless car. Even though 94% of road accidents are due to human errors and AV is a much safer choice in terms of accident prevention and energy consumption [1], in order to ensure public acceptance, stakeholders need real-time explanations of every single decision made by a self-driving car. The social acceptance of autonomous transportation systems is fragile once it's broken, it is strongly hard to regain.

The current literature contains some studies investigating the importance of the explicability in AVs. In [3] the authors demonstrated that providing the "why" information, which explains the cause behind a driverless vehicle's action, created the least anxiety and the highest trust. Thus, explanations in AVs increase the social acceptance of the Intelligent Transportation Systems and help overcome the aforementioned issues existing in the current models. In [8], the authors suggested three dimensions to consider when implementing an AI model in an autonomous vehicle, (i) Explainability, the ability to describe clearly the model to clarify the inner work of the autonomous car, (ii) Interpretability, this feature ensures that the end-user understood the model and made sense of it, (iii) Transparency, the AI model should be transparent for the users, meaning that it exhibits the understandability on its own and without any interface from outside sources.

The problem with the existing XAI models is that the only explanations made are answering the question "how the prediction is made?", this kind of explanations are not enough to justify the driverless vehicle decisions. Thus, when mapping explanations, certain criteria should be taken into considerations as mentioned in [4]. In this work, the authors proposed three assumptions when forming explanation models, (i) User needs, the need to understand who the users are and what are their need, (ii) Explanation strategies, the need for making the explanations user friendly (iii) Real-time vs post-process, this assumption should argues whether the explanations should be given in real-time (as the decision accurses) or retrospectively (after the decision is completed).

In [25], the authors discussed the importance of developing scenario-based and question-driven explanations for autonomous vehicles in order to improve their safety and usability. The traditional methods for explaining the actions of an autonomous vehicle, such as providing a linear sequence of events, are not always effective and can lead to confusion or mistrust. Instead, they suggested using a scenario-based approach, which would involve providing explanations that are generated to specific situations and user queries using a question-driven approach, which would involve the vehicle being able to anticipate and answer questions that a user might have about its actions. While some drivers feel the need for explanations, others feel that the vehicle's actions should be self-explanatory. This ethical issue was addressed in [19] and [22] where the study summarized how explanations for autonomous vehicles can play a role in achieving trust, yet considering the user's perspectives and examining the explanations provided is also important. Moreover, having a good understanding of the vehicle's capabilities and limitations increase the AV's social acceptance. Following the same approach, the authors in [1] and [16] highlighted how important it is to develop XAI systems that can provide understandable and verifiable explanations for the decisions and actions of autonomous vehicles. The authors also suggested the need for more sophisticated methods for generating explanations, the development of standards for evaluating explanations, and the integration of XAI into real-world autonomous systems. Another key aspect that should be thought of when developing an XAI model for AVs, is providing explanations using natural human language with as much less technical terms as possible, since the users might not be familiar with the technical language associated with such a model. This issue was highlighted in reference [6], it suggests using natural language conversations between the automotive system and the driver as a way to provide explanations, as they can be more intuitive and less technical than other forms of explanation.

In summary, Explainable AI can be used to enhance the functionalities of autonomous vehicles by providing transparency and understanding of the decision-making processes of the AI system. This can help to improve the safety, reliability, and trustworthiness of the vehicle's actions.

11.4 JOINT BLOCKCHAIN AND EXPLAINABLE AI FOR AUTONOMOUS VEHICLES

Based in the explored literature, blockchain and Explainable AI are both exciting technologies with great but as yet unfulfilled potential. While the IoT and smart cities have contributed positively to the growth of AV industry, blockchain technology delivers a significant upgrade that could revolutionize the future of driving. The application of blockchain in AVs and IoV grantees trust, security, safety, and sustainability since the system is deployed in a decentralized way without the need of trusted parties' involvement. On the other side, explanations are a focal human need. When interacting with an AV, the explanations are a guarantee for verifiability and control. Integrating an explainable model in the design chain of AV is important to gain user trust and acceptance. Hence, research in these two areas is regard as the sine qua non for AV industry to continue making steady progress and eventually reach the mainstream.

Moreover, from our standpoint, combining values of blockchain and XAI can considerably enhance trustworthiness, security and efficiency of AV. Indeed, in the AV context, firstly human-understandable explanations are required. Then, these explanations must be stored securely to ensure the privacy of the user and should be accessible in the future if required. Hence, blockchain architectures paired to XAI models can help building decentralized AI solutions with more data traceability and audit trail. To the best of our knowledge, no work in the literature have studied yet the potential of converging XAI and Blockchain in AV. One significant initiative has proposed recently a conceptual framework for achieving a more trustworthy and XAI by leveraging features of blockchain. In this work, [17] conceived a solution to address AI opacity by shifting the trust from a single prediction system to a set of distributed predictors, providing predictions and explanations, in which AI predictions and decision outcomes are recorded and stored in a decentralized secure manner, and aggregated and managed in a unbiased, and trusted manner using smart contract and oracles. The paper also discussed how the proposed framework can be utilized for trustworthy and explainable real-time prediction decisions, which may involve for example a fatal accident by an autonomous vehicle.

The use of XAI and blockchain alongside can create a secure, immutable, and decentralized system for the highly sensitive data that AI models need to gather, save, and make use of. In an intelligent transportation system, utilizing a blockchain and XAI based model would ensure greater confidence and accountability. Also, greater computational efficiency since the system is based on decentralized networking, thus the real-time decision-making process will be more accurate. On the other hand, the use of AI models will help the blockchain system in terms of optimization regarding security and performance by optimizing cryptographic hash functions and reduce the scalability concerns related to transactions approval. It would be also worth to investigate how to put together smart contracts and Explainable AI in one IoV decentralized framework to ensure more accurate and trustful oracles (data providers). The XAI models will act as a filter system to indicate the malicious data provided by transportation related oracles.

The blockchain technology augmented with XAI promises to deliver a trusted and traceable automation for AV. However, in order to make the expected big impact, these two technologies should mature and overcomes their intrinsic issues. Particularly, issues related to evaluation and quality assessment for XAI, and issues related to performance, and scalability for blockchain.

11.5 CONCLUSION

In this chapter we conducted a literature analysis regarding the possible uses of blockchain and Explainable AI for autonomous vehicles. In addition to understanding to what stage these technologies can be used alongside to resolve concerns related to automotive transportation systems. The applications of blockchain in AV as treated in the reviewed articles can be divided into four main categories (i) Trust Management, (ii) Safety Management, (iii) Security Management, and (iv) Sustainability Management. As for Explainable AI in AVs the main repeated objective of most of he reviewed studies is providing high a level of trust, reliability and understandability in order to provide awareness over the environmental situation and ensuring a successful human-machine interaction. We also provided an analysis over the merge of blockchain and XAI yet the number of researches regarding such a convergence is still very limited. Until now, a very small number of studies proposed a blockchain and XAI-based framework to achieve maximum efficiency in any domain and even fewer studies regarding intelligent vehicles. Thus, working on combining the blockchain technology and XAI models into one system might be of breakthrough in technology since these technologies are to be the leaders in the years to come, and it is worth taking advantage of.

Bibliography

[1] Shahin Atakishiyev, Mohammad Salameh, Hengshuai Yao, and Randy Goebel. Explainable artificial intelligence for autonomous driving: a comprehensive overview and field guide for future research directions. *arXiv preprint arXiv:2112.11561*, 2021.

[2] Serkan Ayvaz and Salih Cemil Cetin. Witness of things: Blockchain-based distributed decision record-keeping system for autonomous vehicles. *International Journal of Intelligent Unmanned Systems*, 7(2):72–87, 2019.

[3] Janet Fleetwood. Public health, ethics, and autonomous vehicles. *American journal of public health*, 107(4):532–537, 2017.

[4] Jon Arne Glomsrud, André Ødegårdstuen, Asun Lera St Clair, and Øyvind Smogeli. Trustworthy versus explainable ai in autonomous vessels. In *Proceedings of the International Seminar on Safety and Security of Autonomous Vessels (ISSAV) and European STAMP Workshop and Conference (ESWC)*, volume 37, 2019.

[5] Christian Gorenflo, Lukasz Golab, and Srinivasan Keshav. Mitigating trust issues in electric vehicle charging using a blockchain. In *Proceedings of the Tenth ACM International Conference on Future Energy Systems*, pages 160–164, 2019.

[6] Balint Gyevnar. Cars that explain: Building trust in autonomous vehicles through explanations and conversations.

[7] Md Golam Moula Mehedi Hasan, Amarjit Datta, Mohammad Ashiqur Rahman, and Hossain Shahriar. Chained of things: A secure and dependable design of autonomous vehicle services. In *2018 IEEE 42nd Annual Computer Software and Applications Conference (COMPSAC)*, volume 2, pages 498–503. IEEE, 2018.

[8] Fatima Hussain, Rasheed Hussain, and Ekram Hossain. Explainable artificial intelligence (xai): An engineering perspective. *arXiv preprint arXiv:2101.03613*, 2021.

[9] Shafkat Islam, Shahriar Badsha, and Shamik Sengupta. A light-weight blockchain architecture for v2v knowledge sharing at vehicular edges. In *2020 IEEE International Smart Cities Conference (ISC2)*, pages 1–8. IEEE, 2020.

[10] Uzair Javaid, Muhammad Naveed Aman, and Biplab Sikdar. A scalable protocol for driving trust management in internet of vehicles with blockchain. *IEEE Internet of Things Journal*, 7(12):11815–11829, 2020.

[11] Farah Kandah, Brennan Huber, Anthony Skjellum, and Amani Altarawneh. A blockchain-based trust management approach for connected autonomous vehicles in smart cities. In *2019 IEEE 9th Annual Computing and Communication Workshop and Conference (CCWC)*, pages 0544–0549. IEEE, 2019.

[12] Darius Kianersi, Suraj Uppalapati, Anirudh Bansal, and Jeremy Straub. Evaluation of a reputation management technique for autonomous vehicles. *Future Internet*, 14(2):31, 2022.

[13] Shiho Kim. Blockchain for a trust network among intelligent vehicles. In *Advances in Computers*, volume 111, pages 43–68. Elsevier, 2018.

[14] A Mohan Krishna and Amit Kumar Tyagi. Intrusion detection in intelligent transportation system and its applications using blockchain technology. In *2020 international conference on emerging trends in information technology and engineering (IC-ETITE)*, pages 1–8. IEEE, 2020.

[15] Wanxin Li, Collin Meese, Zijia Gary Zhong, Hao Guo, and Mark Nejad. Location-aware verification for autonomous truck platooning based on blockchain and zero-knowledge proof. In *2021 IEEE International Conference on Blockchain and Cryptocurrency (ICBC)*, pages 1–5. IEEE, 2021.

[16] AV Shreyas Madhav and Amit Kumar Tyagi. Explainable artificial intelligence (xai): connecting artificial decision-making and human trust in autonomous vehicles. In *Proceedings of Third International Conference on Computing, Communications, and Cyber-Security: IC4S 2021*, pages 123–136. Springer, 2022.

[17] Mohamed Nassar, Khaled Salah, Muhammad Habib ur Rehman, and Davor Svetinovic. Blockchain for explainable and trustworthy artificial intelligence. *Wiley Interdisciplinary Reviews: Data Mining and Knowledge Discovery*, 10(1):e1340, 2020.

[18] Huong Nguyen, Tri Nguyen, Teemu Leppanen, Juha Partala, and Susanna Pirttikangas. Situation awareness for autonomous vehicles using blockchain-based service cooperation. In *Advanced Information Systems Engineering: 34th International Conference, CAiSE 2022, Leuven, Belgium, June 6–10, 2022, Proceedings*, pages 501–516. Springer, 2022.

[19] Yuan Shen, Shanduojiao Jiang, Yanlin Chen, Eileen Yang, Xilun Jin, Yuliang Fan, and Katie Driggs Campbell. To explain or not to explain: A study on the necessity of explanations for autonomous vehicles. *arXiv preprint arXiv:2006.11684*, 2020.

[20] Rakesh Shrestha and Seung Yeob Nam. Regional blockchain for vehicular networks to prevent 51% attacks. *IEEE Access*, 7:95033–95045, 2019.

[21] Anastasia Theodouli, Konstantinos Moschou, Konstantinos Votis, Dimitrios Tzovaras, Jan Lauinger, and Sebastian Steinhorst. Towards a blockchain-based identity and trust management framework for the iov ecosystem. In *2020 Global Internet of Things Summit (GIoTS)*, pages 1–6. IEEE, 2020.

[22] Gesa Wiegand, Malin Eiband, Maximilian Haubelt, and Heinrich Hussmann. "i'd like an explanation for that!" exploring reactions to unexpected autonomous driving. In *22nd International Conference on Human-Computer Interaction with Mobile Devices and Services*, pages 1–11, 2020.

[23] Zuobin Ying, Longyang Yi, and Maode Ma. Beht: blockchain-based efficient highway toll paradigm for opportunistic autonomous vehicle platoon. *Wireless Communications and Mobile Computing*, 2020:1–13, 2020.

[24] Chenyue Zhang, Wenjia Li, Yuansheng Luo, and Yupeng Hu. Ait: An ai-enabled trust management system for vehicular networks using blockchain technology. *IEEE Internet of Things Journal*, 8(5):3157–3169, 2020.

[25] Yiwen Zhang, Weiwei Guo, Cheng Chi, Lu Hou, and Xiaohua Sun. Towards scenario-based and question-driven explanations in autonomous vehicles. In *HCI in Mobility, Transport, and Automotive Systems: 4th International Conference, MobiTAS 2022, Held as Part of the 24th HCI International Conference, HCII 2022, Virtual Event, June 26–July 1, 2022, Proceedings*, pages 108–120. Springer, 2022.

[26] Ning Zhao and Hao Wu. Blockchain combined with smart contract to keep safety energy trading for autonomous vehicles. In *2019 IEEE 89th Vehicular Technology Conference (VTC2019-Spring)*, pages 1–5. IEEE, 2019.

III

Ethical, Social, and Legal Implications of XAI in ITS

CHAPTER 12

Ethical Decision-Making under Different Perspective-Taking Scenarios and Demographic Characteristics: The Case of Autonomous Vehicles

Kareem Othman

Civil Engineering Department, University of Toronto, Toronto, Ontario, Canada

CONTENTS

DOI: 10.1201/9781003324140-12

OVER THE LAST few years, autonomous vehicles (AVs) have been intensively studied with regard to their benefits, implications, technological development, and public attitude. Results of previous studies in the literature show that AVs have the potential to offer a large number of benefits. The reduction in the number of road accidents is one of the major benefits of AVs because AVs eliminate human errors that contributes to 90% of the accidents. In addition, one of the main advantages of AVs is the ability to program these vehicles to operate based on some ethical logic that maximizes the social value of AVs. However, previous studies show that people prefer to buy an AV that is self-protective that an AV that follows moral decisions. These studies provide the participants with partial perspective taking as they were asked to imagine themselves as the passengers of the AV ignoring that the respondents might be the pedestrians in some situations. As a result, this study employed a questionnaire survey in order to understand the impact of accessibility to perspective-taking on the moral decisions of respondents from the USA following the approach proposed by [16]. The results show that partial perspective taking results in a biased attitude as the respondents select the action that saves their lives at the cost of the others. Moreover, the responses were analyzed based on the demographic properties of the respondents (age, gender, and prior knowledge about AVs). In general, the results show that female and older respondents tend to select the moral action.

12.1 INTRODUCTION

Over the last few years, a large effort has been devoted to the automation of vehicles (Othman, 2022a). In general, it is anticipated that AVs have the potential to offer multiple benefits starting from the reduction in the energy consumption and emissions to the economic value caused by AVs as the trip time will not be considered as an economic loss anymore as the passengers will be able to get engaged in productive activities during their trips [2,3,19,22]. In addition, one of the main benefits of AVs is the ability of this technology to reduce the number of traffic accidents because of the elimination of the human error that contribute to 90% of these accidents [8, 10, 20]. Furthermore, one of the main advantages of AVs is the ability to program these vehicles to operate based on some ethical logic so that if an accident is unavoidable the vehicle can take the decision that focuses on minimizing the overall impact of the accident.

In many scenarios, the decisions of the individuals can affect the safety of others [23]. Autonomous vehicle (AV) technology is one of the emerging technologies in which the decisions of the individuals can have a substantial impact on the safety of the public. For example, programming AVs in a way that prioritizes the passengers' lives would have a major impact on the lives of other road users such as pedestrians, cyclists, and other drivers [20]. In the literature, large number of studies have investigated the public preference of different AVs that has different decision-making approaches. These studies show that people prefer to purchase an AV that is passenger-protective than an AV that is utilitarian which prioritizes the social value and protects the majority of people [4]. However, the latter option is considered a more moral option. Over the last few years, this contradiction was traditionally studied as the ethical dilemma of AVs because people know that it is better to utilize an AV that maximizes the overall value to the society, but they have strong initiatives to adopt a self-protective approach [12]. In general, people do not accept the idea that they would purchase a vehicle that might choose to sacrifice the life of the passenger in some circumstances [9, 17] in order to protect the largest number of people. Alternatively, the public prefers the self-protective or passenger-protective logic that protects the passengers under all circumstances. In general, previous studies focused on presenting multiple scenarios including the moral scenario that focuses on minimizing the overall impact of the accident. In these studies, the participants were asked to imagine themselves the passengers of an AV that cannot avoid an accident and has one of two options: the first is to swerve, hit the road barriers, and kill the passenger of the AV and the second is to hit a group of pedestrians crossing the street. Then, the participants were asked to select the logic that should be adopted in the programming of AVs. The results show that the majority of the respondents believe that the utilitarian logic that saves the largest number of lives is the one that should be adopted; however, they are willing to buy a passenger-protective AV than a utilitarian AV [5–7, 11, 14, 18, 31]. These results shed light on the ethical issue because people might choose to sacrifice the lives of a group of pedestrians to protect themselves. As a result, previous studies suggest that the enforcement of utilitarian AVs through regulations might be an obstacle in front of the wide spread of AVs [16]. Thus, this enforcement might be counterproductive as this delay in the adoption of AVs might mean that the lives saved by the utilitarian AVs might be outnumbered by the number of deaths caused as a result of the delay in the adoption of AVs.

Previous studies that analyzed the ethical dilemma of AVs focused on engaging the participants in a perspective-taking task as the participants were asked about the moral decision or the preferred decision in case the participant is a passenger of the AV and thus they will be the victims of their decision and moral convictions [16]. However, one of the main issues of these studies is the lack of situational perspectives as the participants were always given the perspective of the passenger of the AV ignoring that the participant might be the pedestrian crossing the street in from of the AV. Thus, previous studies consider only a partial situational perspective that focuses on participants as passengers ignoring that the owner of the AV might be a pedestrian (which happens as soon as the passenger o the AV exits the vehicle) as highlighted by Martin et al. (2021) [16]. Thus, it is critical to give the participants of any survey the full situational perspective (give them access to the two scenarios as a passenger and as a pedestrian) as it should affect their ethical decisions. In addition, the partial accessibility to the perspective-taking might be one of the main factors that caused the behavioral inconsistency in the responses of previous studies that showed that people select the moral action that saves the largest number of lives but they are not motivated to buy an AV that follows that logic [16]. As a result, in this study, a questionnaire survey was conducted to understand the impact of the full and partial situational perspective (as proposed by Martin et al. [16]) on the decision-making of the participants. Accordingly, in this study, four different experiments with different perspective taking and different accessibility to the situational perspective were utilized and the responses of different

TABLE 12.1 Number of Responses Collected for the Different Experiments

Experiment Number	Number of Responses	Frequency
1	250	0.246
2	256	0.252
3	250	0.247
4	256	0.253

respondents with different demographic characteristics were collected and analyzed. In addition, previous studies showed that demographic characteristics have a major impact on the public attitude toward AV technology. For example, previous studies show that males respondents are more optimistic towards AVs [1, 25–30] while older respondents were more pessimistic towards this technology [1, 13, 24, 25, 29]. In addition, experts with previous experience are more positive towards AVs. Thus, it can be stated that the demographic characteristics have an influence on the public attitude towards AVs, and it might have an impact on ethical decision-making. However, the influence of the demographic characteristics on ethical decision-making in the case of AVs has never been studied. Thus, in this study, the impact of the demographic characteristics (gender, age, and previous experience about AVs) on the ethical decision-making is studied in detail.

12.2 METHODOLOGY

12.2.1 Questionnaire Survey

In this study, a questionnaire survey was designed to collect the responses of the public in four different scenarios. The survey was designed in Survey Monkey [32] so that every respondent gets access to one scenario. In addition, the survey was designed to assign the scenarios randomly to the respondents so that 25% of the respondents respond to every scenario. In other words, the survey was designed to collect the same number of responses for every experiment studied. Furthermore, respondents from the USA were recruited through Survey Monkey with a monetary reward for participating in the survey. The survey consists of two main sections. The first section focuses on collecting the demographic characteristics of the respondents such as the age, gender, and prior knowledge about AVs. Then, the second section focuses on collecting the respondents' opinions about the preferred actions in the different experiments (the four experiments). In this section, the experiments were assigned randomly to the respondents with 0.25 probability in order to collect the same number of responses for every experiment. The survey was published between March to August 2022. In total, 1012 completed responses were collected (from respondents from the USA) for the four experiments. The number of responses collected for the different experiments are summarized in Table 12.1. In addition, Table 12.2 summarizes the demographic characteristics of the respondents. It is important to make sure that the demographic

TABLE 12.2 Summary of the Demographic Characteristics of the Respondents Who Participated in the Survey

Demographic characteristics		Count (N)	Frequency (%)
Gender	Male	483	48%
	Female	529	52%
Age	18-29	227	22%
	30-44	311	31%
	45-60	232	23%
	>60	242	24%
Have you ever seen, heard, or read anything about self-driving cars before participating in this survey?	Yes, A lot	243	24%
	Yes, a bit	684	68%
	No	85	8%

characteristics of the respondents of the survey are similar to the entire population of the USA. Thus, the results of the demographic characteristics were compared with the latest characteristics of the population of the USA that was published by the United States Census Bureau (2022) [33]. For the gender representation, the results show a balanced representation of male and female respondents as 48% of the respondents were males and 52% were females. In addition, according to the United States Census Bureau (2022) report, 49% of the USA population are males and 51% are females. Thus, comparing these percentages with the results of the survey shows that the sample used in the study is good in representing the two genders. Similarly, the demographic characteristics of the survey show that the different age groups are well represented in the survey and well repetitive of the US population according to the report of the United States Census Bureau (2022) report.

While the previous discussion focuses on the traditional demographics that are commonly used, priori knowledge about AVs technology should be investigated, in the context of AVs, in order to understand whether prior experience about the technology affects ethical decision-making. Thus, part of the demographic characteristics section (section one) focused on collecting information about the respondents' prior knowledge about AVs. In this part, the respondents were asked whether they have heard, seen, or read about AVs prior to the survey. The results show that most of the respondents have prior knowledge about AVs as 92% of the respondents responded with "yes, a lot" or "yes, a bit", while only 8% have no prior knowledge about AVs. Of the respondents with prior knowledge about AVs, only 24% of them believe they have a strong background about AVs.

12.2.2 Study Design

The main scope of this study is to understand the impact of three main factors on moral decisions in the case of AVs following the approach proposed by Martin et al. [16]. These three factors are the type of involvement, the type of perspective-taking as proposed by [16], and the demographic characteristics. Thus, this is a 3*1 study that investigated the impact of three independent variables on one dependent variable. For the type of involvement, this study focuses on analyzing the impact of two types of involvement on ethical decision-making. The levels of involvement are:

1. No involvement: where the respondents are not involved in the situation at all but just watching.

2. Participant involved: in this case the participant has to imagine that he/she is involved in the situation as a passenger or a pedestrian or both.

The second independent variable analyzed in this study is the level of accessibility to the perspective-taking that is adopted from Martin et al. [16]. In this study, three different perspective-taking were analyzed in order to understand their impact on ethical decision-making. These three levels are:

1. No perspective taking: in this scenario, the respondent is not involved at all and is not taking the perspective of any person who is part of the scenario.

2. Partial perspective taking: in this scenario, the respondents were asked to take the perspective of someone who is involved in the scenario. This level of perspective taking has two levels: (a) the respondent is taking the prescriptive of the passengers of the AV; (b) the respondent is taking the perspective of one of the pedestrians involved in the situation.

3. Full perspective taking: in this scenario, the respondent were informed that they might be passengers in some situations and pedestrians in other situations.

The third, dependent variable investigated in this study is the demographic characteristics. Thus, a detailed analysis was conducted based on the demographic properties of the respondents in order to understand the impact of the age, gender, and prior knowledge about AVs on the ethical decision-making of the respondents. On the other hand, the only dependent variable investigated in this study is the judgment of moral appropriateness. Thus, every respondent was given a specific scenario or experiment and in every scenario the respondent was asked to rank two decisions (swerve, stay) on a scale from 1 to 10 where 1 represents that the action is morally inappropriate and 10 represents that the action is morally appropriate. It must be mentioned that one of the pitfalls of this method, which is traditionally used in previous studies in the literature, is that it cannot establish whether the respondent is fully utilitarian or not when judging on every action separately. In addition, this method does not offer a weight to understand whether the respondent is utilitarian or non-utilitarian [16]. Thus, in this study, a different scale was utilized to avoid this issue. This scale is called the utilitarian weight and can be calculated using equation 12.1. This metric was used to estimate the overall utilitarian judgment of every respondent and was proposed by Martin et al. [16]. It is calculated by subtracting the respondent's rank of the non-utilitarian action from the rank of the utilitarian action that saves the largest number of lives. As a result, a high and positive value for the utilitarian weight means that the respondent has utilitarian judgment

$$UW = rank\ of\ the\ UA - rank\ of\ the\ NMA \tag{12.1}$$

where UW is the utilitarian weight; rank of the UA is the rank selected by the respondent for the utilitarian action or moral action; rank of the NMA is the rank selected by the respondent for the non-utilitarian action or non-moral action.

12.2.3 Design of the Experiments

Every respondent was assigned randomly to one of the four experiments tested in the study. Across the different scenarios, the respondents were given a scenario with an AV with one passenger that is about to crash in a group of pedestrians who are crossing the street and the vehicle that cannot avoid the accident. In addition, the AV has the ability to take one of two actions either to swerve and hit the road barrier and kill the passenger of the AV or to stay on the road to hit and kill the ten pedestrians. In the four experiments, the role of the respondent varies from one experiment to the other as follows:

1. Experiment 1: the respondent is watching the situation from a bridge.
2. Experiment 2: the respondent is the passenger of the AV.
3. Experiment 3: the respondent is one of the pedestrians who are crossing the street.
4. Experiment 4: the respondent was informed that he/she might be the passage of the AV or one of the pedestrians crossing the street.

In addition, the responses were given a brief text explanation of the situation with a graphical visualization to make it easy to understand the situation. Then, the respondents were asked to rank the two actions (swerve or stay) on a scale from 1 to 10 where 1 represents that the action is morally inappropriate and 10 represents that the action is morally appropriate. The text explanation provided to the experts in the four experiments is as follows:

1. Experiment 1: "You are walking on a pedestrian bridge above an urban corridor. You monitored an autonomous vehicle that is driving at the speed limit of a road. Suddenly, ten pedestrians appeared on the road in front of the vehicle that the vehicle cannot stop before crashing into them. However, the vehicle has a second option which is swerving to hit the road barrier and save the ten pedestrians. The vehicle can be programmed to swerve and hit the road barrier and kill the passenger of the vehicle and save the ten pedestrians or stays on its path to kill the ten pedestrians and saves the passenger of the vehicle". In addition to the text, Figure 12.1 was provided to the respondents in order to visually understand the experiment.
2. Experiment 2: "You are the sole driver of an autonomous vehicle that is driving at the speed limit of a road. Suddenly, ten pedestrians appeared on the road in front of the vehicle that the vehicle cannot stop before crashing into them. However, the vehicle has a second option which is swerving to hit the road barrier and save the ten pedestrians. The vehicle can be programmed to swerve

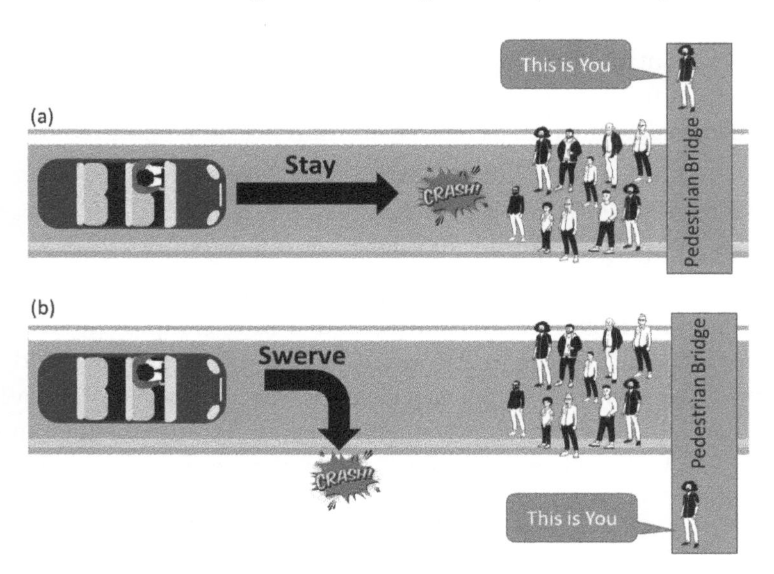

Figure 12.1 Visual illustration of experiment 1 (offered to the respondents of the experiment).

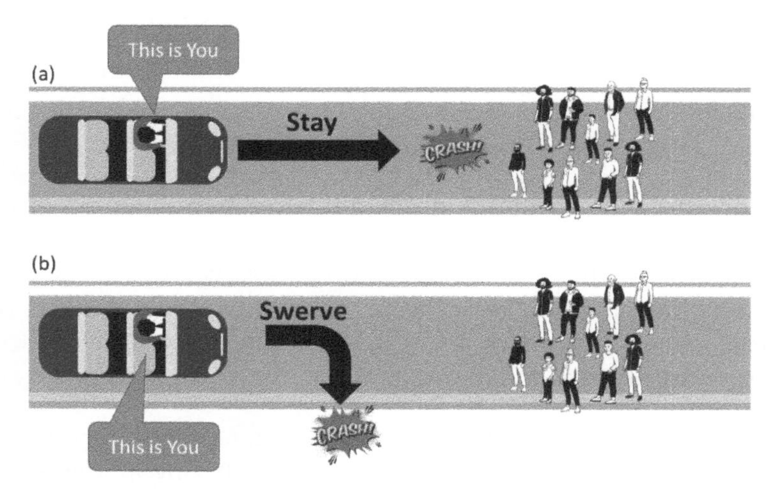

Figure 12.2 Visual illustration of experiment 2 (offered to the respondents of the experiment).

and hit the road barrier and kill you but save the ten pedestrians or stays on its path to kill the ten pedestrians and saves you". In addition to the text, Figure 12.2 was provided to the respondents in order to visually understand the experiment.

3. Experiment 3: "You are one of ten pedestrians who suddenly appeared in front of an autonomous vehicle that travels on the speed limit of the road. In addition, the vehicle cannot stop before hitting the ten pedestrians but it has a second option which is swerving to hit the road barrier. The vehicle can be programmed to swerve and hit the road barrier and kill the passengers of the vehicle and save

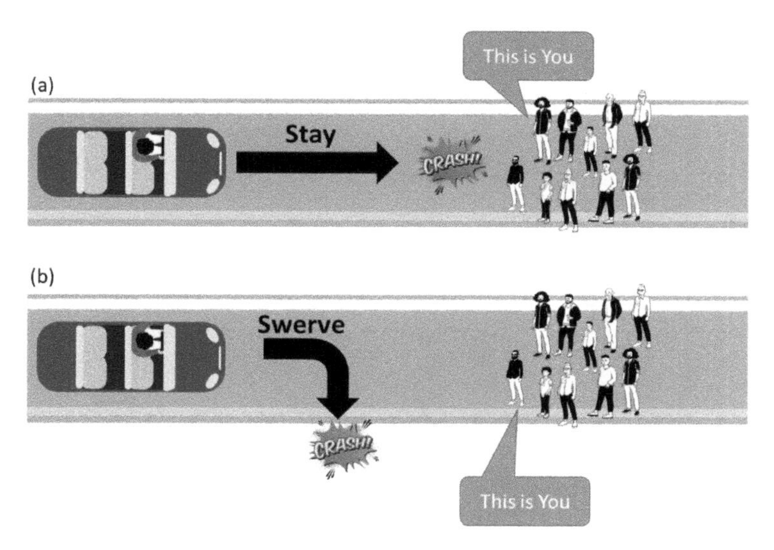

Figure 12.3 Visual illustration of experiment 3 (offered to the respondents of the experiment).

the ten pedestrians or stays on its path to kill the ten pedestrians and saves the passenger of the vehicle". In addition to the text, Figure 12.3 was provided to the respondents in order to visually understand the experiment.

4. Experiment 4: "You could be the sole driver of an autonomous vehicle that is driving at the speed limit of a road. Or you could be one of ten pedestrians who suddenly appeared on the road in front of the vehicle that the vehicle cannot stop before crashing into them. However, the vehicle has a second option which is swerving to hit the road barrier and save the ten pedestrians. The vehicle can be programmed to swerve and hit the road barrier and kill the passenger of the vehicle (that might be you) but save the ten pedestrians or stays on its path to kill the ten pedestrians (that might include you) and saves the passenger of the vehicle". In addition to the text, Figure 12.4 was provided to the respondents in order to visually understand the experiment.

12.3 ANALYSIS AND RESULTS

In this paper, the impact of the level of involvement, perspective-taking, and demographic characteristics on ethical decisions were investigated. The analysis conducted was based on the responses of 1012 individuals who responded to the survey from the USA. In total, this study analyses the responses for the reports in four different experiments. Every respondent has access to only one scenario of the four tested in this study. This scenario was assigned to the respondents randomly with 0.25 probability in order to collect the same number of responses for every experiment.

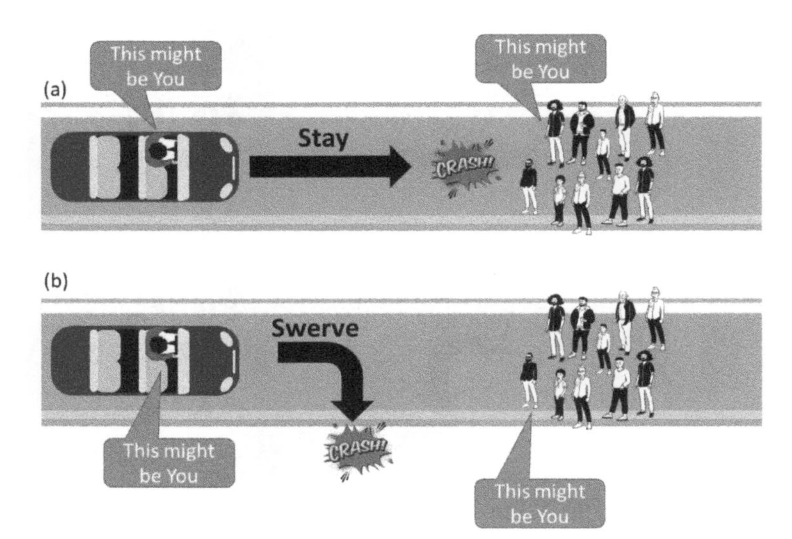

Figure 12.4 Visual illustration of experiment 4 (offered to the respondents of the experiment).

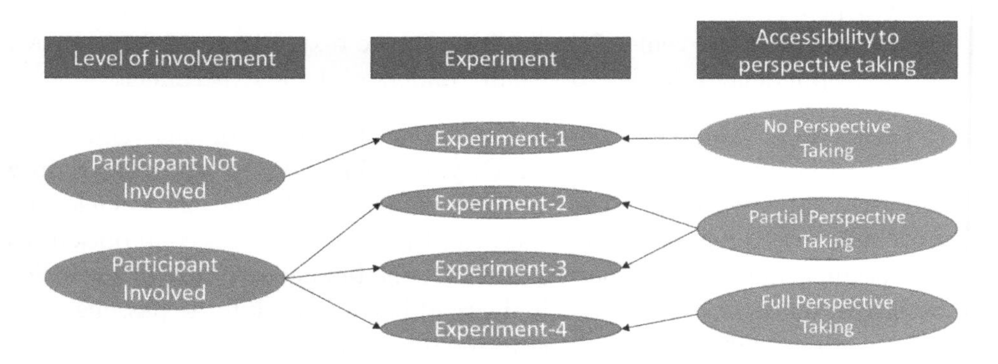

Figure 12.5 Summary of the different levels of involvement and perspective-taking in every experiment tested.

12.3.1 Impact of the Different Levels of Involvement and Perspective Taking on Ethical Decision-Making

In this study, two different levels of involvement were tested: (a) no involvement; (b) participants involved. In addition, for perspective taking, three levels of perspective taking were tested: (a) the participant is not taking the perspective of anyone on the situation; (b) partial perspective taking, and in this case the respondents were asked to imagine themselves as the passenger of the AV or one of the pedestrians crossing the street; (c) full prescriptive taking and in this experiments the respondents were informed that they might be the passage of the AV or one of the pedestrians crossing the street. Every experiment was designed to test a specific scenario for the level of involvement and the perspective-taking accessibility as shown in Figure 12.5. The figure shows that experiment one focuses on evaluating ethical decision-making when the respondents are not involved in the situation at all as the respondents are watching the situation from a pedestrian bridge. In addition, the respondents were not asked to take the perspective of anyone in the scenario so this experiment analyses the ethical decisions for the scenario of no prescriptive taking. For the second level of involvement (participant involved) the respondents were involved in the scenarios of experiments 2, 3, and 4 but with different levels of perspective-taking. In experiment 2, the respondents were asked to imagine themselves as the passengers of an AV, while in experiment 3 the respondents were asked to imagine themselves as one of ten pedestrians crossing the street. While the respondents were involved in the situation in the two experiments (2 and 3), they have partial accessibility to the respective taking as they were asked to imagine themselves as the passengers of an AV or pedestrians who are crossing the street. Finally, in experiment 4, the respondents were informed that they might be the passengers of an AV or a pedestrian crossing the street so that the participants are involved and have full accessibility to the perspective-taking as every person might have an AV but they became pedestrians once they leave the vehicle.

In every scenario, the respondents were asked to rank the two decisions (swerve, stay) on a scale from 1 to 10 where 1 represents that the action is morally inappropriate and 10 represents that the action is morally appropriate. Then, the utilitarian weight was calculated for every respondent using equation (1) as the difference between the utilitarian swerve action and the non-utilitarian stay action in order to understand whether the response is utilitarian or non-utilitarian. Thus, a high and positive value for the utilitarian weight means that the respondent has a utilitarian judgment, while a negative utilitarian weight indicates that the respondent has a non-utilitarian judgment. The average utilitarian weights of the different experiments are shown in Figure 12.6. The results show that the highest and lowest levels of utilitarian weights are observed in experiments 2 and 3 when the respondents got partial accessibility to the perspective-taking. In experiment 2 when the respondents were asked to imagine themselves as the passengers of AVs, this scenario has the lowest average utilitarian weight (−0.5) which indicates that the respondents selected the action that saved themselves at the cost of hitting and killing the ten pedestrians. On the other side, the highest level of utilitarian weight can be observed in experiment 3 when the respondents were asked to imagine themselves as one of the ten pedestrians crossing the street. The results show that the utilitarian weight reached 8.5 indicating that the respondents believe that the vehicle should swerve to sacrifice the passenger of the AV to save the ten pedestrians. Experiment one was designed so that the respondents judge the situation without being involved at all so the results of this scenario can be used as a benchmark for other scenarios. This experiment was selected to be the benchmark as it is expected that the experts will select the ranks that represent their ethical decision as the action selected will not affect them. Thus, comparing the results of experiments 2 and 3 to experiment 1 shows that the respondent became more self-protective in the two scenarios as the utilitarian weights changed in a way that protect the respondents across the two scenarios. As

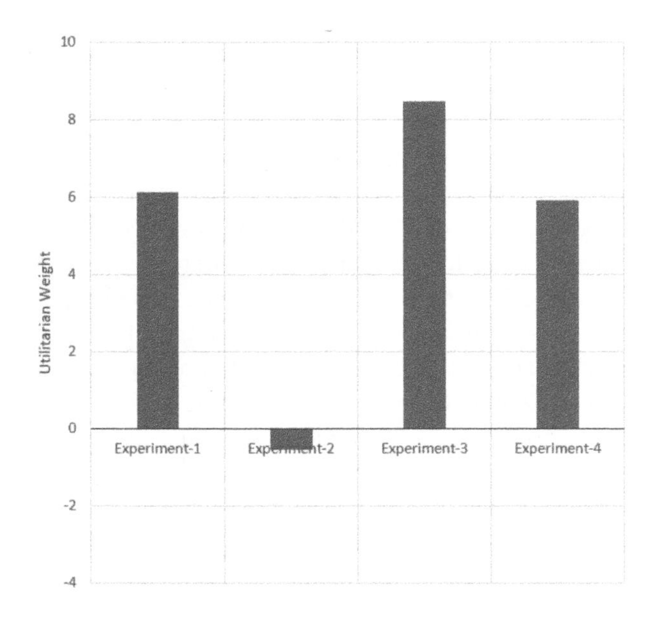

Figure 12.6 Average utilitarian weight across the four experiments.

experiments 2 and 3 provide the respondents with partial perspective taking, it can be concluded that the two scenarios resulted in biased results as the response will try to save their lives at the cost of sacrificing the lives of the others. Thus, it can be concluded that partial perspective taking results in biased results as people will always try to save their lives. Finally, the average utilitarian weight for experiment 4 was similar to the average utilitarian weight for experiment 1. In experiment 4, the respondents were informed that they might be the passengers of an AV or one of the pedestrians crossing the street so the responses have accessibility to full perspective taking as they might be the passengers of the AV or they might be the pedestrian crossing the street. Thus, full accessibility to the perspective-taking resulted in a utilitarian weight that is similar to the no involvement scenario. Thus, the results of experiment 4 show that full accessibility to perspective-taking resulted in moral results that are not biased; similar to the results of experiments 2 and 3.

12.3.2 Impact of the Demographic Characteristics on Ethical Decision-Making

12.3.2.1 Impact of the Gender on Ethical Decision-Making

While the previous discussion in the previous subsection focuses on analyzing the overall responses of the respondents, this subsection focuses on analyzing the responses to the four experiments by gender in order to understand the perspective of male and female respondents. Figure 12.7 shows the average utilitarian weight across the four scenarios for male and female respondents. The results show that the average utilitarian weight is always higher than the average utilitarian weight of male respondents. In addition, conducting the t-test between the responses of the male

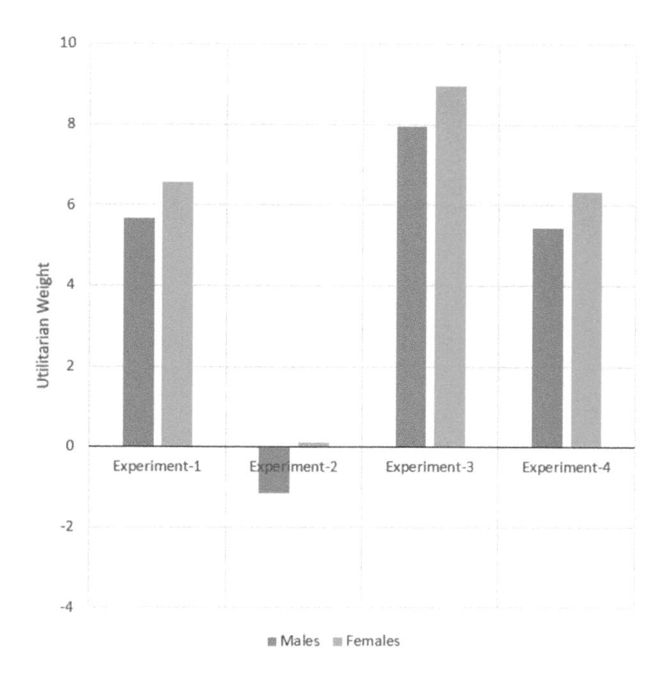

Figure 12.7 average utilitarian weight across the four experiments for male and female respondents.

respondents and female respondents using a 95% level of confidence shows that there is a significant difference between the responses of male and female respondents as the resulting p-value across the four experiments were lower than 0.05. thus, it can be concluded that female respondents will select more utilitarian weights than male respondents, and there is a statically significant difference between the responses of the two categories. However, the results of both males and females have the same pattern across the four experiments as partial accessibility to the preserve taking resulted in biased results as experiments 2 and 3 were the experiments with the highest and lowest utilitarian weights. On the other hand, the utilitarian weights in experiments 1 and 4 are similar for male and female respondents.

12.3.2.2 Impact of the Age on Ethical Decision-Making

This subsection focuses on analyzing the responses to the four experiments by age in order to understand the perspective of the different respondents at different age groups in making the decisions across the different experiments. The results show that the average utilitarian weight increases with the increase in the age of the respondents across the four experiments as shown in Figure 12.8. However, the changes in the utilitarian weights across the different age groups are minor for experiments 1, 3, and 4. On the other side, the changes in the utilitarian weights between the different age groups in experiment 2 are major as the minimum utilitarian weight for the youngest group is −2.45 and the maximum utilitarian weight for the oldest group is 0.678.

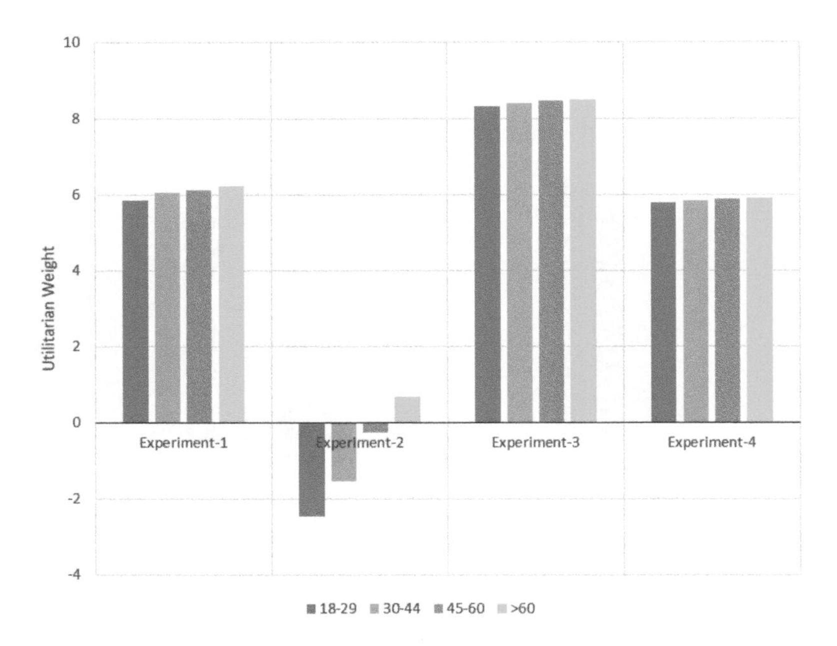

Figure 12.8 average utilitarian weight across the four experiments by age.

thus, the difference in the utilitarian weight between the oldest and youngest groups is 3.128. On the other side, the results of utilitarian weights across the different age groups have the same pattern across the four experiments as partial accessibility to the preserve taking resulted in biased results as experiments 2 and 3 were the experiments with the highest and lowest utilitarian weights. On the other hand, the utilitarian weights in experiments 1 and 4 are similar across the different age groups.

12.3.2.3 Impact of Prior Knowledge about AVs on Ethical Decision-Making

This subsection focuses on understanding how respondents with different prior knowledge about AVs make ethical decisions across the four experiments tested. Figure 12.9 shows the average utilitarian weight for respondents with different prior knowledge about AVs across the four experiments. The results show that the utilitarian weight increases with the increase in the level of knowledge about AVs; however, the changes in the utilitarian weights are minor across the four experiments. In addition, the results show that the responses of the different respondents with different levels of knowledge about AVs have the same pattern across the different experiments. Furthermore, the results show that partial accessibility to the preserve taking resulted in biased results as experiments 2 and 3 were the experiments with the highest and lowest utilitarian weights. On the other hand, the utilitarian weights in experiments 1 and 4 are similar for respondents with different prior knowledge about AVs.

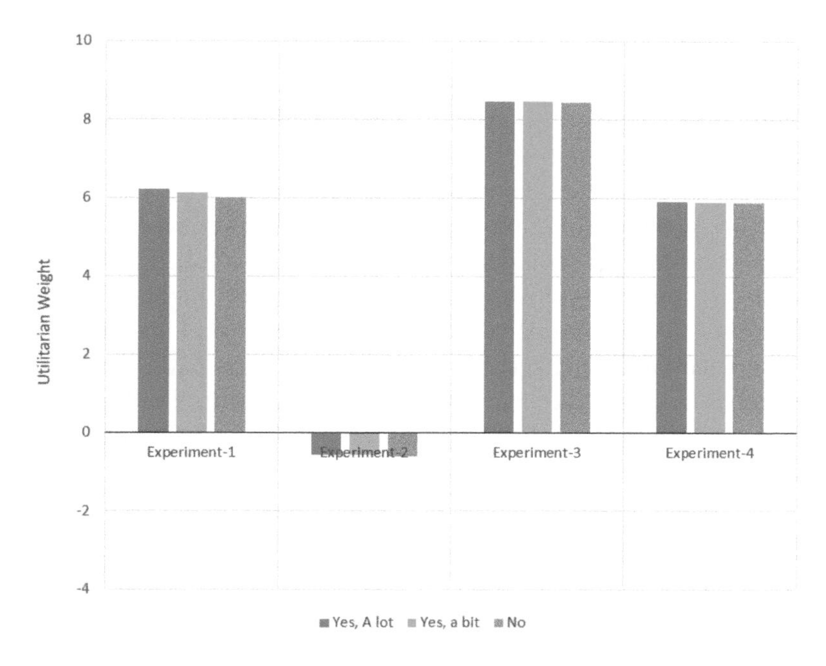

Figure 12.9 Average utilitarian weight across the four experiments for respondents with different background knowledge about AVs.

12.4 CONCLUSION

This study focuses on understanding the impact of three independent factors on the ethical decision-making of the public in the case of AVs. These factors are the level of involvement, accessibility to perspective-taking, and demographic characteristics. Thus, a questionnaire survey was designed and conducted in order to evaluate the moral responses of respondents from the USA. This survey focuses on understanding the public decision in four different scenarios or experiments. Every respondent got access to one of these experiments and the experiments were randomly assigned to the respondents with a probability of 0.25 in order to collect the same number of responses for every experiment. The survey was published between March to August 2022 and a total of 1012 complete responses were collected for the four experiments. 250, 256, 250, and 256 responses were collected for experiments 1, 2, 3, and 4 respectively. The results show that partial accessibility to the perceptive talking, which is the traditional method used in previous studies, results in biased results as the respondents were asked to imagine themselves as the passengers of the AV or pedestrians who are crossing the street in front of the AV (experiments 2 and 3). Thus, in these two experiments, the respondents selected the action that saves their lives at the cost of the others. However, people will not always be passengers of AVs or pedestrians. Thus, experiment 4 focused on providing the respondents with full perspective taking as the respondents were informed that they might be the passengers of an AV or might be pedestrians who are crossing the street in front of the AV, which is the real-life scenario. Thus, full accessibility to the perspective-taking resulted in a utilitarian weight that is similar to the no involvement scenario (experiment 1) during which the respondents were not involved in the situation at all. Thus, the results of experiment 4 show that full accessibility to perspective-taking resulted in moral results that are not biased similar to the results of experiments 2 and 3. In addition, in experiment 4 the respondents were not following a self-protective logic but were selecting the moral action that saves the largest number of lives. As a result, full perspective-taking should be provided to the public in order to avoid the public resistance to buying AVs or buying a vehicle that might take the decision of sacrificing the passenger of the AV to save the pedestrians. Thus, it is expected that the different perspective taking scenarios would have an impact in the public willingness to buy and ride in a utilitarian AV. As a result, future studies should focus on understanding the influence of the different perspective taking scenarios on the level of willingness to buy a self-protective and utilitarian AVs.

Bibliography

[1] Abraham, H., et al.: Autonomous vehicles and alternatives to driving: trust, preferences, and effects of age. In: Transportation Research Board 96th Annual Meeting (2017).

[2] Awad, E., Dsouza, S., Kim, R., Schulz, J., Henrich, J., Shariff, A. et al. (2018). The moral machine experiment. Nature, 563, 59–64.

[3] Bigman, Y. E., & Gray, K. (2018). People are averse to machines making moral decisions. Cognition, 181, 21–34.

[4] Bonnefon, J. F., Shariff, A., & Rahwan, I. (2016). The social dilemma of autonomous vehicles. Science, 352(6293), 1573–1576

[5] Burton, S., Habli, I., Lawton, T., McDermid, J., Morgan, P., & Porter, Z. (2020). Mind the gaps: Assuring the safety of autonomous systems from an engineering, ethical, and legal perspective. Artificial Intelligence, 279, 103201.

[6] Cervantes, J. A., Lopez, S., Rodriguez, L. F., Cervantes, S., Cervantes, F., & Ramos, F. (2020). Artificial moral agents: A survey of the current status. Science and Engineering Ethics, 26(2), 501-532.

[7] De Melo, C. M., Marsella, S., & Gratch, J. (2021). Risk of injury in moral dilemmas with autonomous vehicles. Frontiers in Robotics and AI, 7, 572529.

[8] Fagnant, D. J., & Kockelman, K. (2015). Preparing a nation for autonomous vehicles : Barriers and policy recommendations. Transportation Research Part A: Policy and Practice, 77, 167–181.

[9] Gogoll, J., & Muller, J. F. (2017). Autonomous cars: In favor in mandatory ethics. Science and Engineering Ethics, 23(3), 681–700.

[10] Goodall, N. J. (2014). Ethical decision making during automated vehicle crashes. Transportation Research Record: Journal of the Transportation Research Board, 2424, 58–65.

[11] Kim, R., Kleiman-Weiner, M., Abeliuk, A., Awad, E., Dsouza, S., Tenenbaum, J. B., & Rahwan, I. (2018, December). A computational model of commonsense moral decision making. In Proceedings of the 2018 AAAI/ACM Conference on AI, Ethics, and Society (pp. 197–203). Association for Computing Machinery.

[12] Kollock, P. (1998). Social dilemmas: The anatomy of cooperation. Annual Review Sociology, 24, 183–214.

[13] Lee, C., Ward, C., Raue, M., D'Ambrosio, L. and Coughlin, J.F., 2017, July. Age differences in acceptance of self-driving cars: A survey of perceptions and attitudes. In international conference on Human Aspects of IT for the Aged Population (pp. 3–13). Springer, Cham.

[14] Lucifora, C., Grasso, G. M., Perconti, P., & Plebe, A. (2021). Moral reasoning and automatic risk reaction during driving. Cognition, Technology & Work, 23(4), 705–713.

[15] Martin, R., Kusev, I., Cooke, A., Baranova, V., van Schaik, P., & Kusev, P. (2017). Commentary: The social dilemma of autonomous vehicles. Frontiers in Psychology, 8 (808).

[16] Martin, R., Kusev, P., & Van Schaik, P. (2021). Autonomous vehicles: How perspective-taking accessibility alters moral judgments and consumer purchasing behavior. Cognition, 212, 104666.

[17] Nees, M. A. (2016). Acceptance of self-driving cars: An examination of idealized versus realistic portrayals with a self-driving car acceptance scale. In, Vol. 60(1). Proceedings of the human factors and ergonomics society annual meeting (pp. 1449–1453). Los Angeles, CA: Sage CA, SAGE Publications.

[18] O'Sullivan, S., Nevejans, N., Allen, C., Blyth, A., Leonard, S., Pagallo, U., ... & Ashrafian, H. (2019). Legal, regulatory, and ethical frameworks for development of standards in artificial intelligence (AI) and autonomous robotic surgery. The international journal of medical robotics and computer assisted surgery, 15(1), e1968.

[19] Othman, K. (2021a). Impact of Autonomous Vehicles on the Physical Infrastructure: Changes and Challenges. Designs, 5(3), 40.

[20] Othman, K. (2021b). Public acceptance and perception of autonomous vehicles : a comprehensive review. AI and Ethics, 1(3), 355–387.

[21] Othman, K. (2022a). Cities in the Era of Autonomous Vehicles: A Comparison Between Conventional Vehicles and Autonomous Vehicles. In Resilient and Responsible Smart Cities (pp. 95–108). Springer, Cham.

[22] Othman, K. (2022b). Exploring the implications of autonomous vehicles : A comprehensive review. Innovative Infrastructure Solutions, 7(2), 1-32.

[23] Othman, K. (2022c). Multidimension Analysis of Autonomous Vehicles: The Future of Mobility. Civil Engineering Journal, 7, 71-93.

[24] Park, J., Hong, E. and Le, H.T. (2021). Adopting autonomous vehicles : The moderating effects of demographic variables. Journal of Retailing and Consumer Services, 63, p.102687.

[25] Piao, J., McDonald, M., Hounsell, N., Graindorge, M., Graindorge, T., Malhene, N. (2016). Public views towards implementation of automated vehicles in urban areas. Transp. Res. Procedia 14, 2168–2177.

[26] Pigeon, C., Alauzet, A. and Paire-Ficout, L. (2021). Factors of acceptability, acceptance and usage for non-rail autonomous public transport vehicles: A systematic literature review. Transportation research part F: traffic psychology and behaviour, 81, pp.251-270.

[27] Polydoropoulou, A., Tsouros, I., Thomopoulos, N., Pronello, C., Elvarsson, A., Sigþorsson, H., Dadashzadeh, N., Stojmenova, K., Sodnik, J., Neophytou, S. and Esztergar-Kiss, D., 2021. Who is willing to share their AV? Insights about gender differences among seven countries. Sustainability, 13(9), p.4769.

[28] Rezaei, A., & Caulfield, B. (2020). Examining public acceptance of autonomous mobility. Travel behaviour and society, 21, 235-246.

[29] Richardson, E., Davies, P. (2018). The changing public's perception of self-driving cars.

[30] Schoettle, B. and Sivak, M. (2014). A survey of public opinion about autonomous and self-driving vehicles in the US, the UK, and Australia. University of Michigan, Ann Arbor, Transportation Research Institute.

[31] Seoane-Pardo, A. M. (2016, November). Computational thinking beyond STEM: An introduction to" moral machines" and programming decision making in ethics classroom. In Proceedings of the fourth international conference on technological ecosystems for enhancing multiculturality (pp. 37-44).

[32] SurveyMonkey. https://www.surveymonkey.com/

[33] United States Census Bureau. (2022). Census Bureau Releases New Educational Attainment Data. https://www.census.gov/newsroom/press-releases/2022/educational-attainment.html

Index

For Product Safety Concerns and Information please contact our EU
representative GPSR@taylorandfrancis.com
Taylor & Francis Verlag GmbH, Kaufingerstraße 24, 80331 München, Germany

www.ingramcontent.com/pod-product-compliance
Ingram Content Group UK Ltd.
Pitfield, Milton Keynes, MK11 3LW, UK
UKHW050926180425
457613UK00003B/37

* 9 7 8 1 0 3 2 3 4 4 5 7 7 *